OBJECTIVE AGRIBUSINESS MANAGEMENT

FOR JRF, SRF, ARS, NET, SLET, CIVIL SERVICES & OTHER COMPETITIVE EXAMINATIONS

3rd Edition

Dr Shakti Ranjan Panigrahy
Assistant Professor
IABMI, AAU, Anand

Dr Sanjiv Kumar
Scientist
ICAR-NAARM
Rajendranagar, Hyderabad

Dr Ritambhara Singh
Assistant Professor
IABMI, AAU, Anand

Published by

Competition Tutor
(Imprint of Scientific Publishers, INDIA)
5 A, New Pali Road, P.O. Box 91
Jodhpur 342 001 (INDIA)
E-mail: info@scientificpub.com
Website : https://www.scientificpubonline.com

Print : 2025

ISBN: 978-93-89184-06-8

E-ISBN: 978-93-89184-07-5

Printed in India

Dedication

This book is dedicated to My Parents (Both), My Wife (Arpita) ,My Beautiful Daughter (Gidi) and Son (Aayushman)

SRP

This book is dedicated to my wife (Suprita) and my little daughter (Aashi)

SK

This book is dedicated to my Loving son (Rishi)

RS

PREFACE

When we think about the Indian agriculture, so many things come to our mind. It is the mercy of monsoon that stands on low capital base and that generate low productivity. This is the cause that after sixty years of our independence, we are unable to break the poverty cycle in our grass root level which aggravated the basic socioeconomic problems like migration, unbalanced rural-urban population, less integration and poor innovation in our farmers' community. Why we are more concerned about the Agribusiness, because this states about the philosophy of market, its consumer, overall productivity and the improved shares of farmers in the consumer surplus. This book mainly constitutes more than two thousand five hundred questions that covers all the functional area of management like finance, marketing, production and operation management, human resource management and specialized area like agricultural marketing, agricultural finance etc.

These days we are more concerned about the competent human resource in teaching, research and extension as without them, farmers will not capture right information at right place and time. It is true that information in advance make our farmers' community more competent to succumb any uncertainty and risk. The agribusiness professionals are the front runners who change the agriculture to agribusiness. This book will act as a weapon for the students to put their name in the eligibility list for teaching, research and extension.

The book, "OBJECTIVE AGRIBUSINESS MANAGEMENT" is a complete package for the aspirants for ICAR (NET), UGC (NET), JRF, SRF and Civil Service Examination in the subject category of Agribusiness Management, General Management, Agricultural Economics, Livestock Economics, Dairy economics and Basic Economics Students. These days more than fifteen objective based examinations have being carried out in said subject categories in India and this book will provide customized solution to its stakeholders.

These days the policy makers think for the unison approach in the direction of growth and development and we cannot leave behind our 52.1% farmers' community. If the rule of growth is integration, innovation and inclusiveness then the catalytic approach of each agribusiness professionals cannot be overemphasized. Improve our tomorrow by putting effort on our todays' should be the basic mantra of all the stakeholders' of our farming community and this book may play a little effort in this direction.

Last but not the least time comes, we forget about the mass production, rather think about production by masses. It is only possible if we are more accountable in our agricultural production, its distribution and its consumption. Then we can convince our farmers' community to tilt the toil and set the soil for more production and for more productivity. This is the rule of the agriculture and our todays' agribusiness.

CONTENT

CHAPTER 1

1. Principle of Management, Organizational Behaviour, Human Resource Management & Strategic Management

1. Managers who see the foreign countries and their people as inferior to those of the home country are___________
   (a) Ethnocentric Manager
   (b) Geocentric Manager
   (c) Polycentric Manager
   (d) Regiocentric Manager
2. Managers of which country seem to be more concerned with long term implication and allow subordinates to participate in decision making process____________
   (a) USA (b) South Korea
   (c) Japan (d) India
3. In Japan, managers emphasize group harmony and cohesion expressed in the concept of
   (a) Wa (b) In-wa
   (c) On-wa (d) Off-wa
4. In which country, organizations are quite hierarchical with family members occupying key position
   (a) Japan (b) Korea
   (c) China (d) USA
5. Keiretsu network organization is mainly observed in
   (a) Japan (b) Korea
   (c) China (d) USA
6. A tight collusion model between government and indivisual conglomer ate, called as Chaebol is observed in
   (a) Japan (b) Korea
   (c) China (d) USA
7. Consensus forming group decisions, called as ringsei found in
   (a) Japan (b) Korea
   (c) China (d) USA
8. Six Sigma is the philosophy of which country
   (a) USA (b) China
   (c) Japan (d) Korea
9. Sudden death caused by overworking is called as Karoshi in
   (a) Japan (b) Korea
   (c) China (d) USA
10. JIT systems are called as Kanban (Card/ Signal.)
    (a) Japan (b) Korea
    (c) China (d) USA
11. "A place for everything, everything in its place" is
    (a) American management
    (b) 5S′ Principle
    (c) Peter Principle
    (d) A Chinese principle
12. 5S′ Principle is the concept of
    (a) India
    (b) Japan
    (c) China
    (d) USA
13. In 5S′ principle, sweep is otherwise called as
    (a) Sustain
    (b) Sort
    (c) Standardize
    (d) Shine

14. During learning process, managers copy with the change, is known as
(a) Geometric learning
(b) Generative learning
(c) Adaptive learning
(d) Acceptable learning

15. In which learning process, creativity comes from joint effort among organizational members
(a) Geometric Learning
(b) Generative learning
(c) Adaptive Learning
(d) Acceptable Learning

16. Leaning process where the workers are more directly involved in the actual work and shows greater team spirit, is called as
(a) Single loop Learning
(b) Multiloop Learning
(c) Double loop Learning
(d) Standardize learning

17. Out of frustration with discrimination and the perception that there is little chances for advancement, many minorities simply leave their jobs and start their own business. Situation is called as
(a) Round the door syndrome
(b) Glass ceiling syndrome
(c) Revolving door syndrome
(d) Disguise syndrome

18. The theory that points out , decision maker must cope with inadequate information about the nature of the problem and its possible solutions
(a) Dialectical syndrome
(b) Devil's advocate method
(c) Bounded Rationality
(d) Boundary less organization

19. Bounded rationality theory is stated by
(a) Herzberg
(b) Harminger
(c) Herbert Alexander Simon
(d) Henry Fayol

20. The statement that unveils permanent part of organization's identity, its purpose, competencies and place in the world
(a) Policy (b) Rule
(c) Mission (d) Procedure

21. A single use plan that covers relatively large set of activities
(a) Procedure (b) Program
(c) Project (d) Rule

22. Framework that managers devise for dividing and coordinating the activities of members of organisation
(a) Organizational hierarchy
(b) Organizational culture
(c) Organizational structure
(d) Organizational climate

23. Continuous line of authority that extends from the highest levels in an organisation to the lowest levels and clarifies who reports to whom
(a) Unit of command
(b) Chain of command
(c) Span of control
(d) Organizational structure

24. The process of dividing total workload into tasks for easy and logical work for performance, is called as
(a) Decentralisation
(b) Division of labour
(c) Centralisation
(d) Organiastional Hierarchy

25. The group of employees and tasks is generally referred to as
(a) Decentralization
(b) Division of labour
(c) Departmentation
(d) Centralization

26. The organizational structure that specifies who reports to whom in the organization
(a) Organizational Climate
(b) Organizational Culture
(c) Organizational Hierarchy
(d) Organizational Platform

27. When authority and accountability have been passed down to lower level, it is called as
(a) Centralization
(b) Division of work
(c) Decentralization
(d) Depart mentation

28. Motivational Job Design is stated by
(a) Henry Foyal
(b) J Richard Hackman
(c) Herbert Alexander Simon
(d) Abraham Maslow

29. The combing of various operations at a similar level into one job is
(a) Job enrichment
(b) Job rotation
(c) Job enlargement
(d) Job specification

30. The act of delegating power and authority to a subordinate is called as
(a) Decentralization
(b) Centralization
(c) Empowerment
(d) Enlargement

31. An individual reaction to the difference between high job expectations and the frustrating day to day realities of the workplace
(a) Revolving job syndrome
(b) Reality job syndrome
(c) Reality shock syndrome
(d) Revolving shock syndrome

32. Reality shock syndrome is stated by
(a) Richard Harrison
(b) Linsly-Chittenden Hall
(c) Ernst Wolfgang Hamburger
(d) Milton Friedman

33. State of emotional, mental and physical exhaustion that results from continued exposure to high stress, is known as
(a) Rust out condition
(b) Burnout condition
(c) Burst out condition
(d) Roster syndrome

34. The social approach in which organizations exhibit little environmental sensitivity, is
(a) Dark Green Approach
(b) Light Green Approach
(c) Market Approach
(d) Stake holder Approach

35. The social approach where organizations looks for ways to respect and preserve the earth and its natural resources
(a) Dark Green Approach
(b) Light Green Approach
(c) Market Approach
(d) Stake holder Approach

36. Right to inherent in a managerial position, is by
(a) Power
(b) Authority
(c) Administration
(d) Leadership

37. The ability to exert influence is called as
(a) Power
(b) Authority
(c) Administration
(d) Leadership

38. The authority that provide line managers with advice and services is called as
(a) Mixed Authority
(b) Line Authority
(c) Staff Authority
(d) Single Authority

39. The right to control activities of other departments as they relate to specific staff responsibilities
(a) Mixed Authority
(b) Line Authority
(c) Staff Authority
(d) Functional Authority

40. Obligation or expectation to perform by the employee
(a) Reward
(b) Responsiveness
(c) Responsibility
(d) Rectorship

41. Organization that consists of a small core of full time employees and that hires outside specialties temporarily is
(a) Variable Organization
(b) Viscous Organization
(c) Varicose Organization
(d) Virtual Organization

42. Communication that takes place among employees on the same organizational level, are
(a) Diagonal communication
(b) Horizontal communication
(c) Lateral communication
(d) Vertical communication

43. Communication that cuts across both work areas and organizational level, are
(a) Diagonal communication
(b) Horizontal communication
(c) Lateral communication
(d) Vertical communication

44. The process in which suppliers will ship directly to the end consumers rather than to the seller, is known as
(a) Distant shipping
(b) Divert shipping
(c) Drop shipping
(d) Diagonal shipping

45. The act of assigning formal authority and responsibility for completion of activities to a subordinate
(a) Dividend
(b) Accountability
(c) Delegation
(d) Depart mentation

46. Doing the task correctly, is known as
(a) Efficiency
(b) Effectiveness
(c) Management
(d) Productivity

47. Doing the right tasks, is known as
(a) Efficiency
(b) Effectiveness
(c) Management
(d) Productivity

48. Who was known as, 'Father of scientific management'?
(a) Henri Fayol
(b) Frederick Winslow Taylor
(c) Frank Bunker Gilbreth Sr.
(d) Elton Mayo

49. Motion study is first observed by
(a) Henri Fayol
(b) Frederick Winslow Taylor
(c) Frank Bunker Gilbreth Sr.
(d) Elton Mayo

50. The term, therbligs is related to
(a) Time study technique
(b) Method study technique
(c) Motion study technique
(d) Tag study technique

51. Who is not major contributors of scientific management
(a) Henry Gantt
(b) Frank Bunker Gilbreth Sr.
(c) Frederick Winslow Taylor
(d) Henri Fayol

52. 14 principle of management is stated by
(a) Henry Gantt
(b) Frank Bunker Gilbreth Sr.
(c) Frederick Winslow Taylor
(d) Henri Fayol

53. Administrative Management theory is stated by
(a) Max Weber (b) Elton Mayo
(c) Henri Fayol (d) Chester Bernard

54. Bureaucratic Management theory is stated by
(a) Max Weber
(b) Elton Mayo
(c) Henri Fayol
(d) Chester Bernard

55. Hawthorne study is the contribution of
(a) Max Weber (b) Elton Mayo
(c) Henri Fayol (d) Chester Bernard

56. Modern approach of management is based on
(a) System approach
(b) Contingency approach
(c) Human resource approach
(d) Quantitative approach

57. Situational approach is otherwise called as
(a) System approach
(b) Contingency approach
(c) Human resource approach
(d) Quantitative approach

58. Narrow focus in which one sees things solely through one's own view from one's own perspective
(a) Patriotism
(b) Parochialism
(c) Altruism
(d) Anarchism

59. A system of working at home on a computer that is linked to the office, is known as
(a) Outsourcing
(b) Tele counting
(c) Tele commuting
(d) Tele Printing

60. An firm's obligation, beyond that required by the law and economics, to pursue long term goals that are beneficial to society
(a) Social Obligation
(b) Social responsiveness
(c) Social responsibility
(d) Social ethics

61. MBO is stated by
(a) Mc Claren
(b) Frederick Winslow Taylor
(c) Alister McGrath
(d) Peter Drucker

62. Any of the strengths that represent unique skills or resources that can determine the organization's competitive edge
(a) Competitive Advantage
(b) Core competency
(c) Strategic Advantage
(d) USP

63. SWOT is also written as
(a) OWTS (b) TOWS
(c) TSWO (d) OWST

64. Judgmental shortcuts is otherwise called as
(a) Hypnotics
(b) Rationality
(c) Heuristic
(d) Escalation of comment

65. Software that is designed to imitate the structure of brain cells and connections among them
(a) Telepathy
(b) Neural Network
(c) Fuzzy Logic
(d) Hypnotics

66. The bureaucratic structure that is high in specialization, formalization and centralization found in which organizational structure
(a) Mechanistic organization
(b) Organic Organization
(c) Motivational organization
(d) Biological Organisation

67. An advance form of organizational design is
(a) Simple organization
(b) Divisional Organization
(c) Matrix organization
(d) Functional Organization

68. A description of the organization as a small raft, navigating a raging river is known as
(a) Calm water metaphor
(b) Cool water metaphor
(c) White water rapids
(d) Red water rapids

69. Strategising within frame work is an approach of
(a) Management
(b) Governance
(c) Stake holder
(d) Administration

70. Who had advanced the Stewardship theory of Corporate Governance
(a) J H Davis et., al.
(b) D R Dalton et., al.
(c) D C Hambrick et., al.
(d) None

71. A system with feedback and control components is sometimes called as
(a) Balancing system
(b) Active System
(c) Cybernetic system
(d) Interconnected system

72. Most operational information flows through
(a) Diagonally (b) Vertically
(c) Horizontally (d) Circularly

73. Processed data is called as
(a) Facts
(b) Figures
(c) Information
(d) Knowledge

74. Which stage of decision making process, manager locate the problem?
(a) Design stage
(b) Selection stage
(c) Intelligence stage
(d) Implementation stage

75. Decision Support system is used by
(a) Lower level Manager
(b) Middle level Manager
(c) Higher level Manager
(d) All of them

76. Informal internal knowledge is called
(a) Explicit knowledge
(b) Structured knowledge
(c) External knowledge
(d) Tacit knowledge

77. CAD stands for
(a) Computer Added Design
(b) Computer Analyze Decision
(c) Computer Aided Design
(d) Computer Automatic Design

78. In SECI model of knowledge creation, when tacit knowledge transforms to another in tacit form is called
(a) Socialization (b) Externalization
(c) Combination (d) Internalization

79. In SECI model of knowledge creation, when tacit knowledge transforms to another in explicit way is called
(a) Socialization (b) Externalization
(c) Combination (d) Internalization

80. In SECI model of knowledge creation, when explicit knowledge transforms to another in explicit way is called
(a) Socialization
(b) Externalization
(c) Combination
(d) Internalization

81. In SECI model of knowledge creation, when explicit knowledge transforms to another in tacit way is called
(a) Socialization
(b) Externalization
(c) Combination
(d) Internalization

82. The purpose of stating what an organization wishes to achieve in the long run is called as
(a) Vision (b) Mission
(c) Value (d) Objective

83. A broad goal that reflects purpose, competencies and place of the organisation in the world and that is permanent in nature is called as
(a) Vision (b) Mission
(c) Goal (d) Objective

84. A single use plan that covers a larger set of activities is called as
(a) Project (b) Programme
(c) Policies (d) Procedure

85. General guidelines for decision making is called as
(a) Project (b) Policy
(c) Procedure (d) Rules

86. Statements of specific action to be taken in a given situation is called as
(a) Project
(b) Policy
(c) Procedure
(d) Rules

87. An athlete with a grade point average below 2 cannot be a member of university team is an example of
(a) Project
(b) Policy
(c) Procedure
(d) Rules

88. A detailed set of instructions for performing a sequence of actions that occur often or regularly is called as
(a) Project (b) Policy
(c) Procedure (d) Rules

89. Statements of financial resources set aside for specific activities in a given period of time is called as
(a) Rules (b) Policy
(c) Objectives (d) Budgets

90. Ultimate end results which are to be accomplished by the overall plan over a specified period of time is called as
(a) Rules (b) Policy
(c) Objectives (d) Budgets

91. Find the correct statements
(a) The goals are broad while objectives are specific
(b) Goals are not quantified while objectives are quantified
(c) Goals are more influenced by external environment
(d) All are correct

92. The five forces framework for analyzing the competitive environment developed by
(a) Karl Rogers
(b) Peter Drucker
(c) Michael Porter
(d) None

93. According to Michael Porter, which is not the competitive force
(a) The potential entrants
(b) The substitute products
(c) The credibility of the customer
(d) The bargaining power of suppliers

94. Porter's five force Model is also called as
(a) Competitive Model
(b) Industrial Analysis
(c) Generic Analysis
(d) None

95. 'Value chain Network' a term is given by
(a) Karl Rogers
(b) Peter Drucker
(c) Michael Porter
(d) Philip Kotler

96. In Porter's value chain analysis, which elements come in the group of Primary Activities?
(a) Firm Infrastructure
(b) Procurement
(c) Operation
(d) Human Resource Management

97. In Porter's value chain analysis, that elements come in the group of Support Activities was
(a) Marketing and Sales
(b) Inbound Logistics
(c) Technology Development
(d) Service

98. Three generic competitive strategy is given by
(a) Karl Rogers
(b) Peter Drucker
(c) Michael Porter
(d) Philip Kotler

99. According to Porter's three generic competitive strategy, the firm targeting narrow segment by lowering cost is categorized as
(a) Cost leadership
(b) Differentiation
(c) Cost focus
(d) Focused Differentiation

100. According to Porter's three generic competitive strategy, the firm that is going for niche marketing is categorized in
(a) Cost group
(b) Differentiation group
(c) Focus group
(d) All group

101. The stability strategy in which the company continues at its present rate of development is called as
(a) Harvesting strategy
(b) Stable growth
(c) Holding strategy
(d) Divesting strategy

102. The stability strategy in which firm goes for cost cutting and price increase in its product as it has a dominant market share is called as
(a) Harvesting strategy
(b) Stable growth
(c) Holding strategy
(d) Divesting strategy

103. The stability strategy in which firm's old and obsolete product or technology is being replaced by a new one is called as
(a) Harvesting strategy
(b) Stable growth
(c) Holding strategy
(d) Endgame strategy

104. Product-Market expansion grid is otherwise called as
(a) Porter's five forces model
(b) Porter's Generic Strategy
(c) Ansoff Model
(d) None

105. In Ansoff Model, when current product is targeted to current market then the strategy is called as
(a) Diversification Strategy
(b) Product Development Strategy
(c) Market Penetration Strategy
(d) Market Development Strategy

106. In Ansoff Model, when current product is targeted to new market then the strategy is called as
(a) Diversification Strategy
(b) Product Development Strategy
(c) Market Penetration Strategy
(d) Market Development Strategy

107. In Ansoff Model, when new product is targeted to current market then the strategy is called as
(a) Diversification Strategy
(b) Product Development Strategy
(c) Market Penetration Strategy
(d) Market Development Strategy

108. In Ansoff Model, when new product is targeted to new market then the strategy is called as
(a) Diversification Strategy
(b) Product Development Strategy
(c) Market Penetration Strategy
(d) Market Development Strategy

109. The growth strategy in which two firms, often of about the same size, agree to go forward as a single new company rather than remain separately owned and operated is called as
(a) Acquisition
(b) Merger
(c) Joint Venture
(d) Take over

110. The type of growth strategy in which the parties agree to develop, for a finite time, a new entity and new assets by contributing equity is called as
(a) Acquisition
(b) Merger
(c) Joint Venture
(d) Take over

111. The type of strategy in which two or more organizations share resources, capabilities, or distinctive competencies to pursue some business purpose is called as
(a) Merger
(b) Joint Venture
(c) Strategic Alliance
(d) Long term contracts

112. Merger of two company where there is no common business area is called as
(a) Vertical merger
(b) Product extension merger
(c) Conglomeration
(d) Horizontal Merger

113. An organization goes for international expansion through
(a) Direct Investment
(b) Joint Venture and Exporting
(c) Strategic Alliance and Licensing
(d) All

114. when a firm has a strong competitive position but industry attractiveness is low, then the typeof diversification strategy is called as
(a) Conglomerate diversification
(b) Concentric diversification
(c) Horizontal diversification
(d) None

115. The types of diversification strategy in which the firm may seek new opportunities which have no relation with its current technology, product or market is called as
(a) Conglomerate diversification
(b) Concentric diversification
(c) Horizontal diversification
(d) None

116. The type of diversification strategy in which the firm can develop new products that are technologically unrelated to its current product line and could appear to its current customer is called as
(a) Conglomerate diversification
(b) Concentric diversification
(c) Horizontal diversification
(d) None

117. The types of retrenchment strategy in which firm encounters multiple years of declining financial performance subsequent to a period of prosperity is called as
(a) Survival strategy
(b) Turnaround strategy
(c) Liquidation strategy
(d) Divestment strategy

118. The full form of SPACE matrix is
(a) Strategic Performance and Action Evaluation matrix
(b) Strategic Position and Action Evaluation matrix
(c) Strategic Positioning and Action Elaboration matrix
(d) Strategic Point and Action Experimentation matrix

119. The following personality that is not included as a parameter of strategy evaluation in SPACE matrix is
(a) Conservative
(b) Aggressive
(c) Submissive
(d) Defensive

120. According to SPACE matrix, the fourth quadrant is for
(a) Aggressive (b) Conservative
(c) Competitive (d) Defensive

121. Which two are internal strategic position in SPACE matrix
(a) Financial Position and Growth Potential
(b) Financial Position and Competitive Position
(c) Competitive Position and Stability Position
(d) Stability Position and Industry Position

122. Which two are external strategic position in SPACE matrix?
(a) Financial Position and Growth Potential
(b) Financial Position and Competitive Position
(c) Competitive Position and Stability Position
(d) Stability Position and Industry Position

123. In BCG matrix, the quadrant that show high market share and high market growth is called as
(a) Star
(b) Cash cow
(c) Question mark
(d) Dog

124. In BCG matrix, the quadrant that show high market share and low market growth is called as
(a) Star
(b) Cash cow
(c) Question mark
(d) Dog

125. In BCG matrix, the quadrant that show low market share and high market growth is called as
(a) Star
(b) Cash cow
(c) Question mark
(d) Dog

126. In BCG matrix, the quadrant that show low market share and low market growth is called as
(a) Star
(b) Cash cow
(c) Question mark
(d) Dog

127. Similar to BCG matrix, which matrix is also called as Portfolio Matrix
(a) SPACE Matrix
(b) Grand Strategy Matrix
(c) IE Matrix
(d) Quantitative Strategic Planning Matrix

128. Grand Strategy Matrix is based on the parameter like
(a) Market Share and Market Growth
(b) Market Growth and Competitive Position
(c) Financial Position and Competitive Position
(d) Stability Position and Industry Position

129. Organisational Behaviour is popularized by
(a) F W Taylor (b) Elton Mayo
(c) Keith Davis (d) David Cock

130. Organizational behavior models like autocratic, custodial, supportive and collegial is given by
(a) Frederick Winslow Taylor
(b) Elton Mayo
(c) Keith Davis
(d) David Cock

131. Ability is
(a) Knowledge + Skill
(b) Knowledge – Skill
(c) Knowledge x Skill
(d) Knowledge / Skill

132. Theory of 'need hierarchy' is stated by
(a) Herzberg (b) Maslow
(c) Alderfer (d) McClelland

133. Two factor theory is given by
(a) Herzberg (b) Maslow
(c) Alderfer (d) McClelland

134. Who states that power, affiliation and achievement has a significant role in motivation?
(a) Herzberg (b) Maslow
(c) Alderfer (d) McClelland

135. ERG model of motivation is stated by
(a) Herzberg
(b) Maslow
(c) Alderfer
(d) McClelland

136. Immaturity- Maturity theory is stated by
(a) Blake and Mouton
(b) Bennis
(c) Argyris
(d) Fiedler

137. "Valence and Expectancy" theory of motivation is stated by
(a) Mc Gregor
(b) Vroom
(c) Adam
(d) Alderfer

138. Valence and expectancy relates with motivation in Vroom model as
(a) Motivation = Valence + Expectancy
(b) Motivation = Valence x Expectancy
(c) Motivation = Valence – Expectancy
(d) Motivation = Valence / Expectancy

139. Which need is at the top of the hierarchy according to its priority?
(a) Physiological need
(b) Safety need
(c) Esteem need
(d) Self-actualization need

140. According to Maslow, self-respect, self-confidence, recognition comes in
(a) Safety need
(b) Social need
(c) Self-actualization need
(d) Esteem need

141. 'Motivation-Hygiene' theory is stated by
(a) Abraham Maslow
(b) Frederick Irving Herzberg
(c) Douglas McGregor
(d) Victor H. Vroom

142. That element which is not included in hygienic factor of motivation by Hertzberg is called as
(a) Status
(b) Company policy and Administration
(c) Recognition
(d) Working condition

143. That element which is not included in Motivational factor by Hertzberg is called as
(a) Work itself
(b) Possibility of growth
(c) Interpersonal relationship
(d) Achievement

144. Which need of Maslow was simulates with the Motivational factor by Hertzberg?
(a) Physiological need
(b) Safety need
(c) Social need
(d) Esteem need

145. Alderfer's relatedness need are simulates in which need of Maslow?
(a) Self-actualization
(b) Physiological need
(c) Safety need
(d) Social need

146. Porter and Lawler Models states about
(a) Leadership
(b) Learning Capability
(c) Motivation
(d) Perception

147. Theory X and Theory Y model of motivation is stated by
(a) Abraham Maslow
(b) McGregor
(c) David Mc Clelland
(d) Alderfer

148. The Theory X of McGregor simulates to
(a) Participative leadership
(b) Autocratic leadership
(c) Democratic leadership
(d) Supportive leadership

149. All the responsible and active persons are included in which category by McGregor?
(a) Theory X (b) Theory Y
(c) Both (d) None

150. Theory Z of motivation is stated by
(a) Mc Gregor
(b) Carrot and Stick
(c) Ouchi
(d) Skinner

151. Which theory of motivation places a premium on the effort-reward relationship?
(a) Vroom's (b) McClelland's
(c) Equity (d) Hertzberg

152. Transactional Analysis is a model for knowing the quality of
(a) Motivation (b) Leadership
(c) Personality (d) Attitude

153. Transactional Analysis model was established by
(a) Peter Drucker
(b) Eric Berne
(c) McClelland
(d) Joseph Luft and Harrington Ingham

154. In Transactional Analysis the 'I am Ok, you are not OK' life position is also referred to as
(a) Bossing
(b) Avoidant
(c) Diffident
(d) Bohemian

155. Concept of Johari Window for personality evaluation is stated by
(a) Peter Drucker
(b) Eric Berne
(c) McClelland
(d) Joseph Luft and Harrington Ingham

156. In Johari window, the quadrant where information known to self but information is not known to others is found, is called as
(a) Open (b) Hidden
(c) Blind (d) Unknown

157. In Johari window, the quadrant where information known to self and information is also known to others is found, is called as
(a) Open (b) Hidden
(c) Blind (d) Unknown

158. In Johari window, the quadrant where information not known to self but information is known to others is found, is called as
(a) Open (b) Hidden
(c) Blind (d) Unknown

159. In Johari window, the quadrant where information not known to self and information is also not known to others is found, is called as
(a) Open (b) Hidden
(c) Blind (d) Unknown

160. Which types of ego states are not found in the society?
(a) Parent ego (b) Child ego
(c) Youth ego (d) Adult ego

161. The ego stage is based on rationality is included in
(a) Parent ego
(b) Child ego
(c) Adult ego
(d) All

162. According to ego stage, which types of child ego is not taken into account
(a) Natural child
(b) Adaptive child
(c) Rebellion Child
(d) Sympathetic Child

163. The ideal complementary transaction is
(a) Adult-Parent transaction
(b) Adult-Adult transaction
(c) Parent-Parent transaction
(d) Child-Adult transaction

164. When the communication has double meaning then the transaction is called as
(a) Complementary transaction
(b) Non complementary transaction
(c) Ulterior Transaction
(d) None

165. Any act implying recognition of another's presence is called as
(a) Transactions
(b) Life script
(c) Stroking
(d) Psychological Games

166. The group in which subordinates report to a common superior is called as
(a) Primary group
(b) Reference group
(c) Command group
(d) Secondary group

167. The group to which an indivisual likes to belong is
(a) Secondary group
(b) Reference group
(c) Task group
(d) Out-group

168. The first stage of group development in which indivisual is very cautious is called as
(a) Storming
(b) Forming
(c) Norming
(d) Adjourning

169. The second stage of group development in which indivisual experience a varying degree of tension is called as
(a) Storming
(b) Forming
(c) Norming
(d) Adjourning

170. End phase of group development stage is called as
(a) Performing
(b) Adjourning
(c) Norming
(d) Storming

171. Homans Interaction Theory of group formation states that
(a) Indivisual affiliate one another because of spatial proximity
(b) The more activities person share, more will be the interaction
(c) Similar attitude towards a common object, basis of group formation
(d) Group formation is on the basis of social exchange

172. Propinquity theory of group formation states that
(a) Indivisual affiliate one another because of spatial proximity
(b) The more activities person share, more will be the interaction
(c) Similar attitude towards a common object, basis of group formation
(d) Group formation is on the basis of social exchange

173. Exchange theory of group formation states that
(a) Indivisual affiliate one another because of spatial proximity
(b) The more activities person share, more will be the interaction
(c) Similar attitude towards a common object, basis of group formation
(d) Group formation is on the basis of social exchange

174. A team that meets electronically is called as
(a) Quality Circle
(b) Functional team
(c) Virtual team
(d) Self-managed work team

175. The person who establishes the concept of brain storming is
(a) Louis Kelso (b) Kaplan
(c) Osborn (d) J L Marino

176. The group decision making process in which quantity of ideas are more emphasized and quality follows later is
(a) Delphi technique
(b) Nominal group technique
(c) Brain storming
(d) Dialectic Decision method

177. A creativity technique based on 'the fitting together the diverse elements' principle for the generation of alternative solutions is referred to as
(a) Synectics
(b) Brain-storming
(c) Nominal grouping
(d) Creative thinking

178. The concept of synectics is developed by
(a) Alex F. Osborn
(b) J. J. Gordon
(c) Andra Delbecq & Andrew van de ven
(d) Rand corporation

179. The concept of Brain Storming is developed by
(a) Alex F. Osborn
(b) J. J. Gordon
(c) Andra Delbecq & Andrew van de ven
(d) Rand corporation

180. The concept of Nominal Group Technique is developed by
(a) Alex F. Osborn
(b) J. J. Gordon
(c) Andra Delbecq & Andrew van de ven
(d) Rand corporation

181. The concept 'Delphi Technique' is developed by
(a) Alex F. Osborn
(b) J. J. Gordon
(c) Andra Delbecq & Andrew van de ven
(d) Rand corporation

182. The group decision making in which structured ideas and restricted verbal communication is there is
(a) Delphi technique
(b) Nominal group technique
(c) Brain storming
(d) Dialectic Decision method

183. The group decision making process in which mainly subjective collection of data are occurred through mail is
(a) Delphi technique
(b) Nominal group technique
(c) Brain storming
(d) Dialectic Decision method

184. The group decision making in which only two alternative has been proposed is
(a) Delphi technique
(b) Nominal group technique
(c) Brain storming
(d) Dialectic Decision method

185. If mental age is 50, Chronological age is 40 then Intelligence quotient is
(a) 100 (b) 125
(c) 150 (d) 200

186. Latin word of communication is
(a) De commune (b) La commis
(c) Communis (d) Commecent

187. The formal communication model in which an indivisual is allowed to communicate with his immediate superior as well as his superior's superior is
(a) Wheel model
(b) Free flow model
(c) Inverted V Model
(d) Circular model

188. Which types of communication channel are not formal?
(a) Wheel type (b) Inverted V type
(c) Cluster type (d) Circular type

189. The informal communication channels in which an indivisual communicates non selectively is
(a) Single Strand
(b) Gossip
(c) Probability
(d) Cluster

190. The informal communication channel in which the indivisual communicates with only those indivisuals whom he trusts is
(a) Single Strand
(b) Gossip
(c) Probability
(d) Cluster

191. Grapevine channel of communication is also called as
(a) Formal communication
(b) Informal communication
(c) Both formal and informal
(d) None

192. Type of grapevine information which is communicated without any standard of evidence being present is called as
(a) Gossip (b) Noise
(c) Jargon (d) Rumour

193. An additional idea accompanying every idea that is expressed is called as
(a) Gossip
(b) Meta communication
(c) Jargon
(d) Rumour

194. Unintelligible or meaningless talk or writing, familiar only to a group of person is called as
(a) Gossip (b) Noise
(c) Jargon (d) Rumour

195. The person who passes information to someone is
(a) Liaisons (b) Gatekeeper
(c) Isolates (d) Cosmopolites

196. An indivisual who serves as a communication link between groups but is not a member of either group is known as
(a) Liaisons (b) Gatekeeper
(c) Isolates (d) Cosmopolites

197. Someone who has very little or no conflict with other members of the organisation is called as
(a) Liaisons (b) Gatekeeper
(c) Isolates (d) Cosmopolites

198. A person whose communication network frequently extends into the organisation's external environment is called as
(a) Liaisons
(b) Gatekeeper
(c) Isolates
(d) Cosmopolites

199. The tendency of categorizing people into a single class on the basis of some trait is called as
(a) Projection
(b) Halo effect
(c) Stereotyping
(d) Meta communication

200. A perceptual process by which we try to attribute our own thoughts and feeling to others is called as
(a) Projection
(b) Halo effect
(c) Stereotyping
(d) Meta communication

201. The process of forming opinions based on one element from a group elements, and generalizing that perceptions to all other elements is called as
(a) Projection
(b) Halo effect
(c) Stereotyping
(d) Meta communication

202. The obstruction in communication caused due to differences in indivisual's interpretation of words and symbols, is known as
(a) Psychological barrier
(b) Social barrier
(c) Organisational barrier
(d) Semantic barrier

203. The Balance theory of group formation states that
(a) Indivisual affiliate one another because of spatial proximity
(b) The more activities person share, more will be the interaction
(c) Similar attitude towards a common object, basis of group formation
(d) Group formation is on the basis of social exchange

204. The conditioned learning theory was propounded by
(a) Thiradike
(b) Maslow
(c) Pavlov
(d) Luther Gullick

205. Where anchors are illustration that helps to define as superior, average or below average, behavior of the employee, the rating scale is known as
(a) Behavioural expectation scale
(b) Behavioural observation scale
(c) Forced choice description
(d) Grading scale

206. Joseff Tiffin established a rating scale that mainly used in military is
(a) Behavioural expectation scale
(b) Behavioural observation scale
(c) Forced choice description
(d) Grading scale

207. Leaders who provide indivisualised consideration and intellectual stimulation and who possess a charisma is known as
(a) Transactional leader
(b) Transformational leader
(c) Situational leader
(d) Visionary leader

208. The drive to become what one is
(a) Self-awareness
(b) Self-actualization
(c) Self-efficiency
(d) Self esteem

209. 20+ need theory for personality judgment is stated by
(a) Eysenck
(b) Raymond Cattel
(c) Murray
(d) Costa and Murree

210. 16 trait theory for personality judgment is stated
(a) Eysenck
(b) Raymond Cattel
(c) Murray
(d) Costa and Murree

211. 3 traits theory for personality judgment is stated by
(a) Eysenck
(b) Raymond Cattel
(c) Murray
(d) Costa and Murree

212. 5 traits theory for personality judgment is stated by
(a) Eysenck
(b) Raymond Cattel
(c) Murray
(d) Costa and Murree

213. "Organisational learning" and "Learning organization" are
(a) Same
(b) Different
(c) Reverse to each other
(d) Complementary to each other

214. 'A leader is born and is not made' is the concept of
(a) Trait theory of leadership
(b) Behaviourial theory of leadership
(c) Charismatic theory of leadership
(d) Situational theory of leadership

215. "FIRM is a responsibility of all those who manage people as well as bring a description of the work of those who are employed as specialists. It is that part of management which is concerned with people at work and with their relationships within an enterprise. It applies not only to industry and commerce but to all fields of employment." Known as ________.
(a) Performance Management
(b) Human Resource Management
(c) Personnel Management
(d) Functional Management

216. 'Leadership is shown by a person's acts than by his traits' is the concept of
(a) Trait theory of leadership
(b) Behaviourial theory of leadership
(c) Charismatic theory of leadership
(d) Situational theory of leadership

217. 'Leadership is acquired through learning and experience' is the concept of
(a) Trait theory of leadership
(b) Behaviourial theory of leadership
(c) Charismatic theory of leadership
(d) Situational theory of leadership

218. 'Contingency theory of leadership' is also called as
(a) Trait theory of leadership
(b) Behaviourial theory of leadership
(c) Charismatic theory of leadership
(d) Situational theory of leadership

219. The leadership style in which leader completely gives freedom to subordinates in decision making is called as
(a) Benevolent Autocratic Leadership
(b) Participative Leadership
(c) Strict Autocratic Leadership
(d) Free rein Leadership

220. The leadership style in which leader decentralizes his decision making process is called as
(a) Benevolent Autocratic Leadership
(b) Participative Leadership
(c) Strict Autocratic Leadership
(d) Free rein Leadership

221. The leadership style in which centralized decision making is there but motivation style is positive is called as
(a) Benevolent Autocratic Leadership
(b) Strict Autocratic Leadership
(c) Participative Leadership
(d) Free Rein Leadership

222. Exploitative autocratic leadership style is seen in
(a) Managerial grid System
(b) Employee- Production Orientation System
(c) Likert's Management System
(d) Tridimensional Grid System

223. Employee-Production Orientation model of leadership determination is a
(a) Attitude concerned model
(b) Behaviour concerned model
(c) Relationship orientation model
(d) Task orientation model

224. Managerial grid type of leadership determination model is given by
(a) Renis Likert (b) Reddin
(c) Fiedler (d) Blake and Mouton

225. Who developed the Two-dimensional Managerial Grid?
(a) Blake and Mouton
(b) McCarthy
(c) Howthrone
(d) W A Taylor

226. What is the most effective attribute of an effective leader?
(a) Behaviour (b) Character
(c) Drive (d) Influence

227. Tridimensional Grid model of leadership determination is given by
(a) Renis Likert (b) Reddin
(c) Fiedler (d) Robert House Evan

228. The concept of task oriented and people oriented leadership was developed by
(a) John Berry
(b) Maxwell
(c) Blake and Mouton
(d) Hertzberg

229. Which of the following is not considered as a leadership style?
(a) Task oriented leadership style
(b) Society oriented leadership style
(c) Employee oriented leadership style
(d) Autocratic leadership style

230. The condition of objective incompatibility between values and goals is called as
(a) Altercation
(b) Difference
(c) Conflict
(d) Counter action

231. The conflict episode in which conflict is in the sub conscious mind is called as
(a) Manifest conflict
(b) Felt conflict
(c) Perceived conflict
(d) Latent conflict

232. Conflict behavior like sabotage, aggression, withdrawal is observed in
(a) Manifest conflict
(b) Felt conflict
(c) Perceived conflict
(d) Latent conflict

233. The conflict resulting into anxiety, tension frustration or hostility due to one's emotional involvement, may be termed as
(a) Perceived conflict
(b) Felt conflict
(c) Manifest conflict
(d) Conflict aftermath

234. "Conflict has some positive impact in a group' is the concept of
(a) Traditional theorists
(b) Human Relation theorists
(c) Interactionist theorists
(d) None

235. Approach-Avoidance conflict is a type of
(a) Role conflict
(b) Goal Conflict
(c) Vertical conflict
(d) Intra group conflict

236. The type of goal conflict is very relevant in organizational behaviour
(a) Approach-Approach conflict
(b) Approach-Avoidance conflict
(c) Avoidance–Avoidance conflict
(d) All

237. Types of role conflict, in which role requirements violate the needs, values or capacities of the person is called as
(a) Intra sender role conflict
(b) Inter sender role conflict
(c) Inter role conflict
(d) Role-self conflict

238. Feeling arising out of inadequate definition of roles is known as
(a) Role ambiguity
(b) Role perception
(c) Role conflict
(d) Role expectation

239. The process of conflict resolution in which emphasis is given to common interests that exist between indivisual or groups is
(a) Avoidance
(b) Compromise
(c) Confrontation
(d) Smoothing

240. The negotiation process in which gain to one party is similar amount of loss to another is
(a) Integrative Bargaining
(b) Collective Bargaining
(c) Distributive Bargaining
(d) None

241. The syndrome wherein a person breaks down physically and emotionally due to continuous over-work over a long period of time is called as
(a) Rust out (b) Karoshi
(c) Burn out (d) Burst Out

242. When a person is chronically under-worked and his skills are under utilized, syndrome is called as
(a) Rust out (b) Karoshi
(c) Burn out (d) Burst Out

243. A change touches a sequence of related and supporting changes is called as
(a) Chain effect (b) Domino effect
(c) Both (d) None

244. Any strong stimulus that impels action is called as
(a) Cue (b) Drive
(c) Response (d) Reinforcement

245. Any objects existing in the environment as perceived by the indivisual is called as
(a) Cue
(b) Drive
(c) Generalisation
(d) None

246. Operant Conditioning implies that behaviour is
(a) Involuntary in nature
(b) Voluntary in nature
(c) Both
(d) None

247. The conditioned learning theory was propounded by
(a) Thoradike
(b) Abraham Maslow
(c) Ivan Pevlov
(d) Luther Gullick

248. Anything that increases the strength of response and tends to induce repetitions of the behavior is called as
(a) Drive
(b) Conditioning
(c) Response
(d) Reinforcement

249. In an organisation point of view, loss of memory is called as
(a) Amnesia
(b) Extinction
(c) Implicitness
(d) Spontaneous recovery

250. Person's tendency to perceive familiar stimuli rather than unfamiliar one is called as
(a) Response silence
(b) Response disposition
(c) Perceptual defence
(d) Maintenance of cognitive consistency

251. Set of predispositions which are determined not by the familiarity of the stimulus situations but by the person's own cognitive predisposition is called as
(a) Response silence
(b) Response disposition
(c) Perceptual defence
(d) Maintenance of cognitive consistency

252. Screening of those elements which create conflict and threatening situation in people is called as
(a) Response silence
(b) Response disposition
(c) Perceptual defence
(d) Maintenance of cognitive consistency

253. A predisposition to behave in a certain way and to take particular point of view
(a) Motivation (b) Behaviour
(c) Perception (d) Attitude

254. Sometimes your behavior result in a congruent response from other person, such behaviour is referred to as
(a) Perceptual defense
(b) Misperception
(c) Perceptual bias
(d) Response evocation

255. When faced with incomplete information, people fill up the gaps themselves to make the information meaningful is called as
(a) Figure-Ground (b) Grouping
(c) Simplification (d) Closure

256. The tendency one has to react in a certain way to a given situation is called as
(a) Mental set
(b) Halo effect
(c) Stereotyping
(d) Attribution

257. The process by which indivisuals interpret events as being caused by particular aspects in the setting around them is called as
(a) Mental set
(b) Halo effect
(c) Stereotyping
(d) Attribution

258. The way individuals define themselves as who they are and derive their sense of identity is called as
(a) Self esteem
(b) Self-concept
(c) Machiavellianism
(d) Locus of control

259. Manipulation of others as a primary way of achieving one's goals is called as
(a) Self esteem
(b) Self-concept
(c) Machiavellianism
(d) Locus of control

260. The person who is always showing time urgency and impatient when things are delay, have
(a) Type A personality
(b) Type B personality
(c) External locus personality
(d) Internal locus personality

261. Easy going person donot have time urgency, they have
(a) Type A personality
(b) Type B personality
(c) External locus personality
(d) Internal locus personality

262. A personality characteristic indicating one's willingness to do whatever it takes to get one's own way is known as
(a) Locus of control
(b) Machiavellism
(c) Job characteristic Model
(d) Personality legitimacy

263. Personality theory that states that man is motivated more by unseen forces than he is controlled by conscious and rational thoughts, is
(a) Psychoanalytic Theory
(b) Trait Theory
(c) Socio-psychological Theory
(d) Self-Theory

264. 'The Id', 'The ego' and 'The super ego' comes under
(a) Psychoanalytic Theory
(b) Trait Theory
(c) Socio-psychological Theory
(d) Self-Theory

265. Psychoanalytic Theory of personality is stated by
(a) Cattell
(b) Carl Roger
(c) Adler
(d) Freud

266. Who has not come under the contributor of socio-psychological theory?
(a) Adler
(b) Cattell
(c) Horney
(d) Sullivan

267. The theory that reveals the quantitative approach to the study of personality is
(a) Psychoanalytic Theory
(b) Trait Theory
(c) Socio-psychological Theory
(d) Self-Theory

268. Organismic or field theory of personality is also called as
(a) Trait Theory
(b) Socio-psychological Theory
(c) Psychoanalytic Theory
(d) Self-Theory

269. 'The perception of a person about how others are perceiving his qualities and characteristics' is
(a) Real self
(b) Looking Glass-self
(c) Ideal-self
(d) Self-image

270. In which of the following inter personal styles, people are critical of other's behavior, develop rules and regulations and impose them on others
(a) Rescuing style
(b) Normative style
(c) Prescriptive style
(d) Supportive style

271. When the result of the group cannot be attributed to any person, the relationship between an indivisual's input and the group's output is clouded, is called as
(a) Cognitive dissonance
(b) Organizational citizenship
(c) Social loafing
(d) Organizational climate

272. Five stages of adoption process is given by
(a) Peter Drucker
(b) Rogers
(c) F. W. Taylor
(d) Porter

273. According to Rogers, the most difficult stage of adoption process is
(a) Knowledge
(b) Persuation
(c) Decision
(d) Implementation

274. The change related phenomenon resisted by the organisations in order to maintain an equilibrium, is known as
(a) Conformity to norms
(b) Insecurity and regressions
(c) Homoeostasis
(d) Self-distrust

275. Who is the third person acting as a catalyst, whose assets in the achievement of the smooth introduction and implementation of change
(a) Change Agent
(b) Change Supervisor
(c) Change Manager
(d) Change Player

276. Those function in an organization, which are perceived to be directly contributing to the organization's objectives, are known as
(a) Staff function
(b) Corporate function
(c) Integration function
(d) Line function

277. An organizational memory includes the following except
(a) Information that employees possess
(b) Knowledge embedded in the organisation's systems and structures
(c) Documents
(d) Knowledge that is not documented

278. Which pattern reflects a pure executive form of management
(a) Functional
(b) Line
(c) Line and staff
(d) Committee

279. The amount of satisfaction an individual anticipates from a particular outcome is known as
(a) Theory Y
(b) Valence
(c) Theory Z
(d) Stress

280. In which of the following processes a person reject his own feelings about the other person
(a) Self-analysis (b) Projection
(c) Empathy (d) Denial

281. In the decision tree model of vroom and yetton the style when the leaders share the problem with subordinates as a group, obtains collective ideas, and then makes the decision is defined as
(a) A I (b) A II
(c) C I (d) C II

282. Which of the following is not one of the career anchors?
(a) Managerial competency
(b) Personal competency
(c) Security
(d) Technical-functional competency

283. Which is the oldest and simplest form of organization
(a) Functional organisation
(b) Committee organisation
(c) Line organisation
(d) Line and Staff organisation

284. The process of forecasting a firm's future demand for, and supply of the right type of people in the right number is called as
(a) Human resource management
(b) Human resource development
(c) Human resource planning
(d) Personnel management

285. What is the exclusiveness of Human Resource Management over Personnel Management
(a) Administration Activity
(b) Operational Activity
(c) Strategic Activity
(d) All

286. Which pay is one of the most crucial pay given to the employee & also shown in the pay structure?
(a) Performance
(b) Strategic
(c) Bonus
(d) Commission

287. When jobs are becoming more amorphous and more difficult to define, is called as
(a) Job lysis
(b) De jobbed
(c) Job desection
(d) Job fragmentation

288. When title, tasks, duties and responsibilities involved in a job, it is called as
(a) Job analysis
(b) Job description
(c) Job specification
(d) Job rotation

289. The listing of employee skills, abilities and qualification for a job is called as
(a) Job analysis
(b) Job description
(c) Job specification
(d) Job evaluation

290. Assessment of actual performance of an employee is called as
(a) Performance evaluation
(b) Performance analysis
(c) Performance appraisal
(d) Performance negotiation

291. The logical sequence of job analysis is called as
(a) Job enrichment
(b) Job design
(c) Job engineering
(d) Departmentation

292. The process of moving of employees from job to job to add variety and reduce boredom is called as
(a) Job engineering
(b) Job rotation
(c) Job enlargement
(d) Job enrichment

293. When organization gives more decision making, planning and controlling power to its employee is called as
(a) Job engineering
(b) Job rotation
(c) Job enlargement
(d) Job enrichment

294. The selection test in which organisation takes handwriting as a parameter of judgment is called as
(a) Interest test
(b) Polygraph test
(c) Graphology test
(d) Medical test

295. The art of judging, character of a person from appearance specially the face based on the assumption that face is an index of mind, is known as
(a) Graphology
(b) Phrenology
(c) Physiognomy
(d) None of the above

296. Standardisation of the procedure of administering and scoring the test result in selection procedure is called as
(a) Validity
(b) Reliability
(c) Synthesis
(d) Predictability

297. More formal interview structure in which interviewers may be two or more is called
(a) One-to-one interview
(b) Sequential interview
(c) Panel interview
(d) Personal interview

298. The use of predesigned set of questions used in an interview process leads to
(a) Behavioural Interview
(b) Non-directional Interview
(c) Situational Interview
(d) Structured Interview

299. When new hires do not have predecessors available to guide them or to model their behavior upon, the orientation is called as
(a) Serial orientation
(b) Collective orientation
(c) Formal orientation
(d) Disjunctive orientation

300. Orientation procedure in which organisation try to dismantle the old mind set of the new member and try to inculcate values or norms is called as
(a) Serial orientation
(b) Disjunctive orientation
(c) Investiture orientation
(d) Divestiture orientation

301. Training methods which come under 'on the job' category are
(a) Coaching, Assistantship
(b) Job rotation
(c) Internship, Apprentice
(d) All

302. Training methods which come under 'off the job' category are
(a) Vestibule, Lecture
(b) Simulation, Role Playing
(c) Laboratory training, Conference
(d) All

303. For Sensitivity training, trainees should not be more than
(a) 20
(b) 15
(c) 12
(d) 30

304. T-group training is a type of
(a) Vestibule training
(b) Simulation type training
(c) Sensitivity training
(d) None

305. Problem of performance appraisal method in which the marks reflect on the degree of closeness between rater and ratee
(a) Recency effect
(b) Rater effect
(c) Primary effect
(d) Perceptual method

306. The condition that refers to allowing past performance appraisal ratings to unjustifiably influence current ratings is called as
(a) Staus effect
(b) Recency effect
(c) Spillover effect
(d) Rater effect

307. Assessment Centres is a
(a) Place
(b) Technique
(c) Both
(d) None

308. The systematic collection of performance data on an indivisual or group, derived from a number of stakeholders' is called as
(a) 180-Degree feedback
(b) 360-Degree feedback
(c) 90-Degree feedback
(d) 270-Degree feedback

309. Examples of incentives plan where earnings vary less proportionately than output is
(a) Bedaux Plan
(b) Rowan Plan
(c) Barth scheme
(d) All

310. In Taylor's differential piece rate scheme
(a) Earning vary in the same proportion as output
(b) Earnings vary less proportionately than output
(c) Earnings vary proportionately more than output
(d) Earnings differ at different levels of output

311. In which incentives scheme, earnings differ at different levels of output
(a) Merrick Differential Piece Rate scheme
(b) Gantt task system scheme
(c) Emerson's efficiency plan
(d) All

312. Incentives scheme in Indian industry came in
(a) 1940 (b) 1945
(c) 1946 (d) 1950

313. Methylisocyanide gas leak tragedy is related with which city
(a) New Delhi (b) Bhopal
(c) Rourkela (d) Panipat

314. Oleum gas leak is connected with which company
(a) Chambal fertilizer
(b) Monsanto India Limited
(c) Britannia Industries Limited
(d) Sriram Foods and Fertilizer Industries

315. What comes under the voluntary separation of employee?
(a) Layoff (b) VRS
(c) Quits (d) Discharge

316. Discharge is also called as
(a) Quits (b) VRS
(c) Layout (d) Termination

317. Temporary separation of employee at the instance of the employer is called as
(a) Layout (b) Lay Off
(c) Discharge (d) VRS

318. A situation where management is unable to provide employment due to non availability of work is called
(a) Lock out (b) Lay off
(c) Termination (d) Closure

319. The termination of the employment at the instance of the employee is called as
(a) Quits (b) Retirement
(c) VRS (d) Resignation

320. The process of reducing the size of workforce or increasing it to maintain the employee strength at most desired level is called as
(a) Down sizing
(b) Up sizing
(c) Right sizing
(d) Organisation reshuffling

321. Dramatic reduction of manpower is called
(a) Termination (b) Downsizing
(c) Rightsizing (d) Lay off

322. Horizontal movement employee without a change in responsibility is called as
(a) Job description (b) Job rotation
(c) Transfer (d) Promotion

323. What is true in case of promotion?
(a) It is the upward movement of employees
(b) It carries increased responsibility and enhanced pay
(c) It is a tool to motivate people towards enhanced productivity
(d) All

324. Bonded labour system (Abolition) Act came in
(a) 1952 (b) 1948
(c) 1976 (d) 1986

325. Child labour (Prohibition and Regulation) Act was enacted in
(a) 1976 (b) 1986
(c) 1960 (d) 1991

326. Contract labour (Regulation and Abolition) Act came in
(a) 1970 (b) 1965
(c) 1972 (d) 1948

327. Employee State Insurance Act came in
(a) 1936 (b) 1948
(c) 1970 (d) 1985

328. Employee's provident fund Act was enacted in
(a) 1948 (b) 1952
(c) 1955 (d) 1960

329. Payment Wages Act came in
(a) 1936 (b) 1948
(c) 1976 (d) 1986

330. Minimum wages Act came in
(a) 1936 (b) 1948
(c) 1976 (d) 1986

331. Equal Remuneration Act came in
(a) 1936 (b) 1948
(c) 1976 (d) 1986

332. Factories Act came in
(a) 1936 (b) 1947
(c) 1948 (d) 1952

333. Industrial Dispute Act came in
(a) 1936 (b) 1947
(c) 1948 (d) 1952

334. Bonus Act came in
(a) 1965 (b) 1952
(c) 1970 (d) 1986

335. Gratuity Act came in
(a) 1975 (b) 1972
(c) 1970 (d) 1986

336. Trade Union Act came in
(a) 1936 (b) 1947
(c) 1955 (d) 1926

337. Workmen's compensation Act came in
(a) 1923 (b) 1936
(c) 1928 (d) 1951

338. National and festival holidays Act came in
(a) 1951 (b) 1955
(c) 1958 (d) 1965

339. Maternity Benefit Act came in
(a) 1951 (b) 1955
(c) 1958 (d) 1961

340. Interstate Migrant Workman Act came in
(a) 1970 (b) 1979
(c) 1985 (d) 1991

341. Dock workers (Regulation of Employment.) Act came in
(a) 1936 (b) 1948
(c) 1972 (d) 1986

342. According to Factories Act, 1948, what should be the minimum number of women in a factory for getting crèche facility?
(a) 50 (b) 30
(c) 25 (d) 20

343. According to Mines Act, 1952, what should be the minimum number of women in a factory for getting crèche facility?
(a) 50 (b) 30
(c) 25 (d) 20

344. According to Plantation Labour Act, 1951, what should be the minimum number of women in a factory for getting crèche facility?
(a) 50 (b) 30
(c) 25 (d) 20

345. According to Factories Act, 1948, what should be the minimum number of workers for getting canteen facility?
(a) 150 (b) 200
(c) 250 (d) 500

346. According to Factories Act, 1948, what should be the minimum number of workers for getting ambulance facility?
(a) 500 (b) 1000
(c) 1500 (d) 2000

347. According to Factories Act, 1948, what should be the minimum number of workers for getting first aid facility?
(a) 150 (b) 200
(c) 250 (d) 500

348. According to Mines Act, 1952, what should be the minimum number of workers for getting canteen facility?
(a) 150 (b) 200
(c) 250 (d) 300

349. According to Mines Act, 1952, what should be the minimum number of workers for an appointment of welfare officer
(a) 250 (b) 500
(c) 1000 (d) 2000

350. Under Factory Act, 1948, the maximum time limit for adult to work in a factory in a week is
(a) 40 hrs (b) 48 hrs
(c) 50 hrs (d) 45 hrs

351. Under Factory Act, 1948, the maximum time period for an adult to work in a day is
(a) 6 hrs (b) 8 hrs
(c) 9 hrs (d) 10 hrs

352. The incentives come under fringe benefit category are
(a) Provident Fund, Gratuity
(b) Medical Care, Accidental Relief
(c) Recreation, Health and Group Insurance
(d) All

353. The elements that have direct impact on remuneration
(a) Performance, Seniority
(b) Experience, Potentiality
(c) Working Ability
(d) All

354. Safety Week is observed in between
(a) 1st to 7th July
(b) 2nd to 8th October
(c) 1st to 7th March
(d) 1st to 7th December

355. White lung disease is occurred to the labour community is due to
(a) Coal (b) Asbestos
(c) Cotton (d) Iron

356. Black Lung disease is occurred to the labour community is due to
(a) Coal (b) Asbestos
(c) Cotton (d) Iron

357. Brown Lung disease is occurred to the labour community is due to
(a) Coal (b) Asbestos
(c) Cotton (d) Iron

358. Strike in which union try to paralyse the industry is called as
(a) Sectional Strike
(b) Bumper Strike
(c) General Strike
(d) Sympathetic Strike

359. Strike in which workers cease to performance their duties but do not leave the place of work is called as
(a) Sit Down Strike
(b) Bumper Strike
(c) General Strike
(d) Sympathetic Strike

360. Wild cat strike is otherwise called as
(a) Pen down Strike
(b) Slow Down Strike
(c) Lightning Strike
(d) Bumper Strike

361. The counterpart of Strike that employer puts against the union / worker is called as
(a) Toll down Strike (b) Gherao
(c) Lock Out (d) Lay Off

362. When the group of workers prevent the management to go out side the office premises is called as
(a) Closure (b) Gherao
(c) Lock Out (d) Lay Off

363. When workers motivates other to join in the strike through display signs, banners and placards is called as
(a) Closure (b) Picketing
(c) Layoff (d) Lock out

364. When the employer suspends all the transaction of his business and shuts down the place of business is called as
(a) Picketing (b) Layoff
(c) Lock out (d) Closure

365. Compulsory arbitration is called as
(a) Jurisdiction (b) Adjudication
(c) Conciliation (d) None

366. Dispute settlement Procedure in which a neutral third party studies the bargaining situation and listen to both the parties is called as
(a) Code of discipline
(b) Arbitration
(c) Conciliation
(d) Adjudication

367. Dispute settlement Procedure in which representatives of workers and employees are brought together before a third party is called as
(a) Collective Bargaining
(b) Arbitration
(c) Conciliation
(d) Adjudication

368. Trade Union Act came in
(a) 1905 (b) 1920
(c) 1926 (d) 1931

369. All India Trade Union Congress came in
(a) 1905 (b) 1920
(c) 1926 (d) 1931

370. Indian National Trade Union Congress came in
(a) 1935 (b) 1947
(c) 1948 (d) 1949

371. Hind Mazdoor Sabha came in
(a) 1935 (b) 1947
(c) 1948 (d) 1949

372. United Trade Union Congress came in
(a) 1935 (b) 1947
(c) 1948 (d) 1949

373. United Trade Union Centre came in
(a) 1920 (b) 1948
(c) 1970 (d) 1955

374. Centre of Indian Trade Union Came in
(a) 1920 (b) 1948
(c) 1970 (d) 1955

375. Bharatiya Mazdoor Sangha came in the year
(a) 1920 (b) 1948
(c) 1970 (d) 1955

376. Centre of Indian Trade Union is under the guidance / control of
(a) SUCI (b) Congress Party
(c) BJP Party (d) CPI (M.)

377. United Trade Union Congress is under the guidance / control of
(a) Congress (b) RSP
(c) CPI (d) BJP

378. All India Trade Union Congress is under the guidance / control of
(a) CPI (b) BJP
(c) SUCI (d) CPI (M.)

379. Bharatiya Mazdoor Sangha is under the guidance / control of
(a) SUCI (b) Congress Party
(c) BJP Party (d) CPI (M.)

380. Indian National Trade Union Congress is under the guidance / control of
(a) SUCI (b) Congress Party
(c) BJP Party (d) CPI (M.)

381. United Trade Union Centre is under the guidance / control of
(a) SUCI (b) Congress Party
(c) BJP Party (d) CPI (M.)

382. Change in condition of service of a worker comes under
(a) Section 9 B of the ID Act
(b) Section 9 A of the ID Act
(c) Section 7 B of the ID Act
(d) Section 10 A of the ID Act

383. Robert Keplan is associated with
(a) A Benchmark Practice
(b) Balance Score card
(c) Kaizen
(d) Quality of Life

384. The book "Human Behaviour at Work" was written by
(a) Mc Gregor (b) Abraham Maslow
(c) Herzberg (d) Keith Davis

385. The father of 'Employees State Insurance' Scheme in India is
(a) B. P. Wadia (b) B. P. Adarkar
(c) V. V. Giri (d) N. M. Lokhande

386. Graphical representation of manpower in the organization is known as
(a) Replacement Chart
(b) Placement Chart
(c) Organisational Structure
(d) Manpower Inventory

387. The process of retaining suitable successor is called as
(a) Alternative Staffing
(b) Succession Planning
(c) Position Replacement
(d) Promotion

388. A wilful disobedience of authority is called as
(a) Insubordination
(b) Wilful disobedience
(c) Indiscipline
(d) Misconduct

389. Interview conducted for knowing why an employee is leaving the organisation is called as
(a) Exit Interview
(b) Separation Interview
(c) Termination Interview
(d) Retrenchment Interview

390. Fair allowance for men fatigue can be
(a) 2% for men (b) 4% for men
(c) 6% for men (d) 10% for men

391. Fair allowance for women fatigue can be
(a) 6% for women (b) 6.5% for women
(c) 9% for women (d) 10% for women

392. Minimum Allowance for men and women for eight hour a day is
(a) 4% for men and 6.5% for women
(b) 8% for men and 12% for women
(c) 10% for men and 12.5% for women
(d) 10% for men and 10% for women

393. The Employee Stock Ownership Plan (ESOP.) was developed by
(a) Michael Porter
(b) Kaplan
(c) P F Drucker
(d) Louis Kelso

394. When monotony in work is reduced by giving a wider variety of duties to employees, this is known as
(a) Job rotation
(b) Job redesign
(c) Job enlargement
(d) Job enrichment

395. The book 'HR Champions' is authored by
(a) Mintzberg (b) J Pfeffer
(c) M Porter (d) Dave Ulrich

396. Out Placement Services are provided to employees who have been
(a) Dismissed
(b) Retired
(c) Render Surplus
(d) Given Voluntary Retirement

397. What notice period is required for going on strike/lockout in public utility services
(a) One week (b) Three week
(c) Six week (d) Nine week

398. Modifying methods of performing a job to reduce the levels of skills or skills demanded is known as
(a) Multi Skilling (b) De Skilling
(c) Multi-tasking (d) None

399. Rage Committee is related to
(a) Working condition in Indian Industry
(b) Social security in Indian industry
(c) Indivisual difference in Indian industry
(d) None

400. The wage which has to be paid by the employer to his workers, irrespective of his ability, is known as
(a) Living wage (b) Minimum wage
(c) Fair wage (d) None of the above

401. The wage providing for sustenance of life plus for preservation of the efficiency of worker is called as
(a) Fair wage (b) Minimum wage
(c) Living wage (d) None

402. The wage rate that is equal to the predominant rate for similar work through out the country is called as
(a) Fair wage (b) Minimum wage
(c) Living wage (d) Higher wage

403. A wage period shall not exceed more than
(a) Forty days (b) One month
(c) One week (d) Thirty-five days

404. Appreciative Enquiry is
(a) Psychological theory
(b) Training Method
(c) HR quality
(d) Business etiquette

405. Agenda 21 relates to
(a) Child Labour
(b) Human Rights
(c) Environmental Protection
(d) None

406. Focusing on core competencies means companies have the possibility of generating?
(a) Human Capital Advantage
(b) Larger Profit Margins
(c) Better Output
(d) New Staff

407. Who has advanced the motivational analysis of organisation
(a) D M. Pestonjee (b) T. V. Rao
(c) Udai Pareek (d) None

408. T-group training method is given by
(a) Louis Kelso (b) Kaplan
(c) J L Marino (d) Porter

409. Which of the following theories states that employees make comparision of their efforts and rewards with those of others in similar work situation is called as
(a) Vroom's expectancy theory
(b) Alderfer's ERG theory
(c) Adam's equity theory
(d) Hertzberg's two factor theory

410. Which is the following is a method of indicating the feeling of acceptance or rejection among members of a group
(a) Mutuality
(b) Sociometry
(c) T-group
(d) Both b and c

411. Sociometry concept of feelings acceptance and rejection among employee is stated by
(a) Louis Kelso (b) Kaplan
(c) J L Marino (d) Porter

412. Career plateau that reflects the end of the promotion is
(a) Content plateau
(b) Structural plateau
(c) Life plateau
(d) None

413. When indivisuals with substantially different levels of experience and performance abilities are being paid wages or salaries that relatively equal is called as
(a) Pay Secrecy
(b) Pay Compression
(c) Maturity Curve
(d) Point Manual

414. What is true about gratuity?
(a) It is payable at the time of termination of service
(b) It is normally paid at the rate of 15 days wages for every completed year of service
(c) The maximum gratuity should not exceed one lakh in any case
(d) All are true

415. Standing labour conference was constituted in
(a) 1935 (b) 1942
(c) 1948 (d) 1951

416. Permanent involuntary termination is called as
(a) Lay off (b) Lay out
(c) Firing (d) Attrition

417. Many an instance, organisation is not filling openings created by voluntary resignations or normal retirements is called as
(a) Lay off (b) Lay out
(c) Firing (d) Attrition

418. An effort by one organisation to win the support of the leaders of other organisation by including them in advisory councils, boards of directors and the like. This is called as
(a) Diplomacy
(b) Mediation
(c) Expert incubation
(d) Co-Optation

419. The term 'BARS' stands for
(a) Behavioural Abnormality Rating Scale
(b) Behavioural Abnormality Response Scale
(c) Behavioural Anchored Response Scale
(d) Behavioural Anchored Rating Scale

420. A level or standard in a scale against which performance can be evaluated in order to structure similar process is described as
(a) Standardisation
(b) Structuring
(c) Benchmarking
(d) Institutionalising

421. One of the following is a training provided to enhance decision making skills
(a) On the job training
(b) Management games
(c) Action learning
(d) Behavioural modelling

422. An incentive plan, where improvements in group productivity determines the total amount of money that is allocated, is described as
(a) Profit sharing
(b) Bonus sharing
(c) Gain sharing
(d) None of the above

423. Labour welfare facilities were incorporated in which legislation
(a) MRTP Act
(b) Essential Commodities Act
(c) Consumer Protection Act
(d) Factories Act

424. Selection of persons in an organisation is the responsibility of
(a) Security department
(b) Personnel department
(c) Trade union
(d) Canteen department

425. Minimum wages are fixed by
(a) Trade union
(b) Employees
(c) State government
(d) Court of law

426. Giving general information regarding organization to the newly appointed personnel is called
(a) Training (b) Development
(c) Induction (d) Promotion

427. Industrial relations consists of the relations between
(a) Employer and Employees
(b) State and Management
(c) Union and Management
(d) All of the above

428. Graphical representation of Manpower in the organization is known as
(a) Replacement Chart
(b) Placement Chart
(c) Organisation structure
(d) Manpower Inventory

429. What is the operative function of HRM?
(a) Organising (b) Directing
(c) Controlling (d) Integrating

430. Which of the following is the most important considering promotion of an employee?
(a) Seniority (b) Competency
(c) Loyalty (d) Both a and b

ANSWER KEY (Principle of Management, Organisational Behaviour, Human Resource Management & Strategic Management)

1. (a)	**2.** (c)	**3.** (a)	**4.** (b)	**5.** (a)	**6.** (b)	**7.** (a)	**8.** (a)	**9.** (a)	**10.** (a)
11. (b)	**12.** (b)	**13.** (d)	**14.** (c)	**15.** (b)	**16.** (c)	**17.** (c)	**18.** (c)	**19.** (c)	**20.** (c)
21. (b)	**22.** (c)	**23.** (b)	**24.** (b)	**25.** (c)	**26.** (c)	**27.** (c)	**28.** (b)	**29.** (c)	**30.** (c)
31. (c)	**32.** (b)	**33.** (b)	**34.** (b)	**35.** (a)	**36.** (b)	**37.** (a)	**38.** (c)	**39.** (d)	**40.** (c)
41. (d)	**42.** (c)	**43.** (a)	**44.** (c)	**45.** (c)	**46.** (a)	**47.** (b)	**48.** (b)	**49.** (c)	**50.** (c)
51. (d)	**52.** (d)	**53.** (c)	**54.** (a)	**55.** (b)	**56.** (b)	**57.** (b)	**58.** (b)	**59.** (c)	**60.** (c)
61. (d)	**62.** (b)	**63.** (b)	**64.** (c)	**65.** (b)	**66.** (a)	**67.** (c)	**68.** (c)	**69.** (b)	**70.** (a)
71. (c)	**72.** (c)	**73.** (c)	**74.** (c)	**75.** (b)	**76.** (d)	**77.** (c)	**78.** (a)	**79.** (b)	**80.** (c)
81. (d)	**82.** (a)	**83.** (b)	**84.** (b)	**85.** (b)	**86.** (d)	**87.** (d)	**88.** (c)	**89.** (d)	**90.** (c)
91. (d)	**92.** (c)	**93.** (c)	**94.** (b)	**95.** (c)	**96.** (c)	**97.** (c)	**98.** (c)	**99.** (c)	**100.** (c)
101. (c)	**102.** (a)	**103.** (d)	**104.** (c)	**105** (c)	**106.** (d)	**107.** (b)	**108.** (a)	**109.** (b)	**103.** (c)
111. (c)	**112.** (c)	**113.** (d)	**114.** (b)	**115.** (a)	**116.** (c)	**117.** (b)	**118.** (b)	**119.** (c)	**120.** (c)
121. (b)	**122.** (d)	**123.** (a)	**124.** (b)	**125.** (c)	**126.** (d)	**127.** (c)	**128.** (b)	**129.** (b)	**130.** (c)
131. (c)	**132.** (b)	**133.** (a)	**134.** (d)	**135.** (c)	**136.** (c)	**137.** (b)	**138.** (b)	**139.** (a)	**140.** (d)
141. (b)	**142.** (c)	**143.** (c)	**144.** (d)	**145.** (d)	**146.** (c)	**147.** (b)	**148.** (b)	**149.** (b)	**150.** (c)
151. (a)	**152.** (c)	**153.** (b)	**154.** (a)	**155.** (d)	**156.** (b)	**157.** (a)	**158.** (c)	**159.** (d)	**160.** (c)
161. (c)	**162.** (d)	**163.** (b)	**164.** (c)	**165.** (c)	**166.** (c)	**167.** (b)	**168.** (b)	**169.** (a)	**170.** (b)
171. (b)	**172.** (a)	**173.** (d)	**174.** (c)	**175.** (c)	**176.** (c)	**177.** (a)	**178.** (b)	**179.** (a)	**180.** (c)
181. (d)	**182.** (b)	**183.** (a)	**184.** (d)	**185.** (b)	**186** (c)	**187.** (c)	**188.** (c)	**189.** (b)	**190.** (d)
191. (b)	**192.** (d)	**193.** (b)	**194.** (c)	**195.** (b)	**196.** (a)	**197.** (c)	**198.** (d)	**199.** (c)	**200.** (a)
201. (b)	**202.** (d)	**203.** (c)	**204.** (c)	**205.** (d)	**206.** (c)	**207.** (b)	**208.** (b)	**209.** (c)	**210.** (b)
211. (a)	**212.** (d)	**213.** (d)	**214.** (c)	**215.** (c)	**216.** (b)	**217.** (a)	**218.** (d)	**219.** (d)	**220.** (b)
221. (a)	**222.** (c)	**223.** (b)	**224.** (d)	**225.** (a)	**226.** (d)	**227.** (b)	**228.** (c)	**229.** (b)	**230.** (c)
231. (d)	**232.** (a)	**233.** (c)	**234.** (c)	**235.** (b)	**236.** (b)	**237.** (d)	**238.** (a)	**239.** (d)	**240.** (c)
241. (c)	**242.** (a)	**243.** (c)	**244.** (b)	**245.** (a)	**246.** (b)	**247.** (c)	**248.** (d)	**249.** (b)	**250.** (b)
251. (a)	**252.** (c)	**253.** (d)	**254.** (a)	**255.** (d)	**256.** (a)	**257.** (d)	**258.** (b)	**259.** (c)	**260.** (a)
261. (b)	**262.** (b)	**263.** (a)	**264.** (a)	**265.** (d)	**266.** (b)	**267.** (b)	**268.** (d)	**269.** (b)	**270.** (c)
271. (c)	**272.** (b)	**273.** (c)	**274.** (c)	**275.** (a)	**276.** (d)	**277.** (d)	**278.** (a)	**279.** (b)	**280.** (b)
281. (d)	**282.** (b)	**283.** (c)	**284.** (c)	**285.** (c)	**286.** (a)	**287.** (b)	**288.** (b)	**289.** (c)	**290.** (c)
291. (b)	**292.** (b)	**293.** (d)	**294.** (c)	**295.** (c)	**296.** (b)	**297.** (c)	**298.** (d)	**299.** (d)	**300.** (d)
301. (d)	**302.** (d)	**303.** (c)	**304.** (c)	**305.** (b)	**306.** (c)	**307.** (c)	**308.** (b)	**309.** (d)	**310.** (d)
311. (d)	**312.** (c)	**313.** (b)	**314.** (d)	**315.** (c)	**316.** (d)	**317.** (b)	**318.** (b)	**319.** (d)	**320.** (c)
321. (b)	**322.** (c)	**323.** (d)	**324.** (c)	**325.** (b)	**326.** (a)	**327.** (b)	**328.** (b)	**329.** (a)	**330.** (b)
331. (c)	**332.** (c)	**333.** (b)	**334.** (a)	**335.** (b)	**336.** (d)	**337.** (a)	**338.** (c)	**339.** (d)	**340.** (b)
341. (b)	**342.** (b)	**343.** (a)	**344.** (a)	**345.** (c)	**346.** (a)	**347.** (a)	**348.** (c)	**349.** (b)	**350.** (b)
351. (c)	**352.** (d)	**353.** (d)	**354.** (c)	**355.** (b)	**356.** (a)	**357.** (c)	**358.** (b)	**359.** (a)	**360.** (c)
361. (c)	**362.** (b)	**363.** (b)	**364.** (d)	**365.** (b)	**366.** (b)	**367.** (c)	**368.** (c)	**369.** (b)	**370.** (b)
371. (c)	**372.** (d)	**373.** (b)	**374.** (c)	**375.** (d)	**376.** (d)	**377.** (b)	**378.** (a)	**379.** (c)	**380.** (b)
381. (a)	**382.** (b)	**383.** (b)	**384.** (d)	**385.** (b)	**386.** (c)	**387.** (b)	**388.** (a)	**389.** (a)	**390.** (b)
391. (b)	**392.** (c)	**393.** (d)	**394.** (c)	**395.** (d)	**396.** (c)	**397.** (c)	**398.** (b)	**399.** (b)	**400.** (b)
401. (b)	**402.** (a)	**403.** (b)	**404.** (a)	**405.** (c)	**406.** (a)	**407.** (c)	**408.** (c)	**409.** (c)	**410.** (d)
411. (c)	**412.** (b)	**413.** (b)	**414.** (d)	**415.** (b)	**416.** (c)	**417.** (d)	**418.** (d)	**419.** (d)	**420.** (c)
421 . (b)	**422 .** (c)	**423.** (d)	**424.** (b)	**425.** (c)	**426.** (c)	**427.** (a)	**428.** (c)	**429.** (d)	**430.** (d)

CHAPTER 2

2. Accounting Control and Financial Management

1. The first step in accounting cycle is
(a) Classifying (b) Recording
(c) Summarizing (d) Analyzing

2. Books of original entry is called as
(a) Ledger (b) Trial balance
(c) Journal (d) Balance sheet

3. Double entry system is developed by
(a) Canadian Institute of Chartered Accountants
(b) International Accounting standard committee
(c) Lucas Pacioli
(d) Confederation of Indian Industry

4. The statement of assets and liabilities of the business on a particular date is called as
(a) Ledger (b) Trial balance
(c) Journal (d) Balance sheet

5. 'A business concern is separate and distinct from its owners' is the accounting principle based on
(a) Business Entity Concept
(b) Going Concern Concept
(c) Accounting Period Concept
(d) Money Measurement Concept

6. Organization used to calculate depreciation and consider historical cost over market price for its asset as it reflects the accounting concept of is observed in
(a) Business Entity Concept
(b) Going Concern Concept
(c) Accounting Period Concept
(d) Money Measurement Concept

7. Organization calculate profit and loss of its performance every year as it is a fundamental accounting concept relates to
(a) Business Entity Concept
(b) Going Concern Concept
(c) Accounting Period Concept
(d) Money Measurement Concept

8. Quality of a product though important, cannot be converted into monetary forms. The accounting concept that reveals on it is
(a) Business Entity Concept
(b) Going Concern Concept
(c) Accounting Period Concept
(d) Money Measurement Concept

9. 'Give in one hand and take on another' is the accounting concept reflects
(a) Dual Aspect Concept
(b) Cost Concept
(c) Matching Concept
(d) Realization Concept

10. Organization always recorded the historical cost of its fixed asset to make the calculation more objective oriented. It is the accounting concept of
(a) Dual Aspect Concept
(b) Cost Concept
(c) Matching Concept
(d) Realization Concept

11. 'Revenue of a period should be matched with its cost' as it reflects the accounting concept of
(a) Dual Aspect Concept
(b) Cost Concept
(c) Matching Concept
(d) Realization Concept

12. 'Revenue is recognized when it is realized' is the accounting concept relates to
(a) Cost Concept
(b) Matching Concept
(c) Realization Concept
(d) Accrual Concept

13. Accounting convention that is also called as Doctrine of Prudence is
(a) Convention of Consistency
(b) Convention of Materiality
(c) Convention of Conservatism
(d) Convention of Full Disclosure

14. Accounting Standard that reveals disclosure of accounting policies is
(a) IAS-1
(b) IAS-2
(c) IAS-3
(d) IAS-4

15. The transaction account that relate to property and Assets comes under
(a) Personal Account
(b) Real Account
(c) Nominal Account
(d) All types of Account

16. Transactions relating to expenses and incomes comes under
(a) Personal Account
(b) Real Account
(c) Nominal Account
(d) None

17. If debit and credit occurs in the name of a club, then that type of personal account is called as
(a) Natural Person's Personal Account
(b) Artificial Person's Personal Account
(c) Representative Personal Account
(d) All

18. In personal account, the person who is receiving benefit is given
(a) Debit
(b) Credit
(c) Both
(d) None

19. If transaction occur through trade mark, copy right then the account that is debited or credited is
(a) Representative Personal Account
(b) Tangible Real Account
(c) Intangible Real Account
(d) Nominal Account

20. When transactions occur in salary account, insurance account, commission account, then the type of account that is debited or credited is
(a) Personal Account
(b) Real Account
(c) Nominal Account
(d) All

21. In Real account, the statement that is true is
(a) Debit- what's comes in
(b) Credit- what goes out
(c) Both
(d) None

22. In nominal account, incomes and gains are to be
(a) Debited (b) Credited
(c) Both (d) None

23. Ledger Folio is mainly seen in
(a) Ledger (b) Journal
(c) Trial balance (d) Balance Sheet

24. Folio is found in
(a) Ledger (b) Journal
(c) Trial balance (d) Balance Sheet

25. The transferring entries from the journal to the ledger is called as
(a) Journalising (b) Ledger Folio
(c) Ledger Posting (d) Folio

26. A statement that summarizes the date recorded in the journal as well as in the ledger is called as
(a) Ledger (b) Journal
(c) Trial balance (d) Balance Sheet

27. Opening stock is always shown in
(a) Credit side of trading account
(b) Debit side of trading account
(c) Both
(d) None

28. Direct wages are shown in
(a) Debit side of trading account
(b) Credit side of trading account
(c) Debit side of Profit and Loss account
(d) Credit side of Profit and Loss account

29. In trading account, expenses like custom duty, fuel power, factory expenses are
(a) Credited (b) Debited
(c) Both (d) None

30. Closing stock is found in
(a) Credit side of the trading account
(b) Asset side of the Balance sheet
(c) Both
(d) None

31. Depreciation is found in
(a) Debit side of profit and loss account
(b) Asset side of the Balance sheet
(c) Both
(d) None

32. The account which is not found in the debit side of the profit and loss account is
(a) Commission
(b) Repair and Maintenance
(c) Advertisement expenses
(d) Royalties

33. Liability is shown in which side of the balance sheet?
(a) Right (b) Left
(c) Middle (d) Any where

34. Asset is shown in which side of the balance sheet?
(a) Right (b) Left
(c) Middle (d) Not found

35. The account that is not shown in the balance sheet is
(a) Personal Account
(b) Real Account
(c) Nominal Account
(d) All the above

36. Where the account is settled on the basis of order of liquidity or order of permanency is?
(a) Profit & Loss Account
(b) Trading Account
(c) Manufacturing Account
(d) Balance Sheet

37. Current Asset is also called as
(a) Liquid Asset (b) Fixed Asset
(c) Floating Asset (d) Wasting Asset

38. The assets like mines, oil-wells are categorized in
(a) Liquid Asset (b) Fixed Asset
(c) Floating Asset (d) Wasting Asset

39. Debit balance of profit and loss account, preliminary expenses are types of
(a) Intangible asset
(b) Tangible asset
(c) Wasting asset
(d) Fictitious asset

40. Cases undecided by the law of court, Bills discounted before maturity are the types of
(a) Long term liability
(b) Short term liability
(c) Contingent liability
(d) None of the above

41. The capital that is needed for the purchase of asset is called as
(a) Fixed capital
(b) Working capital
(c) Positive capital
(d) Trading capital

42. The capital which is needed for the purchase of running the business is
(a) Fixed capital
(b) Working capital
(c) Floating Capital
(d) Trading capital

43. Sundry creditor is shown in which side of the balance sheet?
(a) Asset side
(b) Liability side
(c) Both
(d) None

44. Items given on the debit side of the trial balance are shown either in
(a) Debit side of the trading account
(b) Debit side of the profit and loss account
(c) Assets side of the balance sheet
(d) All of the above

45. Working capital ratio is otherwise called as
(a) Liquidity Ratio
(b) Quick Ratio
(c) Current Ratio
(d) Acid test Ratio

46. Quick ratio is otherwise called asas
(a) Liquid ratio
(b) Acid test ratio
(c) Near Money ratio
(d) All of the above

47. For a sound business, the ratio between current asset to current liability in current ratio should be
(a) 3:4 (b) 4:3
(c) 2:1 (d) 1:2

48. For a sound business, the acid test ratio should be
(a) 3:4 (b) 1:1
(c) 2:1 (d) 1:2

49. To check the long term financial status of a firm, the ratio that should be taken into consideration are
(a) Debt-Equity Ratio.
(b) Total Assets to Debts Ratio.
(c) Proprietary Ratio
(d) Acid test ratio

50. For a sound business planning, the debt-equity ratio should be
(a) 3:4 (b) 4:3
(c) 2:1 (d) 1:2

51. For capital intensive industries like fertilizer and cement, the debt ratio can be acceptable up to
(a) 1:4 (b) 4:1
(c) 2:1 (d) 1:2

52. For capital intensive industries like shipping units, the debt ratio can be acceptable up to
(a) 1:6 (b) 6:1
(c) 2:1 (d) 1:2

53. For a sound business planning, the total asset to debt ratio should be
(a) 3:4 (b) 4:3
(c) 2:1 (d) 1:2

54. Gross Profit Ratio is considered good when ratio is in between
(a) 20-25% (b) 25-30%
(c) 15-20% (d) Less than 15%

55. Turnover ratio is also called as
(a) Activity Ratio
(b) Asset Management Ratio
(c) Both
(d) None

56. Manipulation of accounts in a way so as to conceal vital facts and better position than what it actually is. In management point of view, it is called as
(a) Personal Bias
(b) Window dressing
(c) Unethical
(d) None

57. The financial statement that indicates the changes in working capital is called as
(a) Cash flow statement
(b) Fund flow statement
(c) Comparative financial statement
(d) Common-size statement

58. The financial statement that reveals the causes of changes in cash position of an organization is called as
(a) Cash flow statement
(b) Fund flow statement
(c) Comparative financial statement
(d) Common-size statement

59. The financial statement not only show the absolute figures of various years but also provide for columns to indicate the changes in various figure is called as
(a) Cash flow statement
(b) Fund flow statement
(c) Comparative financial statement
(d) Common-size statement

60. The financial statement in which indivisual figures are converted into percentage to some common base is called as
(a) Cash flow statement
(b) Fund flow statement
(c) Comparative financial statement
(d) Common-size statement

61. How the responsibility centres differs from cost centres?
(a) It deals with cost of the product
(b) It deals with cost of the process
(c) It deals with persons who may be managing
(d) It deals with cost of assembling items

62. The calculation of depreciation under the assumption of constant use of assets over the years is
(a) Straight Line Method
(b) Diminishing Balance Method
(c) Sum of the years digit Method
(d) Annual Revaluation Method

63. The calculation of depreciation where higher amount is charged in the initial years of the asset and less amount in the subsequent year is
(a) Straight Line Method
(b) Diminishing Balance Method
(c) Sum of the years digit Method
(d) Annual Revaluation Method

64. The calculation of depreciation where the value of the asset is estimated both at beginning and end of the year is
(a) Straight Line Method
(b) Diminishing Balance Method
(c) Sum of the years digit Method
(d) Annual Revaluation Method

65. The term that is not included in 3A′ of Financial Management is
(a) Anticipating Financial needs
(b) Ascribe financial need
(c) Allocating funds in business
(d) Acquiring financial resources

66. According to traditional view of financial management, the statement that is completely false
(a) It treats the entire subject of finance from the outsider's point of view
(b) It laid heavy emphasis on long-term financing
(c) It put maximum emphasis on non-corporate enterprise
(d) It ignores routine problems

67. Which is not included in the list of Intangible real Asset?
(a) Patent
(b) Factory
(c) Technical Know-how
(d) Technological collaboration

68. The place where the securities issued by firms are traded by investors is called as
(a) Primary Capital Market
(b) Secondary Capital Market
(c) Tertiary Capital Market
(d) None

69. The firm decision to invest its current funds in the long-term assets in anticipation of an expected flow of benefits over a series of years is called as
(a) Working capital management
(b) Dividend Decision
(c) Capital Budgeting
(d) Capital Structure

70. Acquiring of funds in the form of debt and equity is called as
(a) Working capital management
(b) Dividend Decision
(c) Capital Budgeting
(d) Capital Structure

71. The management of current asset in accordance with risk and return is called as
(a) Working capital management
(b) Dividend Decision
(c) Capital Budgeting
(d) Capital Structure

72. The long term financial security that have no maturity date is called as
(a) Preferential share
(b) Ordinary Share
(c) Debenture
(d) Loan

73. The long term financial security that otherwise considered as hybrid security is
(a) Preferential share
(b) Ordinary Share
(c) Debenture
(d) Loan

74. The long term financial security that have residual earnings claim is
(a) Preferential share
(b) Ordinary Share
(c) Debenture
(d) Loan

75. An alternative form of debenture in India is
(a) Loan
(b) Bond
(c) Cumulative Debenture
(d) Preference Share Holder

76. The correct sequence (first-last) for giving dividend is
(a) Debenture, Preference Share Holder, Ordinary Share Holder
(b) Ordinary Share Holder, Preference Share Holder, Debenture
(c) Debenture, Ordinary Share Holder, Preference Share Holder
(d) Preference Share Holder, Debenture, Ordinary Share Holder

77. The share issue to the existing shareholder without any charge is called as
(a) Right share
(b) Equity share
(c) Bonus share
(d) Participative Preference share

78. The use of the fixed charge sources of funds, such as debt and preference capital along with the owner's equity in the capital structure, is called as
(a) Financial leverage
(b) Gearing
(c) Trading on equity
(d) All

79. The percentage change in the earnings before interest and taxes relative to a given percentage change in sales is called as
(a) Coverage ratio
(b) Degree of Operating leverage
(c) Degree of Financial leverage
(d) Degree of Combined leverage

80. The percentage change in EPS due to a given percentage change in Earnings before interest and tax is called as
(a) Degree of Operating leverage
(b) Degree of Financial leverage
(c) Degree of Combined leverage
(d) None

81. Modigliani and Miller(MM) approach on irrelevance capital structure is based on the principle of
(a) Net Income
(b) Earnings Per Share
(c) Return on Investment
(d) Net Operating Income

82. The preferred capital structure for a mature firm is
(a) Ordinary share
(b) Preferential share
(c) Debt with longer maturity
(d) Debt with shorter maturity

83. Place the floatation cost in an increasing order?
(a) Share, Retained earnings, Debt
(b) Debt, Share, Retained earning
(c) Debt, Retained earnings, share
(d) Retained earnings, Debt, Share

84. The firm's investment on current assets is called as
(a) Gross Working Capital
(b) Net working Capital
(c) Both
(d) None

85. The difference between current assets and current liabilities is called as
(a) Gross Working Capital
(b) Net working Capital
(c) Both
(d) None

86. The time duration from acquisition of resources to manufacture of the product and then to sale of the product either for cash or on credit is called as
(a) Cash conversion cycle
(b) Operating cycle
(c) Manufacturing cycle
(d) All

87. The contract in which owner gives the right to use the asset to the user over an agreed period of time for a consideration is called as
(a) Venture Capital Financing
(b) Project Financing
(c) Hire Purchase Financing
(d) Lease

88. When more lease rental is paid in the initial period, then types of lease is called as
(a) Back-ended lease (b) Up-front lease
(c) Operating lease (d) Financial lease

89. The long term, non-cancellable lease contracts is called as
(a) Operating lease
(b) Financial lease
(c) Sale-and-lease-back
(d) All

90. The type of lease in which the asset gets transferred to the lessor at the end, and the risk of obsolescence, residual value remain with the lessor being the legal owner of the asset
(a) Direct lease (b) Open lease
(c) Closed lease (d) Update lease

91. When the lessee is obliged to take care of maintenance, taxes and insurance, then the lease is called as
(a) Leveraged lease
(b) Update lease
(c) Net net net lease
(d) Closed lease

92. Method of venture capital financing in which loan is repayable in the form of royalty and no interest is paid on such loan is called as
(a) Income Note
(b) Participating debenture
(c) Conventional loan
(d) Conditional loan

93. Method of venture capital financing in which the entrepreneur had to pay both interest and royalty is
(a) Income Note (b) Conventional loan
(c) Equity (d) Conditional loan

94. Who is the following shareholder are much concerned for retained earning policy rather than distribution of dividend?
(a) Retired and Old Person
(b) Institutional investor
(c) Wealthy investor
(d) Mature companies

95. Which is the following long term financing security is not a hybrid one
(a) Preference share
(b) Convertible Debenture
(c) Equity share
(d) Income Note

96. Types of merger in which two or more firms in similar type of production, distribution or area of business is called as
(a) Vertical Merger
(b) Horizontal Merger
(c) Conglomerate Merger
(d) Circular Merger

97. The sale of a company's asset, or product lines, or divisions or brand to the outsider is called as
(a) Acquisition
(b) Leveraged Buy-Outs
(c) Divestment
(d) Joint venture

98. When a company sells part of its business to a third party is called as
(a) Sell-off
(b) Spin-offs
(c) Both
(d) None

99. When a company creates a new from the existing entity, it is called as
(a) Sell-off (b) Spin-offs
(c) Both (d) None

100. Market risk is also known as
(a) Systematic risk or diversifiable risk
(b) Unsystematic risk or diversifiable risk
(c) Systematic risk or non-diversifiable risk
(d) Unsystematic risk or non-diversifiable risk

101. Financial risk arises from
(a) R & D and operations stages of value chain
(b) GNP growth rate and competitive environment
(c) Volatility of interest rates, currency rates, commodities prices and stock prices
(d) Changes in laws and regulations

102. Which of the following is an example for unsystematic risk?
(a) Interest rate risk
(b) Exchange rate risk
(c) International risk
(d) Operating risk

103. Which of the following is an important money market instrument?
(a) Debentures
(b) Commercial Paper
(c) Public Deposits
(d) Loan

104. As per the pecking order theory of capital structure will be
(a) Internal equity is preferred over external debt
(b) External debt is preferred over internal equity
(c) External equity is preferred over external debt
(d) None of the above

105. A decision maker has to remember the proverb 'a bird in hand is worth two in the bush' while he examines
(a) Opportunity cost principle
(b) Discounting and Compounding Principle
(c) Marginal or Incremental Principle
(d) Equi-marginal principle

106. If bond sells at discount, the price is less than per value and rate is
(a) YTM = Coupon Rate
(b) YTM > Coupon Rate
(c) YTM < Coupon Rate
(d) None of the above

107. In case where the investment can be made in stages and is dependent on the future outcomes, the capital budgeting technique that can be adopted will be
(a) Simulation technique
(b) Hiller Model
(c) Decision tree analysis
(d) Scenario Analysis

108. When borrowing is less than the equity, the company is said to be
(a) Trading on equity
(b) Trading on thick equity
(c) Trading on thin equity
(d) None of the above

109. Beta in the context of security analysis indicates
(a) Systematic risk
(b) Unsystematic risk
(c) Total risk
(d) All

110. A high rise, high yield often unsecured bond rated below investment grade is known as
(a) Income bond
(b) Junk bond
(c) Subordinated debenture
(d) Serial bond

111. Type of bond where face value of it is promised to pay the investor is called as
(a) Income bond
(b) Junk bond
(c) Subordinated debenture
(d) Serial bond

112. A class of unsecured bond, that in the event of liquidation is prioritized lower than other classes of debt is called as
(a) Income bond
(b) Junk bond
(c) Subordinated debenture
(d) Serial bond

113. A portion of the outstanding bonds mature at regular intervals until eventually all of the bonds matured is called as
(a) Income bond
(b) Junk bond
(c) Subordinated debenture
(d) Serial bond

114. The option to buy an asset is called as
(a) Call Option (b) Put Option
(c) Both (d) None

115. The option to sell an asset is called as
(a) Call Option (b) Put Option
(c) Both (d) None

116. When an option is allowed to be exercised only on the maturity date, it is called as
(a) American Option
(b) European Option
(c) Indian Option
(d) All

117. Which is not the feature of capital structure?
(a) Rigidity in structure
(b) Risky
(c) Control mechanism
(d) Time bound

118. According to SEBI, the ratio of preference capital to equity capital does not exceed more than
(a) 1:4
(b) 1:2
(c) 1:3
(d) 2:3

119. Less degree of financial leverage (debt) is taken by the organization when
(a) It produces more non-durable goods
(b) If products are more seasonal in nature
(c) If there is no or less competition in market
(d) If the product is in growth stage of its life cycle

120. Credit terms include
(a) Credit period
(b) Cash discount
(c) Cash discount period
(d) All of the above

121. If credit terms is written as 2/10 net 30, then 10 signifies
(a) Credit period
(b) Cash discount
(c) Cash discount period
(d) None

122. If credit terms is written as 2/10 net 30, then 2 signifies
(a) Credit period
(b) Cash discount
(c) Cash discount period
(d) None

123. If credit terms is written as 2/10 net 30, then 30 signifies
(a) Credit period
(b) Cash discount
(c) Cash discount period
(d) None

124. Which is not the decentralized method of cash collection
(a) Payment by wire
(b) Payment in nearest bank
(c) Concentration banking
(d) Lock-box system

125. Slowing down cash payment through
(a) Centralized payment
(b) Paying on the last date
(c) Cheque encashment analysis
(d) All

126. The optimum cash balance model that simulates with EOQ model of inventory is
(a) Baumol Model
(b) Miller and Orr Model
(c) Both of the above
(d) None of the above

127. The market for short term securities is known as
(a) Capital Market
(b) Secondary Market
(c) Money Market
(d) Black Market

128. The short term securities in which a company is lending money to a stock market operator who wishes to carry forward his/her transaction from one settlement period to another is called as
(a) Treasury Bill
(b) Commercial Paper
(c) Badla Financing
(d) Bills Discounting

129. For determination of cost of equity the formula is
(a) Rate of Interest (1-t)
(b) Rate of Dividend (1-t)
(c) Rate of Dividend (Face value- Issue Price)/n
(d) D1/Po +G

130. Overall cost of capital is
(a) Net Operating Income / Value of equity
(b) Net Income / Value of debt
(c) Net Operating Income / Value of the firm
(d) Net Income / Value of the firm

131. According to MM-Approach Proposition-1, value of firm is
(a) Independent of capital structure
(b) Dependent of capital structure
(c) Dependent on tax-rate of the enterprise
(d) Dependent on risk of the enterprise

132. A stable dividend policy indicates
(a) Stable Earning/Price ratio
(b) Stable Pay-Out ratio
(c) Stable Price/Earnings ratio
(d) Stable earnings per share ratio

133. The degree of total leverage is equal to
(a) Percentage change in EPS/percentage change in sales
(b) Percentage change in EBIT/ percentage change in sales
(c) Percentage change in EPS/ Percentage change in EBIT
(d) Percentage change in sales/ Percentage change in EBIT

134. Dividend is paid out of
(a) Capital Profit
(b) Reserves
(c) Current or Accumulated Profit
(d) Cash

135. Share capital of a company may be reduced by
(a) A special resolution
(b) An ordinary resolution
(c) By the company Law Board's permission
(d) By the board of director

136. Which of the following statement is correct?
(a) ARR method makes the use of cash flow after tax
(b) Pay back method considers cash flows only up to the payback period
(c) Lower the cost of capital, lower the chance of acceptance of the project
(d) Profitability index method does not make use of salvage value of the project

137. Which is the most commonly used method for evaluation of larger project?
(a) NPV
(b) IRR
(c) PBP
(d) ARR

138. The amount of dividend that can be legally distributed is governed by
(a) Company law
(b) Contractual restrictions
(c) Judicial pronouncement in leading cases
(d) All of the above

139. According to 'Dividend Growth Model' approach the cost of equity is equal to
(a) Cash dividend per share
(b) Cash dividend per share + bonus per share
(c) Dividend pay out ratio + expected growth rate in dividend
(d) Dividend yield + expected growth rate in dividends

140. Pre-emptive right of equity share holder means that they
(a) Can vote to protect their right
(b) Can maintain their proportional ownership by purchasing additional equity shares issued by firm
(c) Can be elected as directors
(d) None of the above

141. Theoretically, with the issue of right share the value of existing shares will
(a) Increase
(b) Reduce
(c) Not change
(d) May increase or decrease

142. Mutually exclusive investment proposals are those
(a) They have unique features
(b) Wherein the acceptance of one proposal leads to rejection of all other proposals
(c) That require special source of financing
(d) That reduce the total risk complexion of the firm

143. The payback period refers to
(a) The number of years a project takes to recovers its investment (original) cost
(b) The number of installments to make payment
(c) The time lag between investment and commissioning of the project
(d) The time required to attain the breakeven point

144. The amount by which the face value of the bond exceeds its current price is known as
(a) Bond premium
(b) Bond discount
(c) Intrinsic value of the bond
(d) Future value of the bond

145. The amount by which the current price of a bond exceeds its face value
(a) Bond premium
(b) Bond discount
(c) Intrinsic value of the bond
(d) Future value of the bond

146. The project whose acceptance precludes the acceptance of one or more alternative projects is known as
(a) Independent project
(b) Dependent project
(c) Mutually exclusive project
(d) Contingent project

147. Annual cash dividend by annual earnings or alternatively dividend per share divided by earning per share is known as
(a) Cost of equity
(b) Cash profit
(c) Dividend-payout ratio
(d) Retention ratio

148. When merger is in the nature of consolidation, the accounting approach will be
(a) Purchase method
(b) Combination
(c) Pooling of interest method
(d) None of the above

149. Other things being equal, stocks which are highly liquid command is
(a) Higher dividend payout ratio
(b) Lower dividend payout ratio
(c) Higher P/E multiplies
(d) Lower P/E multiplies

ANSWER KEY (Accounting Control and Financial Management)

1. (b)	**2.** (c)	**3.** (c)	**4.** (d)	**5.** (a)	**6.** (b)	**7.** (c)	**8.** (d)	**9.** (a)	**10.** (b)
11. (c)	**12.** (d)	**13.** (c)	**14.** (a)	**15.** (b)	**16.** (c)	**17.** (b)	**18.** (a)	**19.** (c)	**20.** (c)
21. (c)	**22.** (b)	**23.** (b)	**24.** (a)	**25.** (c)	**26.** (c)	**27.** (b)	**28.** (a)	**29.** (b)	**30.** (c)
31. (c)	**32.** (d)	**33.** (b)	**34.** (a)	**35.** (c)	**36.** (d)	**37.** (c)	**38.** (d)	**39.** (d)	**40.** (c)
41. (c)	**42.** (b)	**43.** (b)	**44.** (d)	**45.** (c)	**46.** (d)	**47.** (c)	**48.** (b)	**49.** (d)	**50.** (c)
51. (b)	**52.** (b)	**53.** (c)	**54.** (b)	**55.** (c)	**56.** (b)	**57.** (b)	**58.** (a)	**59.** (c)	**60.** (d)

61. (c)	**62.** (a)	**63.** (b)	**64.** (d)	**65.** (b)	**66.** (c)	**67.** (b)	**68.** (b)	**69.** (c)	**70.** (d)
71. (a)	**72.** (b)	**73.** (a)	**74.** (b)	**75.** (b)	**76.** (a)	**77.** (c)	**78.** (d)	**79.** (b)	**80.** (b)
81. (d)	**82.** (c)	**83.** (d)	**84.** (a)	**85.** (b)	**86.** (b)	**87.** (d)	**88.** (b)	**89.** (b)	**90.** (c)
91. (c)	**92.** (b)	**93.** (a)	**94.** (c)	**95.** (c)	**96.** (b)	**97.** (c)	**98.** (a)	**99.** (b)	**100.** (c)
101. (c)	**102.** (d)	**103.** (b)	**104.** (a)	**105.** (b)	**106.** (b)	**107.** (d)	**108.** (b)	**109.** (a)	**110.** (b)
111. (a)	**112.** (c)	**113.** (d)	**114.** (a)	**115.** (b)	**116.** (b)	**117.** (a)	**118.** (c)	**119.** (b)	**120.** (d)
121. (c)	**122.** (b)	**123.** (a)	**124.** (b)	**125.** (d)	**126.** (a)	**127.** (c)	**128.** (c)	**129.** (d)	**130.** (c)
131. (a)	**132.** (b)	**133.** (a)	**134.** (d)	**135.** (a)	**136.** (b)	**137.** (b)	**138.** (d)	**139.** (d)	**140.** (b)
141. (a)	**142.** (b)	**143.** (a)	**144.** (b)	**145.** (a)	**146.** (d)	**147.** (c)	**148.** (c)	**149.** (c)	

CHAPTER 3

3. Agricultural Finance

1. Production loan is payable within a period of
(a) Within 1 year
(b) Within 6 months to 18 months
(c) Within 12 months to 24 months
(d) There is no time limit

2. Consumption loan is payable within a period of
(a) Within 5 crop season
(b) Within 30 months' time period
(c) Either within 5 crop season or Within 30 months' time period
(d) None

3. Consumption loan is given under the discretionary power of bank manager up to
(a) Rs 500 (b) Rs 1000
(c) Rs 1500 (d) Rs 5000

4. The rate of interest on consumption loan is
(a) 10% (b) 11%
(c) 12% (d) 7%

5. Consumption credit for medical expenses is
(a) Rs 500 (b) Rs 250
(c) Rs 200 (d) Rs 100

6. Consumption credit for marriage need is
(a) Rs 400 (b) Rs 500
(c) Rs 250 (d) Rs 100

7. Consumption credit for educational need is
(a) Rs 100 (b) Rs 200
(c) Rs 250 (d) Rs 500

8. Consumption credit for birth, funeral need is
(a) Rs 100
(b) Rs 150
(c) Rs 200
(d) Rs 250

9. Consumption credit for religious ceremony need is
(a) Rs 100
(b) Rs 150
(c) Rs 200
(d) Rs 250

10. The payback period for short term loan is
(a) Within 6 to 18 months
(b) Within 15 months to 5 years
(c) Within 5 year to 20 years
(d) No time limit

11. The payback period for medium term loan is
(a) Within 6 to 18 months
(b) Within 15 months to 5 years
(c) Within 5 year to 20 years
(d) No time limit

12. For reclamation and conservation of land, farmers' need
(a) Short term loan
(b) Medium term loan
(c) Long term loan
(d) All

13. The loan that partial liquid in nature is
(a) Short term loan
(b) Medium term loan
(c) Long term loan
(d) None

14. Form of security in which borrower stands as the guarantor for loan advancement is
(a) Mortgage
(b) Chattel Loans
(c) Personal Security
(d) Collateral Security

15. Which of the following is offered as personal security for claiming the loan by the farmer-borrower from the financial institution?
(a) Agriculture Produce
(b) Land
(c) Gold
(d) LIC Bonds

16. When the movable property of an individual's like LIC bond, warehouse receipts, livestock are offered as security, then it is called as
(a) Mortgage
(b) Chattel Loans
(c) Personal security
(d) Collateral Security

17. When movable property like jewelry, utensils are deposited as security and loan is advanced for a particular category of lender then the type of loan is called as
(a) Collateral security
(b) Chattel loan
(c) Key loan
(d) Open loan

18. Types of security where loans are advanced against immovable property are called as
(a) Hypothecation
(b) Mortgage
(c) Personal security
(d) Collateral security

19. When the property is ancestral in origin then banking institutions go for registration of the property against their name and then advance loan. This type of security is called as
(a) Equitable mortgage
(b) Simple mortgage
(c) Both
(d) None

20. The type of hypothecation in which agriculture produce are deposited as security against loan is called as
(a) Key loan
(b) Open loan
(c) DIR loan
(d) Chattel Loan

21. An example of hypothecated key loan is
(a) Land
(b) Tractor
(c) Agriculture Produce
(d) LIC Bonds

22. Types of hypothecation in which physical possession of the purchased machinery rests with the borrower but the legal right is with the lending institution is called as
(a) Key loans (b) Open loans
(c) DIR loans (d) None

23. The Deccan Agriculture Relief Act came in
(a) 1875 (b) 1879
(c) 1883 (d) 1900

24. Land Improvement Loan Act came in
(a) 1879 (b) 1883
(c) 1885 (d) 1895

25. Cooperative credit society act came in
(a) 1900 (b) 1904
(c) 1910 (d) 1912

26. Legal protection of marketing of the produce through cooperative came in
(a) 1904 (b) 1912
(c) 1919 (d) 1921

27. Cooperative is a provincial subject established in
(a) 1904 (b) 1919
(c) 1935 (d) 1947

28. Land mortgage bank on cooperative was set up first in
(a) Calcutta (b) Madras
(c) Punjab (d) Bombay

29. First land mortgage bank on cooperative was set up in the year
(a) 1915 (b) 1920
(c) 1921 (d) 1925

30. Land Mortgage bank was changed to Land Development Bank in Andhra Pradesh in
(a) 1965 (b) 1970
(c) 1974 (d) 1982

31. Linking credit with marketing of produce through cooperative was recommended first by
(a) Frederick Nicholson
(b) Sir Edward Law
(c) Prof D R Gadgil
(d) R G Saraiya

32. Vaikuntha Mehta National Institute of Cooperative Management was set up at
(a) Hyderabad (b) Pune
(c) Lucknow (d) Chennai

33. Single tier cooperative is found in
(a) Bihar
(b) Jammu & Kashmir
(c) Maharastra and Uttar Pradesh
(d) All

34. Long term loan through cooperative is given by
(a) Primary Agricultural Cooperative Society
(b) District Central Cooperative bank
(c) Primary Land Development bank
(d) State Cooperative bank

35. Primary Land Development Bank is operated in
(a) Village level (b) Taluka level
(c) District level (d) State level

36. Which cooperative bank acts as a link between State Cooperative bank and Primary Agricultural Cooperative Society
(a) Primary Land Development Bank
(b) Central Land development Bank
(c) District Central Cooperative Bank
(d) Regional Rural Bank

37. Single window system for credit disbursement was recommended by
(a) Narsimhan Committee
(b) Bawa Team
(c) Mohan Kanda Committee
(d) AIRCSC

38. In single window system, which cooperative has the power to disburse both short, medium and long term loan to the beneficiary
(a) Primary Agricultural Cooperative Society
(b) District Central Cooperative bank
(c) Primary Land Development bank
(d) State Cooperative bank

39. Large-Sized Adivasi Multipurpose Co-Operative Societies came in
(a) 1965
(b) 1970
(c) 1971
(d) 1975

40. Large-Sized Adivasi Multipurpose Co-Operative Societies was recommended by
(a) AIRCRC
(b) Bawa Team
(c) Mohan Kanda Committee
(d) AIRCSC

41. Large-Sized Adivasi Multipurpose Co-Operative Societies was operated in
(a) Village level (b) Block level
(c) District level (d) State level

42. Farmers Service Societies are registered cooperative society was recommended by
(a) AIRCRC
(b) Bawa Team
(c) National Commission On Agriculture
(d) AIRCSC

43. Farmers Service Societies are operated in
(a) Village level (b) Block level
(c) District level (d) State level

44. Farmers Service Societies came in the year
(a) 1971
(b) 1972
(c) 1975
(d) 1979

45. The share capital of FSS is
(a) Rs 25,000 to Rs 50,000
(b) Rs 50,000 to Rs 1,00,000
(c) More than 1 lakh
(d) Less than Rs 25,000

46. Nationalization of Banks was considered in
(a) 25th January, 1969
(b) 15th March, 1969
(c) 19th July, 1969
(d) 19th August, 1969

47. Number of Banks were nationalized first in 1969
(a) 10 (b) 14
(c) 15 (d) 20

48. Second phase of Nationalization of Banks were done in
(a) 10th January, 1980
(b) 13th March, 1980
(c) 19th July, 1980
(d) 15th April, 1980

49. How many Number of banks were nationalized first in 1980?
(a) 6 (b) 8
(c) 10 (d) 20

50. State Bank of India was set up in
(a) 1935 (b) 1949
(c) 1955 (d) 1960

51. Head quarter of SBI is at
(a) New Delhi
(b) Mumbai
(c) Chennai
(d) Calcutta

52. Head quarter of State Bank of Saurastra is
(a) Indore
(b) Bangalore
(c) Bhavnagar
(d) Jaipur

53. Rural Credit and Development Division is the specialized branches of
(a) SBI
(b) Bank of Baroda
(c) Indian Overseas Bank
(d) Dena bank

54. Gram Vikas Kendras is the specialized branches of
(a) SBI
(b) Bank of Baroda
(c) Indian Overseas Bank
(d) Dena bank

55. Agricultural Development Branches and Agricultural Banking Division is the specialized branches of
(a) SBI (b) IDBI
(c) ICCI (d) UBI

56. Rural Service Center is the specialized branches of
(a) Bank of Baroda
(b) Dena Bank
(c) Syndicate Bank
(d) Indian Overseas bank

57. Farm Clinic is the specialized branches of
(a) Bank of Baroda
(b) Dena Bank
(c) Syndicate Bank
(d) Indian Overseas bank

58. Agricultural Passbook through formal credit institution is recommended by
(a) NCA
(b) AIRCSC
(c) AIRCRC
(d) C E Kamath Committee

59. Village Adoption Scheme conceived first by
(a) SBI
(b) Bank of Baroda
(c) Indian Overseas Bank
(d) Dena bank

60. Village Adoption Scheme came in the year
(a) 1961 (b) 1969
(c) 1975 (d) 1985

61. Lead Bank Scheme was recommended by
(a) National credit Council
(b) Narsimhan Committee
(c) Nariman Committee
(d) C E Kamath Committee

62. Service Area Approach came in the year
(a) 1959 (b) 1969
(c) 1979 (d) 1989

63. In village adoption scheme, the activities of the bank will be
(a) Within the village
(b) Extended outside the village
(c) Restricted to large farmers only
(d) Restricted to marginal farmers only

64. In Service Area Approach, the activities of the bank will be
(a) Within the village
(b) Extended outside the village
(c) Restricted to large farmers only
(d) Restricted to marginal farmers only

65. In village adoption scheme, the village will be selected
(a) As per the will of the Branch Manager
(b) Based on IRDP criteria
(c) As per the will of District Agricultural Officer
(d) As per the will of the farmers' representative

66. In Service Area Approach, the village will be selected
(a) As per the will of the Branch Manager
(b) Based on IRDP criteria
(c) As per the will of District Agricultural Officer
(d) As per the will of the farmers' representative

67. The official forum for coordination for village adoption scheme is
(a) State level Bankers' committee
(b) Block level Bankers' committee
(c) District Technical Committee
(d) No forum

68. Village Adoption scheme was first started by
(a) ICICI
(b) IDBI
(c) SBI
(d) Punjab National Bank

69. Establishment of Regional Rural Bank is suggested by
(a) Narsimhan Committee
(b) Nariman Committee
(c) C E Kamath Committee
(d) Khusro Committee

70. RRB was set up first in the year
(a) 26th February,1975
(b) 26th April,1975
(c) 26th September,1975
(d) 26th December,1975

71. Prathma Regional Rural bank is the sponsor bank of
(a) Syndicate bank
(b) SBI
(c) PNB
(d) United bank of India

72. The head quarter of Prathma RRB is
(a) Jaipur, Rajasthan
(b) Malda, West Bengal
(c) Moradabad, Uttar Pradesh
(d) Bhiwani, Haryana

73. Gaur Grameena bank is the sponsor bank of
(a) SBI
(b) United Bank of India
(c) Syndicate Bank
(d) PNB

74. The head quarter of Gaur Grameena Bank is
(a) Jaipur, Rajasthan
(b) Malda, West Bengal
(c) Moradabad, Uttar Pradesh
(d) Bhiwani, Haryana

75. The authorized share capital of RRB is
(a) Rs 25 lakh
(b) Rs 50 lakh
(c) Rs 1 crore
(d) Rs 5 crore

76. The contribution of central government, state government and the sponsoring bank in the share capital of RRB is
(a) 50:15:35 (b) 15:50:35
(c) 35:50:15 (d) 50:35:15

77. Statutory Liquidity Ratio is to be maintained in RRB is
(a) 20% (b) 25%
(c) 35% (d) 38%

78. Cash Reserve Ratio in RRB is
(a) 3% (b) 5%
(c) 10% (d) 15%

79. Which of the following sequence is correct regarding jurisdiction of banks
(a) PACS>RRB>LAMPS
(b) LAMPS>RRB> PACS
(c) RRB>LAMPS>PACS
(d) RRB>PACS>LAMPS

80. In 1989, one of the agricultural review committees had given their recommendation for rural credit. The chairman of that committee was
(a) M C Bhandari
(b) Dr A M Khusro
(c) Dr B K Hazare
(d) M Narasimham

81. Differential Rate of Interest Scheme (DIR Scheme) is recommended under the chairmanship of
(a) M Narasimhan
(b) F K F Nariman
(c) M C Bhandari
(d) B K Hazare

82. DIR Scheme came into the mainstream of finance in
(a) 1971 (b) 1975
(c) 1978 (d) 1982

83. The concessional interest rate for the needy person in DIR Scheme is
(a) 7% (b) 4%
(c) 3% (d) 10%

84. The loan advanced through DIR Scheme in rural areas is
(a) Rs 3000 (b) Rs 2500
(c) Rs 2000 (d) Rs 1500

85. The loan advanced through DIR Scheme in urban areas is
(a) Rs 3000 (b) Rs 2500
(c) Rs 2000 (d) Rs 1500

86. Farm Gaduate Scheme is the invention of
(a) IDBI (b) SBI
(c) NABARD (d) SIDBI

87. In Farm graduate scheme, a beneficiary will get a maximum loan amount for purchase of land is
(a) Rs 10000 (b) Rs 20000
(c) Rs 25000 (d) Rs 50000

88. In Farm graduate scheme, the minimum qualification of the beneficiary should be
(a) Graduation in any discipline
(b) Matriculation
(c) Graduation in Agriculture
(d) +2 Science with vocational degree

89. The maximum repayment period for a loan under Farm graduate scheme is
(a) 2 years (b) 5 years
(c) 10 years (d) 20 years

90. Failure of Palai Central bank and Laxmi Bank Ltd gave a momentum for establishment of Deposit Insurance and Credit Guarantee Corporation of India The Palai Central bank belongs to
(a) Andhra Pradesh
(b) Tamil Nadu
(c) Kerala
(d) Karnataka

91. Laxmi Bank Ltd belongs to which state?
(a) Punjab
(b) Gujarat
(c) Maharashtra
d) Karnataka

92. Credit Guarantee Corporation of India was established in
(a) 1961 (b) 1971
(c) 1975 (d) 1981

93. Deposit Insurance and Credit Guarantee Corporation of India came in the year
(a) 1961 (b) 1971
(c) 1978 (d) 1985

94. Crop loan system was recommended by
(a) AIRCSC
(b) V L Mehra Committee
(c) Both
(d) None

95. Credit requirement for crop loan is calculated on the basis of
(a) Market price of the produce
(b) Fixed cost of the farmer
(c) Variable cost of the crop
(d) Miscellaneous expenditure of the farmer

96. 10-1 certificate submitted by the farmer-borrower while acquisition of loan indicates
(a) Cropping pattern
(b) Cropping density
(c) Gross cropped area
(d) Ownership right of the land

97. Adangal certificate submitted by the farmer-borrower while acquisition of loan indicates
(a) Cropping pattern
(b) Cropping density
(c) Gross cropped area
(d) Ownership right of the land

98. Encumbrance certificate submitted by the farmer-borrower while acquisition of loan indicates
(a) Cropping Pattern
(b) Ownership rights of land
(c) Mortgage detail
(d) All

99. The Chairman of District Consultative Committee is
(a) Lead bank Officer
(b) Collector/ Magistrate
(c) PD DRDA
(d) District Agriculture Officer

100. The Convener of District Consultative Committee is
(a) Lead bank Officer
(b) Collector / Magistrate
(c) PD DRDA
(d) District Agriculture Officer

101. The International Bank for Reconstruction and Development was established in
(a) 1935 (b) 1942
(c) 1945 (d) 1947

102. World Bank constitutes
(a) International Development Association
(b) International Finance Corporation
(c) Multilateral Investment Guarantee Agency
(d) All

103. RBI was established in
(a) 1921 (b) 1935
(c) 1949 (d) 1955

104. The rate at which RBI lends money to commercial Bank is
(a) Repo rate (b) Reverse repo rate
(c) Bank rate (d) None

105. The rate at which the central bank of a country borrows money from commercial bank
(a) Repo rate
(b) Reverse repo rate
(c) Bank rate
(d) None

106. The rate at which the central bank of a country lends money to commercial bank in the event of any short fall of funds by
(a) Repo rate
(b) Reverse repo rate
(c) Bank rate
(d) None

107. The procedure in which the central bank circulates government bonds in the open market is
(a) Variable reserve requirement policy
(b) Open market operation policy
(c) Bank rate policy
(d) None

108. The policy that completely under the control of the central bank is
(a) Fiscal policy
(b) Monetary policy
(c) Industrial policy
(d) EXIM policy

109. National Agriculture Credit (stabilization) fund came in the year
(a) 1951
(b) 1956
(c) 1960
(d) 1965

110. NABARD was established under the recommendation of
(a) Narsimhan Committee
(b) Nariman Committee
(c) Sivaraman Committee
(d) Patel Committee

111. NABARD came into existence
(a) 12th June, 1982
(b) 12th April, 1982
(c) 12th July,1982
(d) 12th December, 1982

112. Full form of CRAFICARD
(a) Committee to Rearranging formal Institutional Credit for Agriculture and Rural development
(b) Combine Rearrangement for Institutional Credit on Agriculture and Rural development
(c) Committee to Review Arrangements for Institutional Credit for Agriculture and Rural Development
(d) Council to review Arrangements for Institutional Credit for Agriculture and Rural Development

113. Authorized share capital of NABARD is
(a) Rs 100 crore (b) Rs 250 Crore
(c) Rs 500 crore (d) Rs 1000 crore

114. The extent of refinance for Pilot rainfed farming projects through NABARD are
(a) 50% (b) 25%
(c) 100% (d) 75%

115. The extent of refinance for waste land development schemes where indivisuals are beneficiaries through NABARD are
(a) 100% (b) 75%
(c) 50% (d) 25%

116. The extent of refinance for Agro Processing through NABARD are
(a) 75% (b) 50%
(c) 60% (d) 100%

117. The extent of refinance for Bio-gas through NABARD are
(a) 75% (b) 50%
(c) 60% (d) 100%

118. The extent of refinance for farm Mechanisation through NABARD are
(a) 25% (b) 50%
(c) 75% (d) 100%

119. The extent of refinance for Rural Electrification Corporation through NABARD are
(a) 25% (b) 50%
(c) 75% (d) 100%

120. The time period for short term loan refinance though NABARD is
(a) 6 month (b) 12 month
(c) 15 month (d) 18 month

121. National Bank Staff College is at
(a) Hyderabad (b) Nagpur
(c) Lucknow (d) Pune

122. Bankers Institute for Rural Development is at
(a) Hyderabad (b) Nagpur
(c) Lucknow (d) Pune

123. Vikas Volunteer Vahini is an educational programme to borrowers regarding ethics of repayment is started by
(a) NABARD
(b) SBI
(c) SIDBI
(d) Bank of Baroda

124. Small Farmer's Development Agency and Marginal Farmer's and Agricultural Labourer Development Agency came in
(a) 1970 (b) 1971
(c) 1975 (d) 1978

125. According to SFDA, the small farmers are eligible for getting subsidy upto
(a) 20% (b) 25%
(c) 331 / 3% (d) 50%

126. According to SFDA, the marginal farmers' and Agricultural Laborer are eligible for getting subsidy upto
(a) 20% (b) 25%
(c) 331 / 3% (d) 50%

127. Integrated Rural Development Programme (IRDP) is funded by state and central government in the ratio of
(a) 25:75 (b) 75:25
(c) 50:50 (d) 40:60

128. The subsidy for small farmers' under IRDP is
(a) 20% (b) 25%
(c) 331 / 3% (d) 50%

129. The subsidy for Marginal farmers' and Agricultural Labourer under IRDP is
(a) 20% (b) 25%
(c) 331 / 3% (d) 50%

130. The subsidy for Schedule Caste farmers' under IRDP is
(a) 20% (b) 25%
(c) 331 / 3% (d) 50%

131. The Self Help Group should be active existence for at least a period of
(a) 3 months
(b) 6 months
(c) 12 months
(d) 15 months

132. NABARD started a pilot project for linking SHG with banks in
(a) 1991
(b) 1992
(c) 1995
(d) 2000

133. Most popular SHG model in India is
(a) Direct Linkage Model
(b) Modified Direct Linkage Model II
(c) IFAD Model
(d) Modified Indirect Model

134. Which type of SHG model is found in Tamil Nadu, Maharashtra and Uttar Pradesh?
(a) Modified Direct Linkage Model II
(b) IFAD Model
(c) Indirect Linkage Model
(d) Modified Direct Linkage Model I

135. Ganga Kalyan Yojana has been launched in the year
(a) 1995 (b) 1997
(c) 1999 (d) 2001

136. The subsidy for SC / ST through Ganga Kalyan Yojana is
(a) 25% (b) 33 1/3%
(c) 50% (d) 75%

137. Kisan Credit Card has being launched in
(a) 1988-89 (b) 1998-99
(c) 2000-01 (d) 2005-06

138. The validity of Kisan Credit Card is for the period of
(a) 1 year (b) 2 year
(c) 3 year (d) Life time validity

139. The drawing power for contingent need under Kisan credit card is maximum up to
(a) Rs 5000 (b) Rs 10000
(c) Rs 20000 (d) No Limit

140. The Kisan Credit Card would be deemed to be non-performing asset, if it remains inoperative for the period for
(a) 1 year
(b) One cropping period
(c) 2 year
(d) Two cropping period

141. General insurance Scheme was put forward in the year
(a) 1973 (b) 1979
(c) 1985 (d) 1947

142. GIC was first administered in
(a) Maharashtra
(b) Gujarat
(c) Andhra Pradesh
(d) Punjab

143. Crop Insurance Scheme was applied first for
(a) Sonalika (b) CO 43
(c) H-4 Cotton (d) Co J 88

144. Pilot Crop insurance scheme came in the year
(a) 1973 (b) 1979
(c) 1985 (d) 1989

145. Which committee recommended the Pilot crop insurance scheme in India
(a) A D Gorwala Committee
(b) B Siva Raman Committee
(c) V M Dandekar Committee
(d) D R Gadgil Committee

146. Comprehensive Crop Insurance Scheme was introduced by GIC in
(a) 1979 (b) 1985
(c) 1989 (d) 1973

147. Under Comprehensive Crop Insurance Scheme, the crop insurance is shared by GIC and the state government in the ratio of
(a) 1:2 (b) 2:1
(c) 3:4 (d) 4:3

148. National Agricultural Insurance Scheme came in the year
(a) Kharif 1998-99
(b) Rabi 1999-00
(c) Rabi 2000-01
(d) Kharif 2001-02

149. First weather insurance product has being circulated to market by
(a) IFFCO-Tokio General Insurance company
(b) ICICI Lombard General Insurance company
(c) Agriculture Insurance Company of India
(d) None

150. Which of the following is the first 'R' of credit
(a) Rate of Return
(b) Returns
(c) Repayment Capacity
(d) Risk Bearing Ability

151. Which of the following is the second 'R' of credit?
(a) Rate of Return
(b) Returns
(c) Repayment Capacity
(d) Risk Bearing Ability

152. Which of the following is the third 'R' of credit?
(a) Rate of Return
(b) Returns
(c) Repayment Capacity
(d) Risk Bearing Ability

153. Which of the following is the first 'C' of credit?
(a) Condition (b) Character
(c) Capacity (d) Capital

154. Which of the following is the second 'C' of credit?
(a) Condition (b) Character
(c) Capacity (d) Capital

155. Which of the following is the third 'C' of credit?
(a) Condition (b) Character
(c) Capacity (d) Capital

156. Which of the following is the fourth 'C' of credit?
(a) Commonsense
(b) Capacity
(c) Condition
(d) Character

157. Which of the following is the fifth 'C' of credit?
(a) Commonsense (b) Capacity
(c) Condition (d) Character

158. Which of the following is the first 'P' of credit?
(a) Principle of Payment
(b) Principle of Productivity
(c) Principle of Productive Purpose
(d) Principle of Protection

159. Which of the following is the second 'P' of credit?
(a) Principle of Payment
(b) Principle of Personality
(c) Principle of Productive Purpose
(d) Principle of Protection

160. Which of the following is the third 'P' of credit?
(a) Principle of Payment
(b) Principle of Productivity
(c) Principle of Productive Purpose
(d) Principle of Protection

161. Which of the following is the fourth 'P' of credit?
(a) Principle of Productivity
(b) Principle of Productive Purpose
(c) Principle of Protection
(d) Principle of Phased disbursement

162. Which of the following is the fifth 'P' of credit?
(a) Principle of proper utilization
(b) Principle of Payment
(c) Principle of Protection
(d) Principle of personality

163. Which of the following is the sixth 'P' of credit?
(a) Principle of proper utilization
(b) Principle of Payment
(c) Principle of Protection
(d) Principle of personality

164. Which of the following is the seventh 'P' of credit?
(a) Principle of proper utilization
(b) Principle of Payment
(c) Principle of Protection
(d) Principle of personality

165. Which is the first stage of procedural formalities for disbursing loans to farmer?
(a) Scrutiny of records
(b) Submission of loan application by the farmer
(c) Interview with the farmer
(d) Visit to the farmers' field before sanction of loan

166. When the farmer clears principal amount in lumpsum and the interest amount is paid each year, then that repayment plan is called as
(a) Lumpsum repayment plan
(b) Straight-end repayment plan
(c) Single repayment plan
(d) All

167. The repayment plan in which the installment amount will be decreasing as the year pass by except in maturity year is called as
(a) Partial Repayment Plan
(b) Balloon Repayment Plan
(c) Both
(d) None

168. The repayment plan in which the principal component remains constant over the entire repayment period, while the interest part decreases continuously is called as
(a) Balloon repayment Plan
(b) Amortised Even Repayment Plan
(c) Amortised Decreasing Repayment Plan
(d) Variable Repayment Plan

169. The repayment plan in which the principal portion of the installment increases continuously, while the interest part declines gradually is called as
(a) Balloon repayment Plan
(b) Amortised Even Repayment Plan
(c) Amortised Decreasing Repayment Plan
(d) Variable Repayment Plan

170. The repayment plan in which installment is directly proportional to amount of harvest is called as
(a) Balloon repayment Plan
(b) Amortised Even Repayment Plan
(c) Amortised Decreasing Repayment Plan
(d) Variable Repayment Plan

171. Net Worth Statement is also called as
(a) Cash flow statement
(b) Fund flow statement
(c) Balance sheet
(d) Income statement

172. In Net Worth statement, liability is placed on which side
(a) Right side
(b) Left side
(c) Middle
(d) It is a subjective consideration

173. In Net Worth statement, asset is placed on which side?
(a) Right side (b) Left side
(c) Middle (d) None

174. Crop insurance policy reduces
(a) Production risk (b) Price risk
(c) Financial risk (d) Political risk

ANSWER KEY (Agricultural Finance)

1. (b) **2.** (c) **3.** (a) **4.** (b) **5.** (a) **6.** (b) **7.** (b) **8.** (b) **9.** (b) **10.** (a)
11. (b) **12.** (c) **13.** (b) **14.** (c) **15.** (d) **16.** (d) **17.** (c) **18.** (c) **19.** (c) **20.** (a)
21. (c) **22.** (b) **23.** (b) **24.** (b) **25.** (b) **26.** (b) **27.** (b) **28.** (c) **29.** (b) **30.** (c)
31. (c) **32.** (b) **33.** (d) **34.** (c) **35.** (b) **36.** (c) **37.** (c) **38.** (a) **39.** (c) **40.** (b)
41. (b) **42.** (c) **43.** (b) **44.** (a) **45.** (b) **46.** (c) **47.** (b) **48.** (d) **49.** (a) **50.** (c)

51. (b) **52.** (c) **53.** (c) **54.** (b) **55.** (a) **56.** (b) **57.** (c) **58.** (d) **59.** (a) **60.** (b)
61. (c) **62.** (d) **63.** (b) **64.** (a) **65.** (a) **66.** (b) **67.** (d) **68.** (c) **69.** (a) **70.** (c)
71. (a) **72.** (c) **73.** (b) **74.** (b) **75.** (c) **76.** (a) **77.** (b) **78.** (a) **79.** (c) **80.** (b)
81. (d) **82.** (b) **83.** (b) **84.** (c) **85.** (a) **86.** (b) **87.** (c) **88.** (c) **89.** (c) **90.** (c)
91. (c) **92.** (b) **93.** (c) **94.** (c) **95.** (c) **96.** (d) **97.** (a) **98.** (c) **99.** (b) **100.**(a)
101.(c) **102.**(d) **103.**(b) **104.**(c) **105.**(b) **106.**(a) **107.**(b) **108.**(b) **109.**(b) **110.**(c)
111.(c) **112.**(c) **113.**(c) **114.**(c) **115.**(a) **116.**(a) **117.**(a) **118.**(b) **119.**(b) **120.**(b)
121.(c) **122.**(c) **123.**(a) **124.**(b) **125.**(b) **126.**(c) **127.**(c) **128.**(b) **128.**(c) **130.**(d)
131.(b) **132.**(b) **133.**(b) **134.**(b) **135.**(b) **136.**(d) **137.**(b) **138.**(c) **139.**(b) **140.**(d)
141.(a) **142.**(b) **143.**(c) **144.**(b) **145.**(c) **146.**(b) **147.**(b) **148.**(b) **149.**(b) **150.**(b)
151.(c) **152.**(d) **153.**(b) **154.**(c) **155.**(d) **156.**(c) **157.**(a) **158.**(c) **159.**(b) **160.**(b)
161.(d) **162.**(a) **163.**(b) **164.**(c) **165.**(c) **166.**(d) **167.**(c) **168.**(c) **169.**(b) **170.**(d)
171.(c) **172.**(a) **173.**(b) **174.**(a)

CHAPTER 4

4. Marketing Management

1. Who is associated with the statement that marketing is the process of creation and delivery of standard of living?
(a) Mc Nair (b) Philip Kotler
(c) Peter Drucker (d) Mazur

2. Who is associated with the concept that competency is nothing but 'resourcefulness' means identifying resources when there are no resources?
(a) Gopal C Pati and Rabi K. Kanungo
(b) M. Maurice and F Sellier
(c) Evans and Lorange
(d) A Martin and G Ross

3. The 4P's of marketing mix are given by
(a) Lauterborn (b) E J McCarthy
(c) Rogers (d) Boom and Bitner

4. 4P of marketing mix is extended to 7P's by adding service mix- given by
(a) Lauterborn (b) E J McCarthy
(c) Rogers (d) Boom and Bitner

5. Which of the following is not included in 7 Ps of service marketing?
(a) People
(b) Product
(c) Physical evidence
(d) Public Relation

6. The term marketing may be described as the performance of business activities that
(a) Gives profit to the organisation
(b) Directs flow of goods from producer to consumer
(c) Directs selling
(d) Aims at satisfying the consumer

7. Which of the following is in order?
(a) Product, place, price and promotion.
(b) Price, product, promotion and place.
(c) Promotion, place, product and price.
(d) Product, price, place and promotion

8. In 'Marketing Myopia' the tendency of firms is to take a
(a) Broad view of the market
(b) Narrow view of the MIS
(c) Narrow view of the market
(d) Broad view of the MIS

9. Who is associated with concept of marketing myopia?
(a) Levitt
(b) Drucker
(c) Mc Carthy
(d) Kotler

10. "Customer will buy products only if the company promotes it products", is a marketing philosophy of which concept?
(a) Production Concept
(b) Product Concept
(c) Selling Concept
(d) Marketing Concept

11. "Consumers will prefer products that are widely available and inexpensive", is a philosophy of
(a) Production Concept
(b) Product Concept
(c) Selling Concept
(d) Marketing Concept

12. 4C's Concept, in place of 4P is given by
(a) Philip Kotler
(b) Gary Hamel
(c) Robert Lauterborn
(d) Chris Brogan

13. According to Robert Lauterborn, product is synonym to
(a) Customer cost
(b) Convenience
(c) Customer solution
(d) Communication

14. Which of the following is the sole vehicle for creating and delivering customer value?
(a) Marketing channel
(b) Marketing mix
(c) Marketing communication
(d) Marketing research

15. "Stop Smoking", is a type of
(a) Green Marketing Concept
(b) Cause-related marketing Concept
(c) Social Marketing Concept
(d) Humanistic Marketing Concept

16. Tata Chemicals Lt(d) started its, "Desh Ko Arpan" programme. It is a type of
(a) Green Marketing Concept
(b) Cause-related marketing Concept
(c) Social Marketing Concept
(d) Humanistic Marketing Concept

17. Green Marketing Principle relates to
(a) Agricultural Marketing
(b) Preservation of environment
(c) Social marketing environment
(d) Organic product marketing

18. Marketing Information System is the system of Information that flows to assist
(a) Promotional programmes
(b) Sales personnel development
(c) Marketing planning and decision making
(d) Marketing mix

19. ETOP analysis is done for scanning of environment, stands for
(a) Environmental Threat Opportunity Profile
(b) Environmental Threat Opportunity People
(c) Environmental Threat Opportunity Policy
(d) Environmental Threat Opportunity Paradigm

20. Environment scanning involves
(a) Weather forecasting
(b) Studying depression in the sea
(c) Identifying threats and opportunities
(d) Identifying strength and weaknesses

21. Macro marketing environment refers to
(a) Internal to the company
(b) Employees and shareholder of the company
(c) External to the company
(d) Stake holders

22. SWOT analysis is analytical tool for having
(a) Comparative advantage
(b) Competitive advantage
(c) Comprehension of current overall situation of the firm
(d) None of the above

23. Set up marketing tools the firm uses to pursue its marketing objectives is called as
(a) Marketing Myopia
(b) Marketing mix
(c) Mega marketing
(d) Meta marketing

24. A cluster of complementary products and services that are closely related in the minds of consumers but are spread across a diverse set of industries is called as
(a) Marketing Myopia
(b) Marketing mix
(c) Mega marketing
(d) Meta marketing

25. The strategic coordination of economic, psychological, political and public relations skill in marketing process is called as
(a) Marketing Myopia
(b) Marketing mix
(c) Mega marketing
(d) Meta marketing

26. The creation of a viable image in the minds of customers for a product is called as
(a) Segmentation (b) Targeting
(c) Positioning (d) All

27. Grouping of customers who share a similar set of needs and wants is called as
(a) Segmentation
(b) Targeting
(c) Positioning
(d) All

28. The type of marketing in which customer group seeking a distinctive mix of benefits and customer likes to give premium as well is called as
(a) Local Marketing
(b) Customisation
(c) Niche Marketing
(d) Value Marketing

29. Major segmentation category for products is
(a) Geographical Segmentation
(b) Demographic Segmentation
(c) Behavioural Segmentation
(d) Psychographic Segmentation

30. Segmentation of consumer's based on factors like climatic conditions and regions are known as
(a) Demographic segmentation
(b) Geographic segmentation
(c) Cultural segmentation
(d) Administrative segmentation

31. Segmentation of market on the basis of traits, lifestyles or values is called as
(a) Geographical Segmentation
(b) Demographic Segmentation
(c) Behavioural Segmentation
(d) Psychographic Segmentation

32. The psychographical factors in market segmentation include
(a) Income and occupation
(b) Sex and Age
(c) Hobbies and Activities
(d) Living standard and Income

33. Segmentation of market on the basis of age, sex, income or occupation is called as
(a) Geographical Segmentation
(b) Demographic Segmentation
(c) Behavioural Segmentation
(d) Psychographic Segmentation

34. Which of the demographic variable is not used by marketers for demographic segmentation?
(a) Income
(b) Gender
(c) Family life cycle
(d) Poverty

35. What is the basis for segmenting vitamin market?
(a) Age
(b) Gender
(c) Income
(d) Occupation

36. The loyalty status in which customers who buy only one brand all the time is called as
(a) Hard- core loyals
(b) Split loyals
(c) Shifting loyals
(d) Switchers

37. The loyalty status in which customers who are loyal to two or three brands is called as
(a) Hard- core loyals
(b) Split loyals
(c) Shifting loyals
(d) Switchers

38. The loyalty status in which customers who shift loyalty from one brand to another is called as
(a) Hard- core loyals
(b) Split loyals
(c) Shifting loyals
(d) Switchers

39. Consumers who show no loyalty to any brand is called as
(a) Hard- core loyals
(b) Split loyals
(c) Shifting loyals
(d) Switchers

40. A set of segments sharing some exploitable similarity is called as
(a) Single segment
(b) Mega segment
(c) Super segment
(d) Cluster segment

41. When a firm targets several market with its single product is called as
(a) Selective Specialization
(b) Product Specialization
(c) Market Specialization
(d) Full market Coverage

42. When a firm targets a many needs of a particular customer group is called as
(a) Selective Specialization
(b) Product Specialization
(c) Market Specialization
(d) Full market Coverage

43. The service or benefit the customer is really buying is called as
(a) Basic Product
(b) Expected Product
(c) Core benefit
(d) Potential Product

44. All the possible augmentations and transformations the product or offering might go in the future for the customer is called as
(a) Basic Product
(b) Expected Product
(c) Core benefit
(d) Potential Product

45. The goods consumers purchase on regular basis is called as
(a) Impulse goods
(b) Emergency goods
(c) Convenience goods
(d) Staples

46. The goods for which consumer has very little idea and normally do not like buying is called as
(a) Convenience goods
(b) Shopping goods
(c) Specialty goods
(d) Unsought goods

47. Number of brands a company is offering to its customer is called as
(a) Product System
(b) Product Mix
(c) Product Width
(d) Product Length

48. Total number of brands offered by a company in a particular product line is called as
(a) Product System
(b) Product Mix
(c) Product Width
(d) Product Length

49. Total number of variants of each products is called as
(a) Product System
(b) Product Mix
(c) Product Width
(d) Product Length

50. A group of diverse but related items that function in a compatible manner like in camera is called as
(a) Product System
(b) Product Mix
(c) Product Width
(d) Product Length

51. Increasing profits will most likely occur at which stage of the PLC?
(a) Maturity
(b) Growth
(c) Decline
(d) Introduction

52. Which of the following cannot be described by the PLC concept?
(a) product form
(b) product class
(c) product image
(d) Brand

53. The PLC concept can be applied by marketers as a useful framework for describing how?
(a) to develop marketing strategies
(b) to forecast product performance
(c) concept testing is conducted
(d) products and markets work

54. In which stage of the PLC will promotional expenditures be high in an attempt to react to increasing competition?
(a) Maturity (b) Growth
(c) Decline (d) Introduction

55. Which stage in the PLC normally lasts longer and poses strong challenges to the marketing managers?
(a) Maturity (b) Growth
(c) Decline (d) Introduction

56. Most products in the marketplace are in the stage of the product life cycle
(a) Maturity (b) Growth
(c) Decline (d) Introduction

57. Market Modification, Product modification, Market Program modification strategy come in which stage of PLC?
(a) Maturity (b) Growth
(c) Decline (d) Introduction

58. Harvesting and Divesting strategy is mainly observed in which stage of PLC?
(a) Maturity (b) Growth
(c) Decline (d) Introduction

59. Person who first suggests the idea of buying the particular product or service is called as
(a) Influencer (b) Initiator
(c) Decider (d) User

60. In organizational buying process, the person acts as a filter in the flow of information from buyer to seller or vice versa is called as
(a) Buyer (b) User
(c) Gate Keeper (d) Manager

61. People or Organizations who might conceivably have an interest in buying the company's products or services but may not have the means or real intention to buy is called as
(a) Prospect
(b) Suspect
(c) Clients
(d) Advocates

62. People or customers with the motivation, ability and opportunity to make a purchase is called as
(a) Prospect
(b) Suspect
(c) Clients
(d) Advocates

63. The person who duplicates the leader's product and packages and sells it on the black market is called as
(a) Cloner (b) Counterfeiter
(c) Imitator (d) Adapter

64. The person who emulates the leader's products, name and packaging, with slight variation is called as
(a) Cloner (b) Counterfeiter
(c) Imitator (d) Adapter

65. The person who copies somethings from the leader but maintains differentiation in terms of packaging, advertising, pricing or location is called as
(a) Cloner (b) Counterfeiter
(c) Imitator (d) Adapter

66. The person who takes the leader's products and adapts or improves them is called as
(a) Cloner (b) Counterfeiter
(c) Imitator (d) Adapter

67. In which types of promotional tools impersonality is observed?
(a) Sales Promotion
(b) Publicity
(c) Advertising
(d) Personal selling

68. In which type of promotional tools, per unit cost is very low or absolutely zero?
(a) Sales Promotion
(b) Publicity
(c) Advertising
(d) Personal selling

69. Which is the best tool to promote your product in the later stage of buying process?
(a) Sales Promotion
(b) Publicity
(c) Advertising
(d) Personal selling

70. Which promotional tool is mainly preferred in business market?
(a) Sales Promotion
(b) Publicity
(c) Advertising
(d) Personal selling

71. Which promotional tool has maximum impact on decline phase of PLC?
(a) Sales Promotion (b) Publicity
(c) Advertising (d) Personal selling

72. One time mailing comes under
(a) Single vehicle single stage campaign
(b) Single vehicle multi stage campaign
(c) Multi vehicle multi stage campaign
(d) Multi vehicle single stage campaign

73. When an advertiser is scheduling exposures evenly throughout a given period is called as?
(a) Concentration (b) Continuity
(c) Flighting (d) Pulsing

74. When an advertiser is spending all the advertising budget in a single period, it is called as?
(a) Concentration (b) Continuity
(c) Flighting (d) Pulsing

75. When a company is using one brand to advertise another non competing brand is called as?
(a) Tie in Promotion
(b) Cross Promotion
(c) Dual Promotion
(d) Double Promotion

76. When two or more brands or companies team up on coupons, refunds and contests to increase pulling power is called as?
(a) Tie in Promotion
(b) Cross Promotion
(c) Dual Promotion
(d) Double Promotion

77. The level of selective demand for the company's product is called as
(a) Market demand
(b) Market Share
(c) Market Penetration Index
(d) Share Penetration Index

78. The comparison between the current level of market demand to the potential demand level is called as
(a) Market demand
(b) Market Share
(c) Market Penetration Index
(d) Share Penetration Index

79. The attacking strategy in which attacker matches its opponent's products, advertising, price and distribution is called as
(a) Flank Attack Strategy
(b) Encirclement Attack Strategy
(c) Frontal Attack Strategy
(d) By Pass Attack Strategy

80. An enemy's weak spots are natural targets- is the attacking strategy in
(a) Flank Attack Strategy
(b) Encirclement Attack Strategy
(c) Frontal Attack Strategy
(d) By Pass Attack Strategy

81. Defensive strategy in which, company involves itself for building a superior brand power and making the brand almost impregnable is called as
(a) Flank defensive strategy
(b) Position defensive strategy
(c) Pre-emptive defensive strategy
(d) Counter offensive defensive strategy

82. Guerrilla action is mainly observed in which defensive strategy?
(a) Flank defensive strategy
(b) Position defensive strategy
(c) Pre-emptive defensive strategy
(d) Counter offensive defensive strategy

83. Many buyers want to offer other items in payment, a practice is known as
(a) Allowance (b) Auction
(c) Counter trade (d) Discount

84. When the seller receives some items in cash and the rest in products is called as
(a) Barter System
(b) Compensation deal
(c) Buyback Arrangement
(d) Offset

85. When the seller sells a plant, equipment or technology to another country and agrees to accept as partial payment products is called as
(a) Barter System
(b) Compensation deal
(c) Buyback Arrangement
(d) Offset

86. The condition of entry and exit barrier to a new entrants for an attractive segment is
(a) Low Entry barrier, High Exit Barrier
(b) High Entry Barrier, Low Exit Barrier
(c) Low Entry Barrier, Low Exit Barrier
(d) High Entry Barrier, High Exit Barrier

87. The condition of entry and exit barrier to a new entrants for a segment where profit potential is high is
(a) Low Entry barrier, High Exit Barrier
(b) High Entry Barrier, Low Exit Barrier
(c) Low Entry Barrier, Low Exit Barrier
(d) High Entry Barrier, High Exit Barrier

88. The condition of entry and exit barrier to a new entrants for a segment where returns are stable and low is
(a) Low Entry barrier, High Exit Barrier
(b) High Entry Barrier, Low Exit Barrier
(c) Low Entry Barrier, Low Exit Barrier
(d) High Entry Barrier, High Exit Barrier

89. The condition of entry and exit barrier to a new entrants for a segment where overcapacity may be a headache in future is
(a) Low Entry barrier, High Exit Barrier
(b) High Entry Barrier, Low Exit Barrier
(c) Low Entry Barrier, Low Exit Barrier
(d) High Entry Barrier, High Exit Barrier

90. Which of the following book is not written by Michel E Porter
(a) Fortune at the Bottom of the Pyramid
(b) Competitive Strategy
(c) Competitive Advantage
(d) Competitive Advantage of Nation

91. Which of the following factors does not influence the pricing strategy of the firm?
(a) Market (b) Cost
(c) Competition (d) R & D

92. A company using high price and high promotion policy is adopting which of the following strategy
(a) Slow Skimming Strategy
(b) Rapid Penetration Strategy
(c) Slow Penetration Strategy
(d) Rapid Penetration Strategy

93. A pricing strategy in which company sets a higher price for its product during initial entry to the market is called as
(a) Market Penetration Price
(b) Market Skimming Price
(c) Upper Bound Price
(d) Maximum Current Profit Price

94. A pricing strategy in which company sets a lowest price for its product during initial entry to the market is called as
(a) Market Penetration Price
(b) Market Skimming Price
(c) Upper Bound Price
(d) Maximum Current Profit Price

95. Oligopolistic Industry mainly go for
(a) Perceived value pricing Strategy
(b) Auction type pricing Strategy
(c) Going –Rate Pricing Strategy
(d) Mark Up Pricing Strategy

96. Trade discount is also called as
(a) Seasonal Discount
(b) Cash Discount
(c) Functional Discount
(d) Quantity Discount

97. Pricing strategy when a discount stores often drop the price on well-known brands is called as
(a) Psychological Discount Price
(b) Special-event Price
(c) Loss-leader Pricing
(d) Image Pricing

98. For destroying competition, company even sells its product below its cost. This pricing strategy is called as
(a) Yield pricing
(b) Predatory pricing
(c) Auction type pricing
(d) Geographical pricing

99. Who is not the facilitator in marketing channels?
(a) Manufacturer Representative
(b) Bank
(c) Transportation Companies
(d) Advertising agency

100. Consumer's are aware of damages caused by packaging, product process, so they prefer
(a) Social marketing concept
(b) Direct marketing concept
(c) Green marketing concept
(d) Organic marketing concept

101. When the producer wants to maintain control over the services level and outputs offered by the reseller is called as
(a) Selective Distribution
(b) Intensive Distribution
(c) Exclusive Distribution
(d) None

102. When a company launched a web site without any previous existence as a firm is called as
(a) Brick company
(b) Brick and Click company
(c) Pure click company
(d) Pure Brick company

103. DSA means
(a) Directly Selling Authority
(b) Directly Selling Action
(c) Direct Selling Authority
(d) Direct Selling Agent

104. Which stages of new-product development, products often undergo rigorous tests to make sure that they perform safely and effectively or that consumers will find value in them?
(a) Product development
(b) Concept development and testing
(c) Marketing mix
(d) Business analysis

105. During which stage of new-product development is management most likely to estimate minimum and maximum sales to assess the range of risk in launching a new product
(a) Marketing strategy development
(b) Concept testing
(c) Business analysis
(d) Product development

106. After concept testing, a firm would engage in which stage in developing and marketing a new product?
(a) Test marketing
(b) Marketing strategy development
(c) Product development
(d) Business analysis

107. Go error and drop error steps are mainly found in which stage of new product development?
(a) Idea generation
(b) Idea screening
(c) Concept development
(d) Concept testing

108. To ensure success of new products, which market segment should be targeted initially
(a) Innovators
(b) Laggards
(c) Early majority
(d) Early adopters

109. The concept of Core Competency is advanced by
(a) Ghoshla and Nohria
(b) Prahallad and Hammel
(c) Friedman and Philips
(d) Evans and Lorange

110. "Positioning – the Battle for the Mind", statement is given by
(a) Philip Kotler
(b) Peter Drucker
(c) AlRice and Jack trout
(d) Friedman and Philips

111. In channel of distribution of Industrial products, which level is popular?
(a) 0-1 Level
(b) 1-2 Level
(c) 2 Only
(d) Multi Level

112. When the seller requires that these dealers not handle competitors' product, this is called as
(a) Intensive Distribution
(b) Exclusive Distribution
(c) Exclusive dealing
(d) Full line forcing

113. Producers of a strong brand sometimes sell it to dealers only if they will some or all of the rest of the line. This practice is called as
(a) Intensive Distribution
(b) Exclusive Distribution
(c) Exclusive dealing
(d) Full line forcing

114. The customer who purchase from the same place in the same manner is categorized in
(a) High involvement shopper
(b) Variety loving shopper
(c) Habitual shopper
(d) High value deal seeker

115. Buyer who know their needs and channel surf a great deal before buying at the lowest possible price is categorized in
(a) High involvement shopper
(b) Variety loving shopper
(c) Habitual shopper
(d) High value deal seeker

116. Vertical marketing system where from production to distribution is under single ownership, is called as
(a) Administrative VMS
(b) Corporate VMS
(c) Contractual VMS
(d) None

117. The source of competitive advantage that has a significant perceived customer benefit and it is very difficult to imitate by the competitor is called as
(a) Strategic Marketing Plan
(b) Distinct Capability
(c) Core Competency
(d) Core Value

118. Which one of the following is not coming under principle of personal selling?
(a) Prospecting
(b) Promoting
(c) Presentation and Demonstration
(d) Follow up and Maintenance

119. Impulse buying implies
(a) Purchasing with little or no advance planning
(b) Purchasing involving all family members
(c) Purchasing involving all relatives
(d) None of the above

120. A brand is a means of
(a) Naming the products
(b) Packaging of the products
(c) Contents of the product
(d) Identifying and differentiating the company's product

121. Which of the following indicates the distinction between Advertising and Publicity?
(a) Advertising is expensive, publicity is less expensive.
(b) Advertising is through T.V., publicity is through print media
(c) Advertising is a paid form of communication, Publicity is not a paid form of communication.
(d) Advertising is through film stars, Publicity is through cricketers

122. Companies that attack other firms including market leader in an attempt to build market share are called
(a) Market followers
(b) Market challengers
(c) Market penetrators
(d) Market Nichers

123. Distribution channel aims at moving products from producer to
(a) Ultimate consumer
(b) Stockists
(c) Retailer
(d) Broker

124. In business situations, personal selling is
(a) Inside selling
(b) Outside selling
(c) Both
(d) None

125. Mass communication tools available to marketers include
(a) Advertising
(b) Sales Promotion
(c) Public relation
(d) All of the above

126. The financial results of marketing endeavours should be analysed in terms of
(a) Sales volume
(b) Market share
(c) Marketing costs
(d) All of the above

127. In channel distribution of Industrial products which level is popular?
(a) Zero level and one level
(b) Two level only
(c) One level and two level
(d) Multi level

128. Product management do not cover one of the following aspects
(a) Product line
(b) Product mix
(c) Product coordination
(d) Product life cycle

129. Industrial marketing involves
(a) Business to Business
(b) Customers to Customers
(c) Online marketing
(d) Customers to Business

130. In the diffusion of innovation who are the people not willing to try out the product?
(a) Innovators
(b) Progressive people
(c) Early Adopter
(d) Early majority

131. Marketing research does not normally
(a) Gather environmental information
(b) Provide a continuous source of information
(c) Relates to all aspects of marketing information
(d) Describe the current situation

132. Emerging market economies are
(a) A part of developed countries
(b) Newly industrialised countries
(c) A part of developing countries
(d) A part of third world countries

133. Marketing practices that reduce the number of Organisational levels to get closure to the customer is
(a) Thinning (b) Sliding
(c) Flattening (d) Focusing

134. Like Chinese products are cheap, Japanese products are
(a) Prestigeous (b) Innovative
(c) Standardised (d) Hassel free

135. If a store owner is an unauthorized dealer of a certain high-end electronics brand, the product is considered to be sold in the
(a) Black market (b) Gray market
(c) White market (d) None

136. The global entry marketing strategy in which a company's gradually entered sequentially is
(a) Waterfall strategy
(b) Sprinkler strategy
(c) Both
(d) None

137. The marketing strategy when a company mainly search uncontested market space and try to make the competition irrelevant
(a) Red ocean strategy
(b) Blue ocean strategy
(c) Both
(d) None

138. USP in marketing stands for
(a) United States Pharmacopeia
(b) Unique Sell Parity
(c) Unique Sell Positioning
(d) Unique Selling Proposition

139. Whereas advertising offers a reason to buy, sales promotion offers
(a) Answer to buy
(b) Reward to buy
(c) Incentive to buy
(d) Energy to buy

140. The last 'M' of advertising is
(a) Mission (b) Money
(c) Measurement (d) Media

141. The marketing principle when communicators can use their good image to reduce some negative feelings toward a brand is called
(a) Principle of personality
(b) Principle of parity
(c) Principle of congruity
(d) Principle of clarity

142. Putting the goods in boxes and transport through two transporting medium like water and truck is called
(a) Piggyback (b) Fishyback
(c) Trainship (d) Airtruck

143. Putting the goods in boxes and transport through two transporting medium like rail and truck is called
(a) Piggyback (b) Fishyback
(c) Trainship (d) Airtruck

144. Putting the goods in boxes and transport through two transporting medium like rail and water is called
(a) Piggyback (b) Fishyback
(c) Trainship (d) Airtruck

145. Number of channels in direct marketing is
(a) 0 (b) 1
(c) 2 (d) More than 2

ANSWER KEY (Marketing Management)

1. (b)	**2.** (a)	**3.** (b)	**4.** (d)	**5.** (d)	**6.** (d)	**7.** (d)	**8.** (c)	**9.** (a)	**10.** (c)
11. (a)	**12.** (c)	**13.** (c)	**14.** (c)	**15.** (c)	**16.** (b)	**17.** (b)	**18.** (c)	**19.** (a)	**20.** (c)
21. (c)	**22.** (c)	**23.** (b)	**24.** (b)	**25.** (c)	**26.** (c)	**27.** (a)	**28.** (c)	**29.** (b)	**30.** (b)
31. (d)	**32.** (c)	**33.** (b)	**34.** (d)	**35.** (b)	**36.** (a)	**37.** (b)	**38.** (c)	**39.** (d)	**40.** (c)
41. (b)	**42.** (c)	**43.** (c)	**44.** (d)	**45.** (d)	**46.** (d)	**47.** (b)	**48.** (d)	**49.** (c)	**50.** (a)
51. (b)	**52.** (c)	**53.** (d)	**54.** (b)	**55.** (a)	**56.** (a)	**57.** (a)	**58.** (c)	**59.** (b)	**60.** (c)
61. (b)	**62.** (a)	**63.** (b)	**64.** (a)	**65.** (c)	**66.** (d)	**67.** (c)	**68.** (b)	**69.** (d)	**70.** (d)
71. (a)	**72.** (a)	**73.** (b)	**74.** (a)	**75.** (b)	**76.** (a)	**77.** (b)	**78.** (c)	**79.** (c)	**80.** (a)
81. (b)	**82.** (c)	**83.** (c)	**84.** (b)	**85.** (c)	**86.** (b)	**87.** (d)	**88.** (c)	**89.** (a)	**90.** (a)
91. (d)	**92.** (b)	**93.** (b)	**94.** (a)	**95.** (c)	**96.** (c)	**97.** (c)	**98.** (b)	**99.** (a)	**100.** (b)
101. (c)	**102.** (c)	**103.** (d)	**104.** (a)	**105.** (c)	**106.** (b)	**107.** (b)	**108.** (a)	**109.** (b)	**110.** (c)
111. (d)	**112.** (c)	**113.** (d)	**114.** (c)	**115.** (d)	**116.** (b)	**117.** (c)	**118.** (b)	**119.** (a)	**120.** (a)
121. (c)	**122.** (b)	**123.** (a)	**124.** (c)	**125.** (d)	**126.** (d)	**127.** (d)	**128.** (d)	**129.** (a)	**130.** (b)
131. (b)	**132.** (d)	**133.** (c)	**134.** (b)	**135.** (b)	**136.** (a)	**137.** (b)	**138.** (d)	**139.** (c)	**140.** (c)
141. (c)	**142.** (b)	**143.** (a)	**144.** (c)	**145.** (a)					

CHAPTER 5

5. Agricultural and Rural Marketing

1. Market comes from the word
(a) Mercatus
(b) Murcatus
(c) Muscatus
(d) Marcatus

2. In Marketing, Third "P" represents as
(a) Product
(b) Price
(c) Promotion
(d) Place

3. Market defines as
(a) A Place or A Space
(b) An Area
(c) A social institution
(d) Combines all of the above

4. Under modern terminal market, subsidy on capital investment in Agriculture is
(a) 20%
(b) 25%
(c) 331/3%
(d) 50%

5. Modern terminal market is under the purview of
(a) Ministry of Commerce
(b) Ministry of Finance
(c) Ministry of Agriculture & Cooperaion
(d) None of the above

6. Spontaneous Market of developed country simulates
(a) Secular Market
(b) Long Period Market
(c) Periodic Market
(d) Short Period Market

7. When a large number of sellers deal with heterogeneous and differentiated commodity is called as
(a) Perfect Competition
(b) Monopoly
(c) Oligopoly
(d) Monopolistic Competition

8. Unzha market deals in
(a) Pepper (b) Tamarind
(c) Cumin (d) Sugar

9. Meerut commodity market is famous for
(a) Hides & Skin (b) Wheat
(c) Rice (d) Sugar

10. In Input market, farmers mainly face
(a) Perfect Competition
(b) Monopoly
(c) Oligopoly
(d) Monopolistic Competition

11. Hapur and Khanna commodity market deals in
(a) Rice (b) Sugar
(c) Wheat (d) Mesta

12. Which crop is not grown in Kharif season?
(a) Cotton
(b) Maize
(c) Mustard
(d) Sorghum

13. In Agriculture, demand of food grain is mostly
(a) Elastic
(b) Inelastic
(c) Perfectly inelastic
(d) Perfectly elastic

14. Market Intervention Scheme is under the control of
(a) State Govenment
(b) Central Government
(c) Both
(d) Either of the two

15. Numbers of commodity included under MSP
(a) 23 (b) 24
(c) 25 (d) 26

16. The common relationship between Marketed Surplus and Marketable surplus for perishable commodity in agriculture is
(a) Marketable surplus is greater than Marketed surplus
(b) Marketable surplus is less than Marketed surplus
(c) Both are equal
(d) Cannot say

17. Divesting strategy is observed in which stage of PLC?
(a) Introduction Stage
(b) Growth Stage
(c) Maturity Stage
(d) Decline Stage

18. As on date 30/06/2011, the number of regulated market in India was
(a) 21238 (b) 20238
(c) 7246 (d) 7146

19. As on date 30/06/2011, The number of Rural periodic market in India was
(a) 21238 (b) 20238
(c) 7246 (d) 7146

20. Marketing research and training of agricultural commodity is under the guidance of
(a) Director of Marketing
(b) Agricultural Marketing Board
(c) FARD
(d) State Agricultural Marketing Department

21. All India average area served by a regulated market in India is
(a) 103 sq. km (b) 454 sq. km
(c) 11,215 sq. km (d) All of the above

22. The full form of APMC is
(a) Agro Processing Marketing Committee
(b) Agricultural Produce Marketing Center
(c) Agricultural Produce Market Committee
(d) Agricultural Prepared Market Committee

23. The state that has permitted private players to purchase agro-commodity but has not given permission to set up private market of their own is
(a) Maharashtra
(b) Madhya Pradesh
(c) Odisha
(d) Gujarat

24. Only one private market has being set up till yet in
(a) Madhya Pradesh
(b) Gujarat
(c) Karnataka
(d) Maharashtra

25. The states that have issued a single license to the organisation who is going for direct marketing of agro-commodity are
(a) Gujarat (b) Maharashtra
(c) Karnataka (d) All of the above

26. The state that has exempted 30% of market fee on purchase under contract agreement is
(a) Gujarat
(b) Maharashtra
(c) Madhya Pradesh
(d) Karnataka

27. The commission charges for fruit and vegetable is
(a) 1-2.5% (b) 2-5%
(c) 4-8% (d) > 8%

28. The state that abolished the commission agent for agro-commodity is
(a) Maharashtra
(b) Madhya Pradesh
(c) Gujarat
(d) Karnataka

29. The Farmer's market in Odisha is known by
(a) Apni Mandi (b) Krushak Bazar
(c) Shetkari Bazar (d) Kisan Mandi

30. National Horticultural Mission approved a modern terminal market in Perundurai of
(a) Andhra Pradesh (b) Kerala
(c) Tamil Nadu (d) Bihar

31. For contract farming of agricultural products the buyer should render a bank guarantee is the rule of
(a) Madhya Pradesh
(b) Andhra Pradesh
(c) Tamil Nadu
(d) Gujarat

32. For direct marketing of agricultural products Rs 50000/- has being deposited by the player in
(a) Madhya Pradesh
(b) Andhra Pradesh
(c) Tamil Nadu
(d) Gujarat

33. A minimum cost of Rs 10 cr is required by the organisation to set up a private market in
(a) Madhya Pradesh
(b) Andhra Pradesh
(c) Tamil Nadu
(d) Gujarat

34. Kisan Mandi is the direct marketing model of
(a) Rajasthan (b) Gujarat
(c) Haryana (d) Punjab

35. The commission is payable by the seller in
(a) Delhi (b) AP
(c) Tamil Nadu (d) All of the above

36. For fruits and vegetables, the commission is charged from
(a) Buyer
(b) Seller
(c) Both Buyer and Seller
(d) Sometimes by Buyer, Sometimes by seller

37. Which of the following is incorrect option?
(a) APMC bring transparency and provide market led extension
(b) It is mandatory to sell your product within the market area
(c) Grower and legal person are eligible to set up new markets for agro-commodity
(d) Single point levy has being enacted for the agro commodity

38. The number of states that have implemented a policy for single point levy till yet is
(a) 10 (b) 13
(c) 20 (d) 23

39. Only one market committee is seen in
(a) Andhra Pradesh (b) Tamil Nadu
(c) Kerala (d) Punjab

40. Agricultural Marketing Board is advisory in nature in state like
(a) Odisha
(b) Andhra Pradesh
(c) Tamil Nadu
(d) All of the above

41. Apni Mandi / Kisan Mandi is found in
(a) Maharashtra (b) Tamil Nadu
(c) Punjab (d) Odisha

42. Hadaspar Vegetable Market is seen in
(a) Pune (b) Hyderabad
(c) Chennai (d) Lucknow

43. Rythu Bazar is a direct marketing concept of fruits and vegetables is found in
(a) Kanpur (b) Bhubaneswar
(c) Hyderabad (d) Bangalore

44. Uzhavar Sandies is seen in
(a) Maharashtra (b) Tamil Nadu
(c) Haryana (d) Uttar Pradesh

45. Krushak Bazar is a marketing platform between farmer and consumer is seen in
(a) Andhra Pradesh (b) Bihar
(c) Odisha (d) West Bengal

46. Safal Market is the innovative concept of
(a) GCMMF (b) NDDB
(c) ITC (d) Nestle

47. Vegetable and fruit promotion council is an innovative method of marketing of fruits and vegetables through integration of Self Help Group is found in
(a) Meghalaya
(b) Madhya Pradesh
(c) Kerala
(d) Rajasthan

48. LOOP means
(a) Law Of Only Price
(b) Law Of Once Price
(c) Law Of One Price
(d) None

49. LOOP is the concept of
(a) Monopolistic Competition
(b) Duopolistic competition
(c) Perfect Market Competition
(d) Monopoly Competition

50. Central Institute of Agricultural Engineering is at
(a) Barelli (b) Bhopal
(c) Bhubaneswar (d) Baroda

51. Central Storage Institute is at
(a) Pune (b) Kanpur
(c) Chandigarh (d) Palanpur

52. Indian Institute of Packaging is at
(a) New Delhi (b) Mumbai
(c) Chennai (d) Calcutta

53. According to research, transportation cost accounts for about
(a) 25% of total cost
(b) 75% of total cost
(c) 50% of total cost
(d) 60% of total cost

54. Central AGMARK laboratory is at
(a) New Delhi (b) Pune
(c) Hyderabad (d) Nagpur

55. The grading and marking of goods was first started in
(a) 1935 (b) 1938
(c) 1945 (d) 1950

56. The grading and marking of goods was first started for
(a) Vegetable Oil (b) Butter
(c) Rice (d) Ghee

57. The grading and marking of goods for potato was started in
(a) 1940 (b) 1950
(c) 1955 (d) 1957

58. The grading and marking of goods for Rice was started in
(a) 1945 (b) 1950
(c) 1954 (d) 1957

59. The grading and marking of goods for wheat was started in
(a) 1940 (b) 1950
(c) 1955 (d) 1957

60. Directorate of Marketing and Inspection (DMI) came in
(a) Jan 1, 1935 (b) June 1, 1935
(c) July 1, 1935 (d) April 1, 1935

61. Head quarter of DMI is at
(a) Lucknow (b) Hyderabad
(c) Faridabad (d) New Delhi

62. "AGMARKING" for exportable commodity is compulsory, was first started in
(a) 1942 (b) 1945
(c) 1955 (d) 1965

63. "AGMARKING" for exportable commodity is compulsory, was first started for
(a) Tobacco
(b) Wool
(c) Sunnhemp
(d) Sandal Wood Oil

64. Latin Name of codex alimentarius commission
(a) Food And Code law
(b) Food for Code law
(c) Food Law and Food Code
(d) Code for Food Law

65. The full form of MPL in codex alimentarius commission is
(a) Minimum Permissible Level
(b) Maximum Permitted Level
(c) Both
(d) None

66. Codex alimentarius commission came in
(a) 1955 (b) 1963
(c) 1970 (d) 1975

67. The total members under codex alimentarius commission is
(a) 181 (b) 186
(c) 190 (d) 175

68. International Standard of Organisation came in
(a) 1945 (b) 1947
(c) 1955 (d) 1960

69. Indian Standard Institution(certification marks). Act came in
(a) 1951 (b) 1952
(c) 1961 (d) 1975

70. National Grid of Rural Godown was launched in
(a) 1975 (b) 1979
(c) 1985 (d) 1990

71. ECOMARK scheme is administered by
(a) ISO (b) CAC
(c) BIS (d) FAO

72. Non Vegetarian Packed food is reflected in
(a) Dot in a square in green colour
(b) Dot in a square in red colour
(c) Dot in a square in brown colour
(d) Dot in a square in orange colour

73. HACCP stands for
(a) Heavy Analysis and Critics Comment Point
(b) Hot Agenda and Calculated Control Point
(c) Hazard Analysis and Critical Control Point
(d) None

74. CAC is accepted as standard mainly by
(a) FAO (b) WHO
(c) WTO (d) UNICEF

75. HACCP is accepted as standard mainly by
(a) FAO (b) WHO
(c) WTO (d) Both a and b

76. Food Standard Agency of UK is called as
(a) FII (b) FDA
(c) FSA (d) FPA

77. Food Standard Agency of USA is called as
(a) FII (b) FDA
(c) FSA (d) FPA

78. World Standard day is celebrated annually on
(a) 10th January
(b) 14th October
(c) 16th March
(d) 30th September

79. The ISI mark standard design in a rectangle having width and Height ratio
(a) 3:4 (b) 4:3
(c) 3:2 (d) 5:7

80. Caveat emptor means
(a) Let the buyer aware
(b) Let the seller aware
(c) Let the buyer beware
(d) Let the seller beware

81. Caveat venditor means
(a) Let the buyer beware
(b) Let the seller beware
(c) Both
(d) None

82. Which four states have their own cold storage Acts?
(a) Punjab, Bihar, Uttar Pradesh, Odisha
(b) West Bengal, Uttar Pradesh, Bihar, Haryana
(c) Uttar Pradesh, West Bengal, Punjab, Haryana
(d) Punjab, Bihar, Punjab, Haryana

83. West Bengal enacted their own cold storage Act in
(a) 1955 (b) 1960
(c) 1970 (d) 1980

84. Uttar Pradesh enacted their own cold storage Act in
(a) 1955 (b) 1960
(c) 1975 (d) 1980

85. Punjab and Haryana enacted their own cold storage Act in
(a) 1965 (b) 1960
(c) 1979 (d) 1980

86. Value addition for rice and wheat in India is
(a) 5% (b) 7%
(c) 10% (d) 20%

87. Value addition for cotton in India is
(a) 40% (b) 55%
(c) 79% (d) 81%

88. Value addition for tea in India is
(a) 40% (b) 55%
(c) 79% (d) 86%

89. In India, milk is processed up to
(a) 45% (b) 35%
(c) 25% (d) 65%

90. Ministry of Food Processing was set up in
(a) 1975 (b) 1985
(c) 1988 (d) 1995

91. Central Marketing Department was established in
(a) 1900 (b) 1935
(c) 1955 (d) 1975

92. The first regulated Karanjia Cotton Market was established in
(a) 1900 (b) 1886
(c) 1915 (d) 1875

93. Market Intervention scheme came in
(a) 2000-01 (b) 2010-11
(c) 2005-06 (d) 1995-96

94. The total number of commodity come under Minimum Support Price is
(a) 24 (b) 20
(c) 25 (d) 30

95. Statutory Minimum Price is for
(a) Rice (b) Wheat
(c) Sugar Cane (d) Potato

96. Keeping the rise in demand under check by allotting limited quantity per capita per time period is called as
(a) Balancing (b) Neutralising
(c) Rationing (d) Buffering

97. Cotton Corporation of India was set up in
(a) 1965 (b) 1970
(c) 1971 (d) 1958

98. Jute Corporation of India was set up in
(a) 1965 (b) 1970
(c) 1971 (d) 1958

99. Food Corporation of India was set up in
(a) 1965 (b) 1970
(c) 1971 (d) 1958

100. Council of State Agricultural Marketing Board (COSAMB) came in the year
(a) 1955 (b) 1985
(c) 1988 (d) 1995

101. AGMARKNET came in
(a) 8th five year plan
(b) 9th five year plan
(c) 10th five year plan
(d) 11th five year plan

102. Cotton Corporation of India is responsible for price support operation in all states, except
(a) Madhya Pradesh
(b) Gujarat
(c) Maharashtra
(d) Haryana

103. APEDA stands for
(a) Agricultural and Produce Food Export development Authority
(b) Agricultural and Processed Food Products Export Development Authority
(c) Agricultural and Procure Food Products Export development Authority
(d) None

104. APEDA came in the year
(a) 1975 (b) 1986
(c) 1990 (d) 1995

105. National Meat and Poultry processing Board is at
(a) New Delhi (b) Mumbai
(c) Kolkata (d) Hyderabad

106. Indian Grape Processing Board is at
(a) Hyderabad
(b) Pune
(c) Nasik
(d) Nagpur

107. Indian Institute of Crop Processing Technology is at
(a) Pune, Maharashtra
(b) Cuttack, Odisha
(c) Hyderabad, Andhra Pradesh
(d) Thanjavur, Tamil Nadu

108. Coffee Market Expansion Board came in the year
(a) 1953 (b) 1940
(c) 1947 (d) 1976

109. Tea Board Act, came in the year
(a) 1953 (b) 1940
(c) 1947 (d) 1976

110. Rubber Board Act, came in the year
(a) 1953 (b) 1940
(c) 1947 (d) 1976

111. Tobacco Board was established in
(a) 1953 (b) 1940
(c) 1947 (d) 1976

112. Spices Board was established in
(a) 1975 (b) 1953
(c) 1949 (d) 1987

113. Coir Board was set up in
(a) 1953 (b) 1940
(c) 1947 (d) 1976

114. National Dairy development Board came in
(a) 1960 (b) 1965
(c) 1970 (d) 1995

115. National Horticultural Board was set up in
(a) 1980 (b) 1955
(c) 1984 (d) 1987

116. Marine Products Export Development Authority (MPEDA) was set up in
(a) 1971 (b) 1972
(c) 1975 (d) 1970

117. Silk board was established in
(a) 1953 (b) 1949
(c) 1947 (d) 1976

118. Agricultural produce (Grading and Marking) Act was set up in
(a) 1937 (b) 1945
(c) 1951 (d) 1955

119. Bureau of Indian standard came in the year
(a) 1975 (b) 1986
(c) 1991 (d) 1995

120. Meat Food Products Order came in
(a) 1953 (b) 1955
(c) 1973 (d) 1980

121. Milk and Milk products order came in
(a) 1973 (b) 1992
(c) 1995 (d) 2000

122. Standards of weights and measures Act, came in
(a) 1945 (b) 1958
(c) 1965 (d) 1980

123. Standards of weights and measures (packed goods) Act came in
(a) 1958 (b) 1991
(c) 1997 (d) 2000

124. Prevention of Food Adulteration Act, came in
(a) 1954 (b) 1965
(c) 1980 (d) 1951

125. Trade and Merchandise Marks Act, came in
(a) 1951 (b) 1958
(c) 1965 (d) 1970

126. Monopolies and Restrictive Trade Practices Act, came in
(a) 1960 (b) 1965
(c) 1969 (d) 1970

127. Consumer Protection Act, came in the year
(a) 1975 (b) 1986
(c) 1991 (d) 2000

128. Essential Commodities Act came in
(a) 1955
(b) 1965
(c) 1961
(d) 2000

129. Food Safety and Standards Act, came in
(a) 2000
(b) 2005
(c) 2006
(d) 2011

130. Minimum Support Price is recommended by
(a) CACP
(b) Director of Marketing And Inspection
(c) Ministry of Commerce
(d) Ministry of Agriculture

131. Cotton Corporation of India is under the purview of
(a) Ministry of Agriculture
(b) Ministry of Commerce
(c) APEDA
(d) Ministry of Khadi and Village Commission

132. Head Quarter of Cotton Corporation Of India is at
(a) Navi Mumbai (b) New delhi
(c) Chennai (d) Kolkata

133. Head Quarter of Jute Corporation of India
(a) Navi Mumbai (b) New delhi
(c) Chennai (d) Kolkata

134. FCI has set up a dal mill at
(a) Faridabad (b) Hyderabad
(c) Lucknow (d) Ujjain

135. FCI has set up a maize mill at
(a) Faridabad (b) Hyderabad
(c) Lucknow (d) Ujjain

136. FCI produces a protein rich food called as
(a) Milkmin
(b) Suddhahar
(c) Balahar
(d) Bestfood

137. Purchase method in which government competes with trader is called as
(a) Levy on producer
(b) Pre-emptive Purchase
(c) Open Market Purchase
(d) Monopoly Procurement

138. Purchase method in which government assumes the first right in the open market is
(a) Levy on producer
(b) Pre-emptive Purchase
(c) Open Market Purchase
(d) Levy on Trader and Miller

139. Monopoly purchase of wheat was carried out by the government only in the year
(a) 1970 (b) 1971
(c) 1973 (d) 1975

140. The number one state in rice procurement in India is
(a) Andhra Pradesh
(b) Punjab
(c) Chhattisgarh
(d) Tamil Nadu

141. The number one state in wheat procurement in India is
(a) Uttar Pradesh (b) Haryana
(c) Punjab (d) Gujarat

142. The price at which central government supplies food grain to states for distribution through fair price shop
(a) Minimum Support price
(b) Floor Price
(c) Issue price
(d) Minimum Price

143. Decentralised procurement scheme came in
(a) 1995 (b) 1997
(c) 2000 (d) 2005

144. Revamped Public Distribution System came in
(a) January 1, 1992 (b) June 1, 1992
(c) July 1, 1992 (d) March 1, 1992

145. Targeted Public Distribution System came in
(a) 1995 (b) 1997
(c) 2000 (d) 2005

146. Under the food security Act, each priority household should entitled to get millet at
(a) Rs 1 (b) Rs 2
(c) Rs 3 (d) Rs 4

147. Under the food security Act, each priority household should entitled to get wheat at
(a) Rs 1 (b) Rs 2
(c) Rs 3 (d) Rs 4

148. Under the food security Act, each priority household should entitled to get rice at
(a) Rs 1 (b) Rs 2
(c) Rs 3 (d) Rs 4

149. Minimum age for getting benefit under Annapurna Yojana is
(a) 60 (b) 62
(c) 65 (d) 70

150. Wholesale trade take over for wheat by the government only on
(a) 1965-66
(b) 1972-73
(c) 1980-81
(d) 2000-01

151. First contract farming in India was done by
(a) ITC
(b) Pepsi Co
(c) Balaji
(d) McCaines

152. Contract farming through PEPSI was under the guidance of
(a) Patel committee
(b) Tiwari Committee
(c) Johl committee
(d) Khusro Committee

153. Pepsi set up contract farming first in
(a) Andhra Pradesh
(b) Maharashtra
(c) Gujarat
(d) Punjab

154. The model of contract farming in which the central packer/processor buying from a large number of farmers' is called as
(a) Nucleus estate contract farming model
(b) Centralised contract farming model
(c) Multipartite contract far\ming model
(d) Intermediary contract farming model

155. The contract farming model in which the sponsorer manage a large farm and provide raw material to processing plant is called as
(a) Nucleus estate contract farming model
(b) Centralised contract farming model
(c) Multipartite contract farming model
(d) Intermediary contract farming model

156. The contract farming model in which various organizations like cooperative go for contract is called as
(a) Nucleus estate contract farming model
(b) Centralised contract farming model
(c) Multipartite contract farming model
(d) Intermediary contract farming model

157. Contract model in which, sponsorer is involved in sub-contracting with farmer's is called as
(a) Nucleus estate contract farming model
(b) Centralised contract farming model
(c) Multipartite contract farming model
(d) Intermediary contract farming model

158. Petty merchant who moves from village to village is called as
(a) Broker
(b) Class A trader
(c) Itinerant trader
(d) Kachha Arhatias

159. Big retailer or small whole seller is called as
(a) Class A trader
(b) Pacca Arhatiya
(c) Mashakhore
(d) Kachha Arhatiya

160. The person primarily acts for the seller, including farmers' is called as
(a) Class A trader
(b) Pacca Arhatiya
(c) Mashakhore
(d) Kachha Arhatiya

161. The person acts on behalf of the trader is called as
(a) Class A trader
(b) Pacca Arhatiya
(c) Mashakhore
(d) Kachha Arhatiya

162. The market where we find only one seller is called as
(a) Monopoly Market
(b) Duopoly market
(c) Oligopoly Market
(d) Monopolistic Market

163. The market where we find only two seller is called as
(a) Monopoly Market
(b) Duopoly market
(c) Oligopoly Market
(d) Monopolistic Market

164. The market where we find few sellers is called as
(a) Monopoly Market
(b) Duopoly market
(c) Oligopoly Market
(d) Monopolistic Market

165. The market where a large number of seller deal in heterogeneous and differentiated form of a commodity is called as
(a) Monopoly Market
(b) Duopoly market
(c) Oligopoly Market
(d) Monopolistic Market

166. The difference between the value of exports and imports is called as
(a) Balance of payment
(b) Balance of Trade
(c) Both
(d) None

167. Quantity of the produce which can made available to the non-farm population of the country is called as
(a) Marketed Surplus
(b) Marketable Surplus
(c) Both
(d) None

168. That quantity of the produce which the producer farmer actually sells in the market is called as
(a) Marketed Surplus
(b) Marketable Surplus
(c) Both
(d) None

169. In distress sale, the relationship between marketed surplus and marketable surplus is
(a) Marketed Surplus > Marketable surplus
(b) Marketed Surplus < Marketable surplus
(c) Marketed Surplus = Marketable surplus
(d) None

170. In perishable good, the relationship between marketed surplus and marketable surplus is
(a) Marketed Surplus > Marketable surplus
(b) Marketed Surplus < Marketable surplus
(c) Marketed Surplus = Marketable surplus
(d) None

171. Assembling of the produce comes under, which category of marketing function?
(a) Primary function of marketing
(b) Secondary function of marketing
(c) Tertiary function of marketing
(d) Multi-function of marketing

172. Banking, Insurance comes under, which category of marketing function?
(a) Primary function of marketing
(b) Secondary function of marketing
(c) Tertiary function of marketing
(d) Multi-function of marketing

173. The transportation cost carry by Bullock or camel cart is
(a) 50% of the total produce
(b) 25% of the total produce
(c) 35% of the total produce
(d) 10% of the total produce

174. The transportation cost account for tractor-trolley is
(a) 50% of the total produce
(b) 25% of the total produce
(c) 65% of the total produce
(d) 10% of the total produce

175. The transportation cost account for truck is
(a) 50% of the total produce
(b) 25% of the total produce
(c) 65% of the total produce
(d) 10% of the total produce

176. Find out the incorrect in the following giving options
(a) Standisation follows grading
(b) Grading follows standisation
(c) Grading is otherwise called as Grade Standards
(d) Grade standards have been fixed by the Agricultural marketing Advisor

177. Which commodity comes under centralized grading system?
(a) Ghee (b) Butter
(c) Vegetable oil (d) All

178. Which commodity comes under decentralized grading system?
(a) Egg (b) Ghee
(c) Butter (d) Vegetable oil

179. Central AGMARK Laboratory is at
(a) Lucknow (b) New Delhi
(c) Nagpur (d) Chandigarh

180. Central Nodal agency for food grain purchase is by
(a) FCI
(b) NAFED
(c) CCI
(d) Jute Corporation of India

181. Central Nodal agency for Pulses and oilseeds including Copra purchase is by
(a) FCI
(b) NAFED
(c) CCI
(d) Jute Corporation of India

182. Which are small grain storage structure
(a) PAU Bin (b) PUSA Bin
(c) Hapur Tekka (d) All

183. Which are large grain storage structures?
(a) CAP storage
(b) Ware House
(c) Silo
(d) All

184. Indian Grain Storage Institute is at
(a) Nagpur, Maharashtra
(b) Hapur, Uttar Pradesh
(c) Chandigarh, Punjab
(d) Anand, Gujarat

185. Entrepot trade means
(a) Export of goods
(b) Import of goods
(c) Re-import of exported goods
(d) Re-export of imported goods

186. The first state warehouse was set up in
(a) Uttar Pradesh (b) Odisha
(c) Bihar (d) West Bengal

187. National Grid of Rural Godown was launched in
(a) 1963 (b) 1965
(c) 1979 (d) 1985

188. The first cold storage was established in India in 1892 at
(a) New Delhi (b) Mumbai
(c) Kolkata (d) Chennai

189. Maximum percentage of which commodity is stored in cold storage?
(a) Onion (b) Tomato
(c) Potato (d) Cauliflower

190. The government programme that deals with construction of godown in rural areas is
(a) MGNREGA
(b) SGRY
(c) Gramin Bhandaran Yojana
(d) All of the above

191. Bulletin of Agricultural Prices is published
(a) Monthly (b) Weekly
(c) Annual (d) Daily

192. Agricultural Situation in India is published
(a) Monthly (b) Weekly
(c) Annually (d) Daily

193. Agricultural Prices in India is published
(a) Monthly (b) Weekly
(c) Annually (d) Daily

194. Bulletin in Food Statistics is published
(a) Monthly
(b) Weekly
(c) Annually
(d) Daily

195. The monthly situation and outlook reports are published by
(a) Ministry of Agriculture
(b) ICAR
(c) Ministry of Commerce
(d) DMI, GOI

196. National Agricultural Cooperative Marketing Federation (NAFED) came in the year
(a) 1955 (b) 1958
(c) 1961 (d) 1965

197. Value addition in food in India is around
(a) 5% (b) 10%
(c) 12% (d) 15%

198. Value addition in food in China is around
(a) 20% (b) 23%
(c) 25% (d) 30%

199. Value addition for food grain like rice and wheat is around
(a) 5% (b) 7%
(c) 10% (d) 15%

200. Value addition for cotton is around
(a) 65% (b) 79%
(c) 86% (d) 95%

201. Value addition for tea is around
(a) 65% (b) 79%
(c) 86% (d) 95%

202. Processing of products add
(a) Possession utility
(b) Time utility
(c) Space utility
(d) Form utility

203. Storage of products add
(a) Possession utility
(b) Time utility
(c) Space utility
(d) Form utility

204. Method of selling in which produce in different lots is mixed and then sold as one lot is
(a) Moghum sale method
(b) Dara sale method
(c) Open auction method
(d) Close tender method

205. Method where sale of the produce is effected on the basis of verbal understanding between the buyer and seller without any pre-settlement of price is
(a) Moghum sale method
(b) Dara sale method
(c) Open auction method
(d) Close tender method

206. These days the market charges that are observed in regulated market is
(a) Darmada (b) Muddat
(c) Hamali (d) Karda

207. AGMARKNET came in
(a) Eighth five year plan
(b) Ninth five year plan
(c) Tenth five year plan
(d) Eleventh five year plan

208. The post-harvest food grain loss is around
(a) 6% (b) 10%
(c) 15% (d) 18%

209. The post-harvest loss of fruits and vegetables is around
(a) 6% (b) 10%
(c) 18% (d) 30%

210. Market information in which we try to reveal the prices that prevailed in the past is
(a) Market News
(b) Market Intelligence
(c) Market Outlook
(d) Market Research

211. Market information in which we try to reveal the current information of the price is
(a) Market News
(b) Market Intelligence
(c) Market Outlook
(d) Market Research

212. Market information in which we try to estimate the future price of the produce is
(a) Market News
(b) Market Intelligence
(c) Market Outlook
(d) Market Research

213. Cooperative credit society advances, how much percentage of credit on the value of the produce they store
(a) 50% (b) 25%
(c) 75% (d) 100%

214. Maximum cooperative agro processing units in India is in
(a) Sugar processing factories
(b) Food processing sector
(c) Cotton growing and processing societies
(d) Fruits and vegetable processing units

215. Which states are marked for sugarcane in cooperative sector
(a) Gujarat and Maharashtra
(b) Maharashtra and Karnataka
(c) Maharashtra and Uttar Pradesh
(d) Karnataka and Uttar Pradesh

216. Which states are marked for cotton in cooperative sector?
(a) Gujarat and Maharashtra
(b) Maharashtra and Karnataka
(c) Maharashtra and Uttar Pradesh
(d) Karnataka and Uttar Pradesh

217. Which state is marked for plantation crop in cooperative sector?
(a) Gujarat
(b) Maharashtra
(c) Uttar Pradesh
(d) Karnataka

218. Cooperative sector handles how much percentage of agricultural commodity approximately in India?
(a) 10% (b) 20%
(c) 25% (d) 50%

219. The head quarter of NAFED is at
(a) New Delhi (b) Mumbai
(c) Nagpur (d) Faridabad

220. In 1976-77, NAFED undertook price support operation for
(a) Moong (b) Groundnut
(c) Toria (d) Sunflower

221. NAFED became the central agency for price support policy of oilseeds and coarse grain designated in
(a) Sixth five Year Plan
(b) Seventh five Year Plan
(c) Eighth five Year Plan
(d) Ninth five Year Plan

222. National Cooperative Development Corporation came in the year
(a) January, 1963
(b) March, 1963
(c) July, 1963
(d) October, 1963

223. Tribal Cooperative Marketing Federation was established in the year
(a) 1985 (b) 1987
(c) 1990 (d) 1991

224. 'Sangam' and 'Vijaya' is the milk cooperative of
(a) Tamil Nadu (b) Andhra Pradesh
(c) Uttar Pradesh (d) Odisha

225. The milk cooperative of Uttar Pradesh is
(a) OMFED (b) Saras
(c) Parag (d) Bhagirath

226. Sneha is the milk cooperative of
(a) Himachal Pradesh
(b) Madhya Pradesh
(c) Bihar
(d) West Bengal

227. The milk cooperative of Tamil Nadu is
(a) AAVIN (b) MAAVIN
(c) Both (d) None

228. 'Vita' is the milk cooperative of
(a) Himachal Pradesh
(b) Punjab
(c) Haryana
(d) Rajasthan

229. Tilam Sangh' is the state level cooperative of oilseed in
(a) Punjab (b) Rajasthan
(c) Haryana (d) Gujarat

230. 'Mahagrape', 'Mahabanana', 'Mahamango' is the marketing entity of state
(a) Madhya Pradesh
(b) Maharashtra
(c) Gujarat
(d) Odisha

231. Spending on food accounting how much percentage of India's GDP is spend?
(a) 15% (b) 21%
(c) 30% (d) 50%

232. Major segment in food processing in India is in
(a) Bread and bakery
(b) Dairy products
(c) Food grain milling
(d) Meat and Poultry

233. Unorganised sector contribute how much percentage of food processing in India?
(a) 25% (b) 42%
(c) 50% (d) 75%

234. Country which has maximum share in world processed food market is?
(a) India (b) Netherland
(c) Germany (d) USA

235. Edible oil that is produced maximum in world is
(a) Soya oil (b) Palm Oil
(c) Rape Oil (d) Rapeseed Oil

236. India's rank in silk production is
(a) First (b) Second
(c) Third (d) Fourth

237. Types of market integration where two or more functions have being carried out under a single ownership is
(a) Horizontal integration
(b) Vertical integration
(c) Conglomeration
(d) None

238. A combination of agencies or activities not directly related with each other is
(a) Horizontal integration
(b) Vertical integration
(c) Conglomeration
(d) None

239. Shepherd Approach of marketing efficiency measurement mainly deals with
(a) Ratio of value added output and input cost
(b) Ratio of total value of goods marketed to marketing cost
(c) Comparision of the efficiency of alternate markets/channels
(d) All

240. Acharya approach of marketing efficiency measurement mainly deals with
(a) Ratio of value added output and input cost
(b) Ratio of total value of goods marketed to marketing cost
(c) Comparision of the efficiency of alternate markets/channels
(d) None

241. The difference between the prices prevailing at successive stages of marketing at a given point of time is called
(a) Concurrent margin
(b) Lagged margin
(c) Both
(d) None

242. The difference between the price received by a seller at a particular stage of marketing and the price paid by him at the preceding stage of marketing during an earlier period is called as
(a) Concurrent margin
(b) Lagged margin
(c) Both
(d) None

243. Calculation of laggard margin based on
(a) FIFO method of accounting
(b) LIFO method of accounting
(c) First In Last Out of accounting
(d) Last In First Out of accounting

244. Indian Society of Agricultural Marketing came in the year
(a) 1980 (b) 1986
(c) 1988 (d) 1990

245. Head Quarter of National Institute of Agricultural Marketing is at
(a) Anand (b) Jodhpur
(c) Hyderabad (d) Jaipur

246. Agricultural inputs are showing the characteristics of
(a) Bulky (b) Costly
(c) Used rarely (d) All of the above

247. Which is not a capital agricultural input?
(a) Pumpsets (b) Diesel
(c) Harvestor (d) Tractor

248. Which is not a consumable agricultural input?
(a) Seed (b) Electricity
(c) Thresher (d) Fertiliser

249. Agricultural input product and price shows the behaviour similar to
(a) Consumer product
(b) Industrial product
(c) Both
(d) None

250. Sole Selling Agent is the major distributing channel for
(a) Seed
(b) Fertiliser
(c) Pesticides
(d) Plant Protection Chemical

251. Distributor is the major distributing channel for
(a) Seed
(b) Fertiliser
(c) Pesticides
(d) Plant Protection Chemical

252. National Seed Corporation was set up in
(a) 1960 (b) 1963
(c) 1965 (d) 1970

253. Seed Act came in the year
(a) 1960 (b) 1963
(c) 1966 (d) 1970

254. State farm Corporation of India came in the year
(a) 1965 (b) 1969
(c) 1975 (d) 1979

255. National Seed project came in the year
(a) 1963 (b) 1966
(c) 1974 (d) 1981

256. India rank in seed production in world is
(a) First (b) Second
(c) Fifth (d) Seventh

257. Total number of state seed corporation in India is
(a) 10 (b) 13
(c) 15 (d) 20

258. Tag colour of breeder seed is
(a) Blue (b) Golden yellow
(c) White (d) Black

259. Tag colour of foundation seed is
(a) Blue (b) Golden yellow
(c) White (d) Black

260. Tag colour of certified seed is
(a) Blue (b) Golden yellow
(c) White (d) Black

261. Fertilizer contribution is how much percentage approximately for increase crop production per unit area?
(a) 25% (b) 33%
(c) 50% (d) 66%

262. Optimal consumption of overall NPK in India is
(a) 6:4:1 (b) 4:2:1
(c) 12:4:1 (d) 7:5:2

263. First fertilizer plant was established after Independence of India in
(a) Naya Nangal, Punjab
(b) Jodhpur, Rajasthan
(c) Paradip, Odisha
(d) Sindhi, Bihar

264. Fertiliser corporation of India was set up in
(a) 1951 (b) 1961
(c) 1965 (d) 1973

265. Maximum fertilizer production is in
(a) Northern India
(b) Southern India
(c) Western India
(d) Eastern India

266. Maximum nitrogenous fertilizer consumption is in
(a) Northern India
(b) Southern India
(c) Western India
(d) Eastern India

267. Maximum phosphorus consumption in fertilizer is in
(a) Northern India
(b) Southern India
(c) Western India
(d) Eastern India

268. Maximum consumption of potash in fertilizer is in
(a) Northern India
(b) Southern India
(c) Western India
(d) Eastern India

269. Overall maximum fertilizer consumption is in
(a) Northern India
(b) Southern India
(c) Western India
(d) Eastern India

270. Fertilizer came under Essential Commodity Act, in the year
(a) 1955 (b) 1957
(c) 1960 (d) 1961

271. Maximum fertilizer move by
(a) Road (b) Rail
(c) Water route (d) Air route

272. Retention Price Scheme for fertilizer marketing came in the year
(a) 1965 (b) 1969
(c) 1977 (d) 1982

273. All fertilizer is decontrolled by government except
(a) CAN
(b) Ammonium Sulphate
(c) Urea
(d) DAP

274. Nutrient Based Subsidy Scheme came in the year
(a) 2010 (b) 2011
(c) 2012 (d) 2013

275. IFFCO came in the year
(a) 1963
(b) 1965
(c) 1967
(d) 1969

276. The first plant of IFFCO was commissioned at
(a) Kandla, Gujarat
(b) Kalol, Gujarat
(c) Phulpur, Gujarat
(d) Aonla, UP

277. KRIBHCO came in the year
(a) 1975 (b) 1982
(c) 1991 (d) 1995

278. In India, bio-pesticides currently represent how much percentage of market share?
(a) 3.5% (b) 4.2%
(c) 6.3% (d) 10%

279. In India, crop protection is largely dominated by
(a) Fungicide (b) Insecticide
(c) Herbicide (d) Weedicide

280. Globally, crop protection is dominated by
(a) Fungicide
(b) Insecticide
(c) Herbicide
(d) Weedicide

281. The maximum electricity produced in India from
(a) Hydel source
(b) Thermal source
(c) Nuclear source
(d) None

282. Total electricity consume for agriculture in India is nearly about
(a) 10% (b) 17%
(c) 40% (d) 45%

283. Rural Electrification Corporation was established at the national level in
(a) 1965 (b) 1969
(c) 1992 (d) 2007

284. Rajiv Gandhi Vidyutikaran Yojana came in the year
(a) 1995 (b) 2000
(c) 2005 (d) 2007

285. Which segment is the core of the tractor industry?
(a) Under 20 HP (b) 21-30 HP
(c) 31-40 HP (d) 41-50 HP

286. Approximately how many percentage of population are in the rural India?
(a) 65% (b) 69%
(c) 71% (d) 75%

287. In India, rural population literacy rate is approximately
(a) 74% (b) 69%
(c) 85% (d) 60%

288. Total number of rural village in India is
(a) 6, 30,000 (b) 6,35,000
(c) 6,40,000 (d) 6,45,000

289. Maximum rural population in India is in
(a) Bihar (b) Uttar Pradesh
(c) Odisha (d) Chhattisgarh

290. Least number of rural population in India is in
(a) Sikkim (b) Mizoram
(c) Goa (d) Meghalaya

291. In rural India, how many percentage of population use internet
(a) 30% (b) 50%
(c) 70% (d) 100%

292. In rural India, how many percentage of population use internet only through their mobile
(a) 25% (b) 32%
(c) 40% (d) 45%

293. According to 2011 census , how many population of India have television set?
(a) 35% (b) 47%
(c) 50% (d) 75%

294. State with maximum rural haat in India is in
(a) Odisha (b) Uttar Pradesh
(c) Jharkhand (d) Maharashtra

295. Maximum brand awareness for a particular product in haat in india is for India?
(a) Toothpaste (b) Soap
(c) Shampoo (d) Talcum powder

296. FMCG product preferred in haat in India is maximum for
(a) Toilet soap (b) Grocery
(c) Detergent (d) Tea

297. Hariyali Stores is the retail chain of
(a) Mahindra & Mahindra
(b) DCM Shriram Consolidated Limited
(c) Indian Oil Corporation
(d) Indian Tobacco Company

298. Shubhlabh Stores is the retail chain of
(a) Mahindra & Mahindra
(b) DCM Shriram Consolidated Limited
(c) Indian Oil Corporation
(d) Indian Tobacco Company

299. Kisan Seva Kendra is the retail chain of
(a) Mahindra & Mahindra
(b) DCM Shriram Consolidated Limited
(c) Indian Oil Corporation
(d) Indian Tobacco Company

300. Har Gaon, Har Angan' was the promotional slogan of
(a) Bajaj Auto
(b) Tata Motors
(c) Hero Honda
(d) Mahindra bikes

301. E-choupal is the revolutionized programme of
(a) HUL (b) ITC
(c) IOC (d) DSCL

302. Project Shakti is the rural penetration programme of
(a) HUL (b) ITC
(c) IOC (d) DSCL

303. Sundesh- a sustainable development society of
(a) ICICI
(b) Dabur India
(c) Lafarge Cement
(d) Godrej

304. The main objective of 'Sundesh' programme of Dabur India is
(a) Establish low cost housing in rural areas
(b) Extending educational and medical facility in rural areas
(c) Providing sanitation and fresh drinking water in rural areas
(d) To promote research and welfare activities in rural areas

305. Availability, Affordability and Acceptance are 3A of marketing in rural areas. Another 'A' is
(a) Amicability (b) Applicability
(c) Accessibility (d) Adjustability

306. In Project Disha, HUL empower the rural women by
(a) Marketing their product and providing self-sustainability
(b) Providing skill development in embroidery and garment designing
(c) Assisting career guidance for young girls
(d) Providing scholarships for girls

307. In Project Kaladarshan, HUL empower the rural women by
(a) Marketing their product and providing self-sustainability
(b) Providing skill development in embroidery and garment designing
(c) Assisting career guidance for young girls
(d) Providing scholarships for girls

308. In Project Sanjivani, HUL empower the rural women by
(a) Marketing their product and providing self-sustainability
(b) 3-month nurse training in collaboration with Dr Reddy's lab
(c) Assisting career guidance for young girls
(d) Providing scholarships for girls

309. Project Jagruti, a rural based hygienic programme of
(a) Dabur (b) HUL
(c) ITC (d) Colgate

310. Uttam Bandhan is the community welfare programme in which company trains rural youths as extension workers as Krishi Sewak. What is the name of that company?
(a) HUL (b) DSCL
(c) CFCL (d) BPCL

311. An IT based project for dairy farmers' in Gujarat is
(a) Tarahaat
(b) AkashGanga
(c) Bhoomi
(d) Gyandoot

312. An IT based land records in Karnataka is
(a) Tarahaat (b) AkashGanga
(c) Bhoomi (d) Gyandoot

313. Kutir Jyoti programme in the direction of rural sector electrification came in the year
(a) 1982 (b) 1985
(c) 1989 (d) 1995

314. 'Head Start' is the rural based educational programme of
(a) Andhra Pradesh
(b) Madhya Pradesh
(c) Rajasthan
(d) Kerala

315. 'Project Arrow' is an initiative in the direction of communication implemented by
(a) Telecom Department
(b) Postal Department
(c) Information and Broadcasting Department
(d) Income Tax department

316. Full form of CAPART
(a) Committee for Advancement of People's Action and Rural Technology
(b) Council for Advancement of People's Action and Rural Technology
(c) Committee for Advancement of People's Action and Rural Terminology
(d) Council for Achievement of People's Action and Rural Technology

317. National Institute of Rural Development (NIRD) is at
(a) Pune
(b) Delhi
(c) Hyderabad
(d) Kanpur

318. Consumer evaluation model in which a consumer selects a product that satisfies few attributes that he considers important is
(a) Expectancy value model
(b) Lexicographic model
(c) Conjunctive model
(d) Disjunctive model

319. Consumer evaluation model in which a consumer arrange attributes in an order of their priority is
(a) Expectancy value model
(b) Lexicographic model
(c) Conjunctive model
(d) Disjunctive model

320. Number of choices are more in which sets of consumer decision making process
(a) Awareness set
(b) Decision set
(c) Choice set
(d) Consideration set

321. Cognitive dissonance is earmarked in which stage of consumer buying decision process
(a) Evaluation of alternatives
(b) Purchase decision
(c) Post purchase behavior
(d) Information search process

322. 'Bright Smile, Bright Future' is a promotional programme in which school teachers are acted as opinion leader for
(a) Pepsodent
(b) Mentadent
(c) Colgate
(d) Close Up

323. Storage enhances
(a) Form utility for a product
(b) Time utility for a product
(c) Place utility for a product
(d) Possession utility for a product

324. Through buying and selling we enhances
(a) Form utility for a product
(b) Time utility for a product
(c) Place utility for a product
(d) Possession utility for a product

325. Brand value of a product reduces
(a) Functional risk
(b) Perceived risk
(c) Social risk
(d) All of the above

326. Sankat Haran Policy is a non- crop insurance policy in which company gives free insurance cover worth Rs 4000 per each bag of fertilizer. The name of the company is
(a) Monsanto
(b) IFFCO
(c) KRIBHCO
(d) CFCL

327. Logo of a company differs from its symbol in the way that
(a) Logo do not change
(b) Logo is exclusively created by the company
(c) Both
(d) None

328. Brand elements include
(a) Name, symbol, character
(b) Symbol, logo, package design
(c) Jingles, Unique resource locator
(d) All of the above

329. A group of product possess the same brand name is
(a) Umbrella Brand
(b) Blanket Family Brand
(c) Both
(d) None

330. When a marketer offers one brand with another brand of the same company or another company is called
(a) Bundling
(b) Co-Branding
(c) Combipack
(d) Cross selling

331. Pricing strategy that has taken Return On Investment (ROI) in to consideration is
(a) Cost-Plus strategy
(b) Experience curve strategy
(c) Marginal cost strategy
(d) Target return strategy

332. Climbers, Aspirants, Destitute are
(a) More quality conscious
(b) More value conscious
(c) More price conscious
(d) None of the above

333. Pricing strategy where a service organisation set a fix price for one part and variable price for another part depending on the usage is
(a) Product-Bundling pricing
(b) Two-part pricing
(c) Promotional pricing
(d) Captive product pricing

334. 'Dharti' a distribution strategy in the rural sector by
(a) Emami
(b) Colgate-Palmolive
(c) Godrej
(d) Maruti

335. Aadhaar centre is based on hub and spoke model for distribution of commodity and integrate both rural and urban is the retail chain of
(a) ITC
(b) HUL
(c) Godrej
(d) CFCL

336. Manthan is the agricultural outlet of company
(a) ITC
(b) HUL
(c) Godrej
(d) CFCL

337. Kisan Sansar is the agricultural outlet of company
(a) Godrej
(b) ITC
(c) Tata
(d) Bharti

338. Community Kitchen is the rural marketing strategy of
(a) IOC
(b) BPCL
(c) HPCL
(d) Reliance Industries and Petro Chemicals

339. Kisan Seva Kendras are rural malls set up by
(a) IOC
(b) BPCL
(c) HPCL
(d) Reliance Industries and Petro Chemicals

340. Advanced Sales Training for Retail Ascendance (ASTRA), is a customized training programme for the frontline sales force of company
(a) Colgate Palmolive
(b) Tata
(c) Dabur
(d) Godrej

341. 'Placing the life-size dummy of a cow mounted on a branded van, named Kamdhenu' is a promotional campaign in the direction of cattle loan launched by
(a) SBI (b) HDFC
(c) ICICI (d) Bank of Baroda

342. Self Employed Women Association (SEWA) has launched a community radio station as
(a) 92.1 FM (b) 104 FM
(c) 90.4 FM (d) 106 FM

343. Sonepur mela is famous in
(a) UP (b) MP
(c) Bihar (d) Odisha

344. Chappar Mela is famous in
(a) Rajasthan (b) Punjab
(c) UP (d) Bihar

345. Kartik Mela is famous in
(a) UP (b) MP
(c) Bihar (d) Odisha

346. Kaila Devi mela is famous in
(a) Rajasthan (b) Punjab
(c) UP (d) Bihar

347. Kaila devi mela is hold on in the month of
(a) November (b) March
(c) October (d) July

348. Onam is a famous festival of Kerala is usually hold on in the month of
(a) March
(b) July
(c) August
(d) December

349. Rural teledensity in India is approximately
(a) 25 (b) 40
(c) 50 (d) 55

350. Asha is low cost snacks and beverages- especially for girls age between 12 to 16, is the product of
(a) Coca Cola
(b) Pepsi
(c) Parle Agro
(d) Dabur

351. Samridhi is an initiative in rural farm sector segment is launched by
(a) Tata
(b) Mahindra
(c) John Dheere
(d) HMT

352. Which is the parent scheme of Bringing Green Revolution in Eastern India (BGREI)?
(a) Mid-day Meal Scheme of India
(b) Rashtriya Krishi Vikas Yojana
(c) Vishesh Krishi Upaj Yojana
(d) Krishi Sinchai Yojana

353. Zatak is the male deodorant of which company for its rural market segments?
(a) Marico
(b) ITC
(c) Calvin Klein
(d) Nivea

354. Which company promotes its product as 'Desh Ka Namak'?
(a) TATA Salt
(b) Keya Salt
(c) Sundar Health Salt
(d) Patanjali Namak

355. Which organization promotes its mobile product as 'Behtar Zindegi' for farm communities in Rajasthan?
(a) TATA (b) Reliance
(c) AIRTEL (d) BSNL

356. Who has been credited as multi-media distance education package for Panchayati Raj functionaries?
(a) Dr. M. S. Swaminathan
(b) Dr. Verghese Kurien
(c) Dr. Yoginder K Alagh
(d) Professor M Aslam

357. Which organization develops 'Audiobook Project' for visually challenged students of Odisha?
(a) Being Human
(b) Prakruti Charitable Society
(c) We4You
(d) Animal Welfare

358. Which year is declared as 'Year of Farmer Producer Organization'?
(a) 2012 (b) 2013
(c) 2014 (d) 2015

359. Which company promote 'Chootukool' refrigerator for the rural segments?
(a) Godrej (b) Samsung
(c) Whirpool (d) Alwyn

360. Which company promotes its rural beverages as 'Piyo Josh Mein Jiyo Hosh Mein'?
(a) PEPSI
(b) COCACOLA
(c) AMUL
(d) MOUNTAIN DEW

361. 'PowerLight A Village' campaign is a concept of which company for electrifying villages?
(a) SYSKA
(b) PHILIPS
(c) GARNIER MEN
(d) CROMPTON

362. Which company has the tagline 'Happy Food Happy People' targeting young adults?
(a) AMUL (b) Mother Dairy
(c) Parle Agro (d) NESTLE

363. Which Product related to the tagline 'Pahle istemal karein phir vishwaas karein'?
(a) NIRMA (b) GHADI
(c) FENA (d) WHEEL

364. Which organization entered the rural market as 'Jo diktha hai wo bikta hai'?
(a) Mahindra and Mahindra
(b) NIRMA
(c) GHADI
(d) Regional Banks

365. Mileage Mukabala' campaign for rural market entry strategy related to
(a) Hero Honda (b) Bajaj
(c) Ashok Leyland (d) Maruti

366. Rasoi Ghar' tagline for rural campaign related to
(a) HPCL
(b) BPCL
(c) IOC
(d) Reliance

ANSWER KEY (Agricultural and Rural Marketing)

1. (d) **2.** (d) **3.** (d) **4.** (b) **5.** (c) **6.** (c) **7.** (d) **8.** (c) **9.** (d) **10.** (d)
11. (c) **12.** (c) **13.** (d) **14.** (d) **15.** (c) **16.** (c) **17.** (d) **18.** (c) **19.** (a) **20.** (b)
21. (b) **22.** (c) **23.** (b) **24.** (d) **25.** (d) **26.** (d) **27.** (c) **28.** (d) **28.** (b) **30.** (c)
31. (b) **32.** (b) **33.** (b) **34.** (a) **35.** (d) **36.** (c) **37.** (b) **38.** (b) **39.** (b) **40.** (a)
41. (c) **42.** (a) **43.** (c) **44.** (b) **45.** (c) **46.** (b) **47.** (c) **48.** (c) **49.** (c) **50.** (b)
51. (c) **52.** (b) **53.** (c) **54.** (d) **55.** (b) **56.** (d) **57.** (b) **58.** (c) **59.** (c) **60.** (b)
61. (c) **62.** (a) **63.** (c) **64.** (c) **65.** (b) **66.** (b) **67.** (b) **68.** (b) **69.** (b) **70.** (b)
71. (c) **72.** (c) **73.** (c) **74.** (c) **75.** (d) **76.** (c) **77.** (b) **78.** (b) **79.** (b) **80.** (c)
81. (b) **82.** (c) **83.** (b) **84.** (c) **85.** (c) **86.** (b) **87.** (c) **88.** (d) **89.** (b) **90.** (c)
91. (b) **92.** (b) **93.** (b) **94.** (c) **95.** (c) **96.** (c) **97.** (b) **98.** (c) **99.** (a) **100.** (c)
101. (b) **102.** (c) **103.** (b) **104.** (b) **105.** (a) **106.** (b) **107.** (d) **108.** (b) **109.** (a) **110.** (c)
111. (c) **112.** (a) **113.** (a) **114.** (b) **115.** (c) **116.** (b) **117.** (b) **118.** (a) **119.** (b) **120.** (c)
121. (b) **122.** (b) **123.** (c) **124.** (a) **125.** (b) **126.** (c) **127.** (b) **128.** (a) **128.** (c) **130.** (a)
131. (b) **132.** (a) **133.** (d) **134.** (c) **135.** (a) **136.** (c) **137.** (c) **138.** (b) **139.** (c) **140.** (b)
141. (c) **142.** (c) **143.** (b) **144.** (b) **145.** (b) **146.** (a) **147.** (b) **148.** (c) **149.** (c) **150.** (b)
151. (b) **152.** (c) **153.** (d) **154.** (b) **155.** (a) **156.** (c) **157.** (d) **158.** (c) **159.** (c) **160.** (d)
161. (b) **162.** (a) **163.** (b) **164.** (c) **165.** (d) **166.** (b) **167.** (b) **168.** (a) **169.** (a) **170.** (c)
171. (a) **172.** (c) **173.** (b) **174.** (c) **175.** (d) **176.** (a) **177.** (d) **178.** (a) **179.** (c) **180.** (a)
181. (b) **182.** (d) **183.** (d) **184.** (b) **185.** (d) **186.** (c) **187.** (c) **188.** (c) **189.** (c) **190.** (c)
191. (b) **192.** (a) **193.** (c) **194.** (c) **195.** (d) **196.** (b) **197.** (b) **198.** (b) **199.** (b) **200.** (b)
201. (c) **202.** (d) **203.** (b) **204.** (b) **205.** (a) **206.** (c) **207.** (b) **208.** (a) **209.** (c) **210.** (b)
211. (a) **212.** (c) **213.** (c) **214.** (b) **215.** (c) **216.** (a) **217.** (d) **218.** (a) **219.** (a) **220.** (b)
221. (b) **222.** (b) **223.** (b) **224.** (b) **225.** (c) **226.** (b) **227.** (c) **228.** (c) **228.** (b) **230.** (b)
231. (b) **232.** (c) **233.** (b) **234.** (d) **235.** (b) **236.** (b) **237.** (b) **238.** (c) **239.** (b) **240.** (c)

241. (a) **242.** (b) **243.** (a) **244.** (b) **245.** (d) **246.** (d) **247.** (b) **248.** (c) **249.** (b) **250.** (b)
251. (c) **252.** (b) **253.** (c) **254.** (b) **255.** (c) **256.** (c) **257.** (c) **258.** (b) **259.** (c) **260.** (a)
261. (c) **262.** (b) **263.** (d) **264.** (b) **265.** (c) **266.** (a) **267.** (b) **268.** (b) **269.** (b) **270.** (b)
271. (b) **272.** (c) **273.** (c) **274.** (a) **275.** (b) **276.** (a) **277.** (c) **278.** (b) **279.** (b) **280.** (c)
281. (b) **282.** (b) **283.** (b) **284.** (c) **285.** (c) **286.** (b) **287.** (b) **288.** (c) **289.** (b) **290.** (a)
291. (c) **292.** (b) **293.** (b) **294.** (b) **295.** (c) **296.** (b) **297.** (b) **298.** (a) **299.** (c) **300.** (c)
301. (b) **302.** (a) **303.** (b) **304.** (d) **305.** (c) **306.** (c) **307.** (b) **308.** (b) **309.** (d) **310.** (c)
311. (b) **312.** (c) **313.** (c) **314.** (b) **315.** (b) **316.** (b) **317.** (c) **318.** (c) **319.** (b) **320.** (a)
321. (c) **322.** (c) **323.** (b) **324.** (d) **325.** (d) **326.** (b) **327.** (c) **328.** (d) **328.** (c) **330.** (b)
331. (d) **332.** (c) **333.** (b) **334.** (c) **335.** (c) **336.** (c) **337.** (c) **338.** (a) **339.** (c) **340.** (c)
341. (c) **342.** (c) **343.** (c) **344.** (b) **345.** (b) **346.** (a) **347.** (b) **348.** (c) **349.** (b) **350.** (b)
351. (b) **352.** (b) **353.** (a) **354.** (a) **355.** (c) **356.** (d) **357.** (c) **358.** (c) **359.** (a) **360.** (a)
361. (c) **362.** (b) **363.** (b) **364.** (d) **365.** (c) **366.** (a)

CHAPTER 6

6. Agricultural Supply Chain Management

1. ECR stands for
   (a) Effective Customer Response
   (b) Efficient Consumer Response
   (c) Effective Consumer Response
   (d) Effective Customer Reply

2. The term Supply Chain Management was coined by
   (a) Peter Drucker
   (b) Keith Oliver
   (c) Raphael Kaplinsky
   (d) Michael Porter

3. Different Contract Farming models have been given by
   (a) Clarence Danhof
   (b) Eaton and Shepherd
   (c) Richard Cantillon
   (d) Margret et. al

4. Which of the following is the nodal agency for AEZ?
   (a) APEDA
   (b) ICAR
   (c) NIAM
   (d) DARE

5. Which of the following is the source of revenue in the Supply Chain?
   (a) Wholesaler
   (b) Retailer
   (c) Customer
   (d) All of the above

6. Supply chain includes
   (a) Wholesalers and retailer, Transporters
   (b) Manufacturers, wholesalers and retailers, Warehouses
   (c) Suppliers, manufacturers, wholesalers and retailers
   (d) All of the above

7. Primary purpose of any supply chain is
   (a) To satisfy the customers need
   (b) To maximize the profit
   (c) To minimize the overall cost
   (d) To create link between the intermediaries

8. The book "Competitive advantage: Creating & Sustaining Superior performance" is written by
   (a) Keith Oliver
   (b) Goldberg & Davis
   (c) Michael Porter
   (d) T.L. Friedman

9. Supply chain activities at strategic level does not include
   (a) Benchmarking
   (b) Location & size of warehouse
   (c) Strategic partnership with suppliers & distributors
   (d) Location of facilities

10. Supply chain activities at operational level does not include
    (a) Taking customer order
    (b) Movement of goods from warehouse to point of consumption
    (c) Creating a purchasing plan with the preferred supplier
    (d) Maintenance of inventory record

11. The ownership of inventory in the Vendor Managed Inventory system lies with
    (a) Purchaser
    (b) Transport company
    (c) Financial Institution
    (d) Supplier

12. Supply Chain mapping is the study of
(a) Logistics network
(b) Activities sequence
(c) Value and non-value added activities
(d) All of the above

13. Electronic Data Interchange(EDI) used in the logistics system ensures
(a) Information transfer
(b) Document transfer
(c) Material transfer
(d) Money transfer

14. The design of storage system for perishable goods should consider
(a) First In First Out(FIFO)
(b) Last In First Out(LIFO)
(c) Random order
(d) All of the above.

15. For perishable products transportation, major emphasis is on vehicle scheduling and
(a) Route planning
(b) Inventory planning
(c) Warehouse planning
(d) Handling equipment

16. Contract farming is an example of
(a) Forward integration
(b) Backward integration
(c) Conglomeration
(d) Diversification

17. Central processor or buyer buying from a large number of small farmers is ____________ model of contract farming
(a) Centralized
(b) Nucleus estate
(c) Multipartite
(d) Intermediary

18. In ____________ model of contract farming, in addition to individual contract with farmers, the buyer also owns his own cultivation source to guarantee output for processing purpose.
(a) Centralized
(b) Nucleus estate
(c) Multipartite
d) Intermediary

19. The ____________ model of contract farming usually involves a partnership between government bodies, private companies, lending agencies, NGO and farmers.
(a) Centralized (b) Nucleus estate
(c) Multipartite (d) Intermediary

20. The ____________ model of contract farming involves subcontracting by companies to intermediaries who have their own (informal) arrangements with farmers.
(a) Centralized (b) Nucleus estate
(c) Multipartite (d) Intermediary

21. The number of cold storages in India as of 2014 is
(a) 2307 (b) 1024
(c) 5307 (d) 6891

22. Cold storage capacity of India as of 2014 is
(a) 31.82 MMT (b) 28.22 MMT
(c) 26.11 MMT (d) 24.33 MMT

23. Grameen Bhandaran Yojana was launched as a central sector scheme in the year
(a) 1998 (b) 1999
(c) 2000 (d) 2001

24. The project for construction of rural godowns cannot be taken up by
(a) Individual
(b) NGOs
(c) Companies
(d) None of the above

25. Rural godowns capacity in India till 30/09/2012 is to the tune of
(a) 24.5 MMT (b) 27.5 MMT
(c) 31.5 MMT (d) 33.5 MMT

26. Small Farmers Agribusiness Consortium was set up as a registered society in the year
(a) 1990 (b) 1992
(c) 1994 (d) 1996

27. FDI allowed for cold storage construction in India is
(a) 49% (b) 74%
(c) 100% (d) None of the above

28. Which of the following commodity has maximum share in cold storage?
(a) Meat & Fish
(b) Potato
(c) Milk
(d) Tomato

29. Ripening chambers are used for
(a) Banana
(b) Mango
(c) Both of the above
(d) None of the above

30. Which of the following is not a problem in Indian supply chain?
(a) Numerous stakeholders working in isolation
(b) Lack of demand estimation
(c) Lack of technology applications
(d) All of the above

31. Which of the following is a game of Agribusiness Supply Chain Strategy?
(a) VAG (Virtual Reality Agribusiness World)
(b) MIDAS (Marico Industries Distribution Automation Software)
(c) SCEM (Supply Chain Event Management)
(d) None of the above

32. 3PL stands for
(a) Third product Life
(b) Third Party Logistics
(c) Third Product Logistics
(d) Third Party Life

33. Ratio of Logistics cost with GDP for India is
(a) More than Japan
(b) More than US
(c) Both of the above
(d) None of the above

34. Project Shakti is an initiative of
(a) Procter & Gamble
(b) Hindustan Unilever Ltd.
(c) Marico Industries Ltd.
(d) Hero Industries Ltd.

35. Supply Chain Lead time is the sum of time taken for
(a) Sourcing the materials
(b) Manufacturing the product
(c) Delivering the finished product to the end customer
(d) All of the above

36. The point at which customer enters the supply chain is called
(a) Customer penetration point
(b) Order penetration point
(c) Customer entry point
(d) None of the above

37. If an order is to be fulfilled instantaneously, then the supply chain is in the ________ business
(a) Make to stock (MTS)
(b) Make to Order (MTO)
(c) Configure to Order (CTO)
(d) None of the above

38. If the supplier gives enough time to the firm to assemble the product before delivery, then the supply chain is in the ________ business
(a) Make to stock (MTS)
(b) Make to Order (MTO)
(c) Configure to Order (CTO)
(d) None of the above

39. If the customer gives enough time to the manufacturer to carry out the complete set of operations after placing the order, then the supply chain is in the ________ business
(a) Make to stock (MTS)
(b) Make to Order (MTO)
(c) Configure to Order (CTO)
(d) None of the above

40. Supply Chain Operations Reference (SCOR) model is used to
(a) Measure SCM performance of the organization
(b) Capture the order delivery lead time
(c) Describe every business process or activity
(d) All of the above

41. The action of unloading materials from an incoming trailer or rail car and immediately loading these materials on to outbound trailers or rail cars, thus eliminating the need for warehousing (storage) is known as
(a) Cross Docking
(b) Containerization
(c) Dunnage
(d) None of the above

42. Moving up from the retailers to wholesalers and to manufacturers, each stage in the chain distorts demand and the variability in the demand keeps increasing as one moves up the chain is known as
(a) Bullwhip effect
(b) Decoupling
(c) Speculative Supply Chain
(d) None of the above

43. Bullwhip effect is caused by
(a) Multiple demand forecast update by various supply chain members
(b) Order batching at various stages in the chain
(c) Price fluctuations and long lead time in purchasing and manufacturing within the chain
(d) All of the above

44. Bullwhip effect can be avoided by
(a) Information sharing
(b) Improving operational efficiency in the chain
(c) Both of the above
(d) None of the above

45. Which of the following is a qualitative method of demand forecasting?
(a) Customer Survey
(b) Sales force composite
(c) Executive opinion
(d) All of the above

46. Which of the following is a qualitative method of demand forecasting
(a) Delphi method
(b) Past analogy
(c) Trend analysis
(d) None of the above

47. Which of the following is a quantitative method of demand forecasting
(a) Simple moving average
(b) Simple exponential smoothing
(c) Holt's double exponential smoothing
(d) All of the above

48. Which of the following is a quantitative method of demand forecasting
(a) Winter's exponential smoothing
(b) Linear Regression analysis
(c) Trend analysis
(d) All of the above

49. Which forecasting method should be used in case of no increasing or decreasing trend, only random fluctuations are present
(a) Simple moving average
(b) Simple exponential smoothing
(c) Holt's double exponential smoothing
(d) Both a and b

50. Which forecasting method should be used in case of no increasing or decreasing trend, but there are seasonal variations at regular interval?
(a) Simple moving average
(b) Simple exponential smoothing
(c) Holt's double exponential smoothing
(d) Both a and b

51. Which forecasting method should be used in case of increasing or decreasing trend, but there is no seasonal variation?
(a) Simple moving average
(b) Simple exponential smoothing
(c) Holt's double exponential smoothing
(d) Both a and b

52. Which forecasting method should be used in case of increasing or decreasing trend with seasonal variation?
(a) Simple moving average
(b) Simple exponential smoothing
(c) Holt's double exponential smoothing
(d) Winters' Triple exponential smoothing

53. In case of ____________ method of demand forecasting, a jury of top executives of the company from different functional areas such as marketing, finance, human resources, production etc brought together to give their opinion about the forecast.
(a) Customer Survey
(b) Sales force composite
(c) Executive opinion
(d) None of the above

54. In ____________ method of demand forecasting, a questionnaire is sent to experts from various diversified streams, seeking their opinion on the forecast of a product.
(a) Delphi method
(b) Past analogy
(c) Trend analysis
(d) None of the above

55. For forecasting the sales of a new product, sales growth trend of other existing products may be taken in case of
(a) Delphi method
(b) Past analogy
(c) Trend analysis
(d) None of the above

56. In ____________ method of demand forecasting, past data required is limited to just the last period's actual demand and its forecast.
(a) Simple moving average
(b) Simple exponential smoothing
(c) Holt's double exponential smoothing
(d) None of the above

57. Which of the following is a method for estimating the forecast error?
(a) Mean absolute deviation (MAD)
(b) Mean Square Error (MSE)
(c) Mean Absolute Percentage Error (MAPE)
(d) All of the above

58. Which of the following is the most popular method for estimating the forecasting error for demand forecast of a product?
(a) Mean Error (ME)
(b) Mean absolute deviation (MAD)
(c) Mean Square Error (MSE)
(d) Mean Absolute Percentage Error (MAPE)

59. Which of the following helps in comparing forecasting errors across product categories?
(a) Mean Error (ME)
(b) Mean absolute deviation (MAD)
(c) Mean Square Error (MSE)
(d) Mean Absolute Percentage Error (MAPE)

60. ____________ is a tool which estimates inventory requirements at the stocking locations and ensures that the supply sources are able to meet the demand.
(a) Materials Requirement Planning (MRP)
(b) Enterprise Resource Planning (ERP)
(c) Distribution Requirement Planning (DRP)
(d) None of the above

61. ____________ is a route in which a truck either delivers product from a single supplier to multiple retails or goes from multiple suppliers to single retailer.
(a) Cross Docking
(b) Milk Runs
(c) Direct Shipping
(d) Hub & Spoke model

62. In hub and spoke model of distribution
(a) Hub refers to location holding inventory for a large region
(b) Spoke leads to smaller distribution centre which holds inventory for smaller region
(c) Both of the above
(d) None of the above

63. The primary function of package is
(a) To hold/withstand the product
(b) To protect the product
(c) To communicate the product
(d) All of the above

64. _____________ is the technology that can be used for determining the position of any object on earth with high amount of accuracy.
(a) Geographical Information System (GIS)
(b) Geographical Positioning System (GPS)
(c) Radio Frequency Identification (RFID)
(d) None of the above

65. _________________ is a computerized system for capturing the position of an object and then processing, enhancing, querying, analyzing and storing it with spatially reference data.
(a) Geographical Information System (GIS)
(b) Geographical Positioning System (GPS)
(c) Radio Frequency Identification (RFID)
(d) None of the above

66. On an average how many intermediaries exist between primary producer and consumer in Indian agricultural system?
(a) 2-3
(b) 5-6
(c) 8-10
(d) None of the above

67. Post-Harvest loss percentage in case of cereals (study conducted by CIPHET, 2010) ranges between
(a) 3.9% - 6.0% (b) 4.3% - 6.1%
(c) 2.8% - 10.1% (d) 5.8%-18%

68. Post-Harvest loss percentage in case of pulses (study conducted by CIPHET, 2010) ranges between
(a) 3.9% - 6.0% (b) 4.3% - 6.1%
(c) 2.8% - 10.1% (d) 5.8%-18%

69. Post-Harvest loss percentage in case of oilseeds (study conducted by CIPHET, 2010) ranges between
(a) 3.9% - 6.0% (b) 4.3% - 6.1%
(c) 2.8% - 10.1% (d) 5.8% - 18%

70. Post-Harvest loss percentage in case of fruits (study conducted by CIPHET, 2010) ranges between:
(a) 3.9% - 6.0% (b) 4.3% - 6.1%
(c) 2.8% - 10.1% (d) 5.8% - 18%

71. Post-Harvest loss percentage in case of vegetables (study conducted by CIPHET, 2010) ranges between
(a) 6.8% - 12.4% (b) 3.9% - 7.4%
(c) 0.8% - 6.9% (d) 5.8% - 18%

72. Post-Harvest loss percentage in case of spices (study conducted by CIPHET, 2010) ranges between
(a) 6.8% - 12.4% (b) 3.9% - 7.4%
(c) 0.8% - 6.9% (d) 5.8% - 18%

73. Post-Harvest loss percentage in case of livestocks (study conducted by CIPHET, 2010) ranges between
(a) 6.8% - 12.4% (b) 3.9% - 7.4%
(c) 0.8% - 6.9% (d) 5.8% - 18%

74. Under the current marketing system, the rate of market fee generally varies between
(a) 0.5 – 2.0 % (b) 2.0 – 5 %
(c) 5.0 – 7.5 % (d) 7.5 – 10.0 %

75. The model APMC act, 2003 provides
(a) Contract farming
(b) Direct marketing
(c) Private market
(d) All of the above

76. Which of the following is the direct marketing model in Punjab and Haryana?
(a) Apni Mandi
(b) Kisan Mandi
(c) Krushak Bazaar
(d) Shetkari Bazaar

77. Which of the following is the direct marketing model in Rajasthan?
(a) Apni Mandi
(b) Kisan Mandi
(c) Krushak Bazaar
(d) Shetkari Bazaar

78. Which of the following is the direct marketing model in Orissa?
(a) Apni Mandi
(b) Kisan Mandi
(c) Krushak Bazaar
(d) Shetkari Bazaar

79. Which of the following is the direct marketing model in Maharashtra?
(a) Apni Mandi
(b) Kisan Mandi
(c) Krushak Bazaar
(d) Shetkari Bazaar

80. Which of the following is the direct marketing model in Tamil Nadu?
(a) Apni Mandi
(b) Kisan Mandi
(c) Uzhavar Santhai
(d) Shetkari Bazaar

81. Which of the following is the direct marketing model in Andhra Pradesh?
(a) Apni Mandi
(b) Kisan Mandi
(c) Uzhavar Santhai
(d) Rythu Bazaar

82. Under Model APMC Act, 2003, extent of market fee levied on the farmers in Farmers market (Direct Sale by the Farmer) is
(a) 2% of the total value
(b) 5% of the total value
(c) No levy on farmers
(d) Changes from market to market

83. Total number of regulated markets in India as on 31/03/2014 is
(a) 2356 (b) 7114
(c) 3654 (d) 4672

84. Total number of Wholesale and Rural primary markets in India as on 31/03/2013 is
(a) 28723 (b) 17777
(c) 24444 (d) 14444

85. Which state has maximum number of wholesale & rural primary market as on 31/03/2014?
(a) Maharashtra
(b) Uttar Pradesh
(c) West Bengal
(d) Bihar

86. Which state has maximum number of regulated markets as on 31/03/2014?
(a) Maharashtra
(b) Uttar Pradesh
(c) West Bengal
(d) Andhra Pradesh

87. How much is the storage capacity of Central Warehousing Corporation (in 2016)?
(a) 10.09 MMT
(b) 8.5 MMT
(c) 6.5 MMT
(d) 5.5 MMT

88. How many warehouses does the Central Warehousing Corporation operate (2016)?
(a) 271
(b) 371
(c) 440
(d) 571

89. Central Warehousing Corporation (CWC) offers the services of
(a) Clearing & forwarding
(b) Handling & transportation
(c) Procurement & Distribution
(d) All of the above

90. Which of the following is correct?
(a) Central Warehouse Corporation (CWC) has only owned storage capacity
(b) Central Warehouse Corporation (CWC) has only hired storage capacity
(c) Central Warehouse Corporation (CWC) has both owned & hired storage capacity
(d) Central Warehouse Corporation (CWC) does not store food grains

91. Which of the following is not true with regard to functions of CWC?
(a) To acquire and build godowns and warehouses at suitable places in India or abroad.
(b) To run warehouses for the storage of agricultural produce, seeds, manures, fertilizers, agricultural implements and notified commodities offered by individuals, cooperative societies and other institutions.
(c) To arrange facilities for transport of agricultural produce, seeds, manures, fertilizers, agricultural implements and notified commodities to and from warehouses.
(d) None of the above

92. The warehouses under Central Warehouse Corporation (CWC) include
(a) Custom Bonded Warehouses
(b) Container Freight Stations (CFSs)/ Inland Container Depots (ICDs),
(c) Air Cargo Complexes & Integrated Check Post/Truck Terminal
(d) All of the above

93. CWC provides storage and warehousing facilities to
(a) Public and private institutions, Co-operative societies, traders, farmers, importers/exporters, etc.
(b) Only Public institutions
(c) Public institutions & Co-operative societies
(d) None of the above

94. Total Storage capacity of Food Corporation of India (FCI) as on 01/04/2014 is
(a) 70.75 MMT (b) 60.65 MMT
(c) 55.56 MMT (d) 50.58 MMT

95. Which of the following is correct?
(a) Food Corporation of India (FCI) has only owned storage capacity
(b) Food Corporation of India (FCI) has only hired storage capacity
(c) Food Corporation of India (FCI) has both owned & hired storage capacity
(d) Food Corporation of India (FCI) does not store food grains

96. Total Covered Storage capacity of Food Corporation of India (FCI) as on 01/04/2014 is
(a) 33.86 MMT
(b) 23.86 MMT
(c) 25.46 MMT
(d) 20.58 MMT

97. The objective of Food Corporation of India (FCI) is
(a) Effective price support operations for safeguarding the interests of the farmers.
(b) Distribution of food grains throughout the country for public distribution system.
(c) Maintaining satisfactory level of operational and buffer stocks of food grains to ensure National Food Security.
(d) All of the above

98. The storage capacity with State Warehousing Corporations (SWCs) is to the tune of
(a) 12.29 MMT (b) 15.37 MMT
(c) 17.37 MMT (d) 21.29 MMT

99. The term 'Value Chain' was developed and popularized by
(a) Peter Drucker
(b) Keith Oliver
(c) Raphael Kaplinsky
(d) Michael Porter

100. According to Michael Porter, ____________ is the combination of nine generic value added activities operating within a firm.
(a) Supply Chain (b) Value Chain
(c) Supply Web (d) Value Web

101. ________________ focusses upstream on integrating supplier & producer processes, improving efficiency & reducing waste
(a) Supply Chain (b) Value Chain
(c) Demand Chain (d) Product Chain

102. ____________ focusses downstream on creating value in the eyes of the customer
(a) Supply Chain (b) Value Chain
(c) Demand Chain (d) Product Chain

103. Difference between supply chain and value chain is
(a) Shift in focus from supply base to the customers
(b) Shift in focus from one supplier to other
(c) Shift in focus from one customer to other
(d) No shift in focus, both have same focus, i.e. the suppliers

104. Relationship between firms in different stages of the supply chain such as processors and retailers is refereed as
(a) Horizontal cooperation
(b) Vertical cooperation
(c) Conglomerate relation
(d) Concentric relation

105. Which of the following is an example of horizontal integration?
(a) Corporate farming
(b) Contract farming
(c) Cooperative farming
(d) Tenant farming

106. Which of the following is a benefit of horizontal integration?
(a) Economies of scale & cost sharing
(b) Better access to new market
(c) Increase bargaining power
(d) All of the above

107. NCDEX stand for
(a) National Commodity Derivatives Exchange
(b) National Commodity and Derivatives Exchange Limited
(c) National Commodity Derivatives Extension Limited
(d) National Common Derivatives Exchange

108. NCDEX is a
(a) Online Commodity Exchange
(b) Offline Commodity Exchange
(c) Both
(d) None

109. NCDEX offers trading in
(a) 20 commodities which includes 11 agricultural commodities.
(b) 16 commodities which includes 11 agricultural commodities.
(c) 26 commodities which includes 21 agricultural commodities.
(d) 26 commodities which includes 11 agricultural commodities.

110. Star Agriwarehousing and Collateral Management Ltd. is engaged in
(a) agri-warehousing
(b) origination and procurement
(c) collateral management and logistics
(d) All of the above

111. Forward Market commission was merged with SEBI in the year
(a) 2000 (b) 2010
(c) 2015 (d) 2010

112. The lead promoter of National Agricultural Market is
(a) ISAP
(b) SFAC
(c) FICCI
(d) CII

113. The strategic partner to implement NAM is
(a) Indian Society of Agribusiness Professionals
(b) Directorate of Marketing and Inspection
(c) Coromandal Fertilizers Ltd
(d) Nagarjuna Fertilizers and Chemicals Ltd

114. ATIF stands for
(a) Agri-Tech Infrastructure Fund
(b) Agricultural Technology India Fund
(c) Agri-Technique Infrastructure Fund
(d) Agri-Technique India Fund

115. Government of India approved NAM in the year
(a) 2012
(b) 2014
(c) 2015
(d) 2016

116. Which of the following is true with respect to NAM?
(a) single trading license to be valid across the State
(b) single point levy of market fee across the State
(c) provision for e-auction / e-trading as a mode of price discovery
(d) All the above

117. How many markets are targeted to be covered under NAM by 2017-18?
(a) 585 (b) 485
(c) 535 (d) 435

118. The technical partner for hosting of e-NAM software on Meghraj Cloud is
(a) National Informatics Centre (NIC)
(b) C-DAC
(c) IIT-Mumbai
(d) IIIT-Allahabad

119. ____________ is a strategic tool to analyze internal firm activities.
(a) SWOT analysis
(b) Value Chain Analysis
(c) PESTEL Analysis
(d) None of the above

120. Currently the FDI in trading of food products is at the level of:
(a) 49% (b) 51%
(c) 74% (d) 100%

121. Big Basket and Grofers are examples of
(a) Online Food Retailer
(b) Offline Food Retailer
(c) Both of the above
(d) None of the above

122. Food Industry in India is currently valued at (2016)
(a) US$ 19.71 billion
(b) US$ 29.71 billion
(c) US$ 39.71 billion
(d) US$ 49.71 billion

123. Food Processing Industry in India accounts for ____ percent of the country's total food market (in 2016).
(a) 22% (b) 32%
(c) 42% (d) 52%

124. Organized food business in India is worth (in 2016)
(a) US$ 28 billion (b) US$ 38 billion
(c) US$ 48 billion (d) US$ 58 billion

125. Which of the following is a food delivery start-up?
(a) Swiggy (b) Foodpanda
(c) Zomato (d) All of these

126. Which of the following provide integrated post harvest management solution?
(a) Star AGri
(b) Sohal Lal Commodity Management
(c) Shree Shubham Logistics
(d) All of the above

127. Cold Star Logistics deals with
(a) Only Cold Storage solution
(b) Only Refrigerated transportation solution
(c) Both Cold storage and refrigerated transportation
(d) None of the above

128. Which of the following company connects buyers and sellers of agricultural produce and allied services via a web and mobile based information exchange platform?
(a) eFarm
(b) Reuters Market Light
(c) Fasal Intuit
(d) TCS' mKRISHI

129. Which of the following provides personalized agricultural market information to the farmers through a mobile based service?
(a) Reuters Market Light
(b) Fasal Intuit
(c) Both of the above
(d) None of the above

130. Which of the following offers personalized advisory services to farmers via mobile phones?
(a) eFarm
(b) Reuters Market Light
(c) Fasal Intuit
(d) TCS' mKRISHI

131. Which of the following is a hosted web service for supply chain management, which can be accessed via basic mobile phones and web browsers?
(a) Logistimo
(b) eFarm
(c) Reuters Market Light
(d) Fasal Intuit

132. ______of Agricultural commodities for domestic consumption is transported by road in Indi(a)
(a) 50% (b) 60%
(c) 70% (d) 80%

133. Agri warehousing accounts for ____________ of the warehousing market in India
(a) 10% (b) 15%
(c) 20% (d) 25%

134. Agri warehousing industry in India in FY 2013 is valued at:
(a) 60-65 billion (b) 70-75 billion
(c) 80-85 billion (d) 90-95 billion

135. Agri logistics activities include
(a) Procurement (b) Storage
(c) Distribution (d) All of the above

136. Beginning of the agri logistics value chain comprises of
(a) Farmers (b) Traders
(c) Processors (d) Retail Chains

137. Agri logistics value chain culminates at
(a) Retail Customer
(b) Retail Chains
(c) Processors
(d) Wholesalers

138. Cold stores account for approximately ____% of total warehousing industry
(a) 14% (b) 16%
(c) 18% (d) 20%

139. Logistics costs account for around ____% of average retail prices in India
(a) 1-5% (b) 6-10%
(c) 10-15% (d) 15-20%

140. Global logistics costs towards average retail prices stands at
(a) 4-5% (b) 6-8%
(c) 8-10% (d) 12-15%

141. Organized players contribute ______ of the cold chain industry market in India
(a) 4-6% (b) 8-10%
(c) 12-15% (d) 15-18%

142. Retail Market Size in India (2015) is to the tune of
(a) US $ 400 billion
(b) US $ 500 billion
(c) US $ 600 billion
(d) US $ 700 billion

143. Contribution of Retail Sector to total GDP (2015) is _____
(a) 15% (b) 20%
(c) 25% (d) 30%

144. Share of Organized retail to total retail in India in (2015) is
(a) 5-6% (b) 6-8%
(c) 8-10% (d) 10-11%

145. Food Retail accounts for ______ of total Retail (2015)
(a) 41% (b) 51%
(c) 61% (d) 71%

146. Organized food retail is _____ of total retail (2015)
(a) 1.22% (b) 1.44%
(c) 1.66% (d) 1.88%

147. Overhead cost in Organized retail in India is approximately
(a) 5-10% (b) 10-15%
(c) 17-25% (d) 32-37%

148. Overhead cost in Unorganized retail in India is approximately
(a) 1-2%
(b) 5-8%
(c) 10-12%
(d) 18-20%

149. Which of the following is a non-store base retail model?
(a) Direct marketing
(b) Vending Machine
(c) Internet marketing
(d) All of the above

150. Which of the following E-Grocers follow inventory model?
(a) Mygrahak (b) Zopnow
(c) Both of these (d) None of these

151. Which of the following E-Grocers follow Hyper-local sourcing model?
(a) Grofers (b) Peppertap
(c) Aaramshop (d) All of the above

152. Which of the following E-Grocers follow hybrid model?
(a) BigBasket (b) LocalBanya
(c) Both of these (d) None of these

153. Which of the following is first mover among Brick & Mortar retailers into online retailing?
(a) Godrej
(b) Reliance Retail
(c) Aditya Birla Retail
(d) Future Group

154. Value Chain Financing includes:
(a) Financing for buying inputs & Creating assets
(b) Financing for Risk Mitigation & Commodity Future Market
(c) Warehouse Receipt
(d) All of the above

155. Which of the statement is true?
(a) Nearly half of the farmers in India have no access to credit
(b) 27% of the farmers in India have access to formal sources
(c) Those have no access; 88% are marginal farmers
(d) All of the above

156. Share of processing sector in total agricultural credit is:
(a) 16% (b) 18%
(c) 20% (d) 24%

157. Out of the following, which has received highest FDI inflow?
(a) Food processing
(b) Agriculture Service
(c) Fertilizer industry
(d) Sugar industry

158. Which of the following is a public sector initiative of innovative institutions?
(a) State level Agro-industries Corporation
(b) Safal
(c) Mahagrapes
(d) All of the above

159. Which of the following is a Cooperative sector initiative of innovative institutions?
(a) Safal
(b) Mahagrapes
(c) Both of the above
(d) None of the above

160. Which of the following is a Private sector initiative of innovative institutions?
(a) ITC echoupal
(b) Mahindra Subh Labh
(c) Suguna Poultry
(d) All of the above

161. Which of the following is a constraint in Value chain financing?
(a) Breach of contract
(b) Low bargaining power of smallholders
(c) Monopsony in the long run
(d) All of the above

162. Size of India Seed industry in 2015 stands at:
(a) US $ 2 billion (b) US $ 3 billion
(c) US $ 4 billion (d) US $ 5 billion

163. Which of the following has maximum share in global seed industry?
(a) Monsanto (USA)
(b) Dupont Pioneer (USA)
(c) Syngenta (Switzerland)
(d) Vilmorin (France)

164. Total Fertilizer produced in the country in 2014-15 is:
(a) 3.08 million tonnes
(b) 4.08 million tonnes
(c) 5.08 million tonnes
(d) 6.08 million tonnes

165. Which of the following has maximum share in global fertilizer industry?
(a) Yara (Norway)
(b) Agrium (Canada)
(c) The Mosaic Company (USA)
(d) Potashcorp (Canada)

166. Market Size of Pesticide industry in India stands at (2014)
(a) US $ 2.25 (b) US $ 3.25
(c) US $ 4.25 (d) US $ 5.25

167. Which of the following has maximum share in global Agrochemical industry?
(a) Syngenta (Switzerland)
(b) Bayer Cropsciences (Germany)
(c) BASF (Germany)
(d) DOW Agrosciences (USA)

168. Which of the following company has recently acquired US based monsanto for $64.5 billion?
(a) Bayer Cropsciences (Germany)
(b) BASF (Germany)
(c) DOW Agrosciences (USA)
(d) None of the above

169. Nestle's model of milk procurement in dairy industry is commonly known as
(a) Hub and spoke model
(b) Milk run model
(c) Milk district model
(d) Milk centre model

170. Which of the following has maximum share in Indian pesticide industry?
(a) Syngenta India
(b) Bayer Cropscience
(c) United Phosphorus
(d) Rallis India

ANSWER KEY (Agricultural Supply Chain Management):

1. (b) **2.** (b) **3.** (b) **4.** (a) **5.** (c) **6.** (d) **7.** (a) **8.** (c) **9.** (a) **10.** (c)
11. (d) **12** (c) **13.** (b) **14.** (a) **15.** (a) **16.** (b) **17.** (a) **18.** (b) **19.** (c) **20.** (d)
21. (d) **22.** (a) **23.** (d) **24.** (d) **25.** (d) **26.** (c) **27.** (c) **28.** (b) **29.** (c) **30.** (d)
31. (a) **21.** (b) **33.** (c) **34.** (b) **35.** (d) **36.** (b) **37.** (a) **38.** (c) **39.** (b) **40.** (a)
41. (a) **42.** (a) **43.** (d) **44.** (c) **45.** (d) **46.** (c) **47.** (d) **48.** (d) **49.** (d) **50.** (a)
51. (c) **52.** (d) **53.** (c) **54.** (a) **55.** (b) **56.** (b) **57.** (d) **58.** (b) **59.** (d) **60.** (c)
61. (b) **62.** (c) **63.** (a) **64.** (b) **65.** (a) **66.** (b) **67.** (a) **68.** (b) **69.** (c) **70.** (d)
71. (a) **72.** (b) **73.** (c) **74.** (a) **75.** (d) **76.** (a) **77.** (b) **78.** (c) **79.** (d) **80.** (c)
81. (d) **82.** (c) **83.** (b) **84.** (a) **85.** (a) **86.** (d) **87.** (a) **88.** (c) **89.** (d) **90.** (c)
91. (d) **92.** (d) **93.** (a) **94.** (b) **95.** (c) **96.** (a) **97.** (d) **98.** (d) **99.** (d) **100.**(b)
101.(a) **102.**(b) **103.**(a) **104.**(b) **105.**(c) **106.**(d) **107.**(b) **108.**(a) **109.**(c) **110.**(d)
111.(c) **112.**(b) **113.**(d) **114.**(a) **115.**(c) **116.**(d) **117.**(a) **118.**(a) **119.**(b) **120.**(d)
121.(a) **122.**(c) **123.**(b) **124.**(c) **125.**(d) **126.**(d) **127.**(c) **128.**(a) **129.**(c) **130.**(d)
131.(a) **132.**(d) **133.**(b) **134.**(c) **135.**(d) **136.**(a) **137.**(a) **138.**(b) **139.**(b) **140.**(a)
141.(b) **142.**(c) **143.**(b) **144.**(d) **145.**(c) **146.**(b) **147.**(c) **148.**(b) **149.**(d) **150.**(c)
151.(d) **152.**(c) **153.**(a) **154.**(d) **155.**(d) **156.**(a) **157.**(a) **158.**(a) **159.**(c) **160.**(d)
161.(d) **162.**(b) **163.**(a) **164.**(a) **165.**(a) **166.**(c) **167.**(a) **168.**(a) **169.**(c) **170.**(c)

CHAPTER 7

7. Production & Operations Management

1. Operations Management includes all of these activities except:
(a) Schedule work
(b) Secure financial resources
(c) Maintain quality
(d) Manage inventories

2. Which of the following is not included in "5 Ps" of Operations Management?
(a) Product
(b) Plant
(c) Processes
(d) Price

3. The major difference between goods and services is:
(a) It is difficult to inventory services
(b) Services lack the profitability that goods enjoy
(c) Goods production usually requires more labor than services
(d) Goods are always customized, services are always standardized

4. Technology choices don't have a major effect on:
(a) costs
(b) Employee turnover
(c) productivity
(d) quality

5. Which of the following is not true about a lean production system?
(a) It puts emphasis on quality, flexibility, and time reduction.
(b) It puts emphasis on reducing a company's labor force.
(c) It uses small production batch sizes.
(d) It uses safety stock to protect against uncertainty.

6. Many factors impact productivity, one factor not generally stated as having a negative impact on productivity is:
(a) Government regulations
(b) Liability claims
(c) Increased emphasis on services
(d) Emphasis on short-term performance

7. Which is the correct hierarchy of operations management decisions (highest first)?
(a) Operational, strategic, tactical
(b) Goals, missions, productivity
(c) Strategic, operational, tactical
(d) Strategic, tactical, operational

8. Which of the following has least likely source of input for developing an operations strategy?
(a) Planning personnel
(b) Suppliers
(c) Top Management
(d) Production personnel

9. A measure of productivity which reflects a combination of some or all of the resources used to obtain a certain output is:
(a) machine productivity
(b) multi-factor productivity
(c) materials productivity
(d) overhead productivity

10. Quality, Supply Chains, Inventory and Capacity are strategic decisions generally made by:
(a) Marketing
(b) Human Resources
(c) Project Managers
(d) Operations

11. The ratio of good output to quantity of raw material input is:
(a) Overall good productivity
(b) Process yield
(c) Quality measurement
(d) Quality/Quantity ratio

12. Which of the following is a partial measure of productivity?
(a) Output/Materials
(b) Output/All resources used
(c) Output/Inputs
(d) All of the above

13. Which of the following is a multifactor measure of productivity?
(a) Output/(Labor + Capital + Energy)
(b) Output/All resources used
(c) Output/Inputs
(d) All of the above

14. Which of the following is a total measure of productivity?
(a) Output/Materials
(b) Output/(Labor + Capital + Energy)
(c) Output/Inputs
(d) All of the above

15. If total daily output in a production system is 400 units and the total labor necessary to produce the 400 units is 350 hours, what kind of productivity measure should one use to compute productivity?
(a) Partial measure
(b) Multifactor measure
(c) Total measure
(d) All of the above

16. If total daily output in a production system is 400 units and the total labor necessary to produce the 400 units is 350 hours, and the total materials used is 425 units, what kind of productivity measure should one use to compute productivity?
(a) Partial measure
(b) Multifactor measure
(c) Both A and B
(d) None of the above

17. Labour productivity cannot be measured in terms of
(a) Units of output per labor hour
(b) Units of output per shift
(c) Value-added per labor hour
(d) Units of output per rupee input

18. Forecasts based on Judgment and Opinion include:
(a) Delphi method
(b) Weighted moving average
(c) Linear regression
(d) Naive Methods

19. Which probability distribution is used most extensively in dealing with forecasting errors?
(a) Normal distribution
(b) Poisson distribution
(c) Exponential distribution
(d) Beta distribution

20. The cumulative forecast error is important for determining the:
(a) Mean squared error
(b) Bias in forecast error
(c) Mean absolute deviation
(d) Correlation coefficient

21. Which of the following smoothing constants would make an exponential smoothing forecast equivalent to a naive forecast?
(a) 0.0 (b) .01
(c) 0.10 (d) 1.0

22. Simple exponential smoothing is being used to forecast demand. The previous forecast of 66 turned out to be six units less than actual demand. The next forecast is 69, implying a smoothing constant, alpha, equal to:
(a) 0.01 (b) 0.10
(c) 0.15 (d) 0.20

23. The demand of a product on Tuesday was 60 kg, on Wednesday it was 82 kg, on Thursday it was 77 kg. A naive forecast for the demand on Friday would be:
(a) 73 (b) 77
(c) 80 (d) 82

24. Product design that uses computers is referred to as:
(a) QFD (b) CAM
(c) CAD (d) TQM

25. Which of the following isn't a dimension of quality?
(a) Conformance
(b) Reliability
(c) Durability
(d) Profitability

26. Service design generally differs from product design in which of the following ways?
(a) Service design tends to focus on tangible factors
(b) There is less freedom in detecting and correcting errors prior to delivery
(c) There is less visibility to customers
(d) Once designed, service tends to remain constant while product design will change often.

27. Efficiency, in capacity terms, is the ratio of:
(a) actual output to effective capacity
(b) effective capacity to actual output
(c) design capacity to effective capacity
(d) design capacity to actual capacity

28. A bottlenecked operation will result in
(a) Unbalanced systems
(b) Labor unrest
(c) Increased capacity
(d) Job shop efficiency

29. Given the following information: Effective capacity = 20 units per day; Design capacity = 60 units per day; Actual output = 15 units per day, what is the utilization?
(a) 25% (b) 33%
(c) 50% (d) 75%

30. The maximum output rate or service capacity an operation, process or facility is designed for is the
(a) Effective capacity
(b) Theoretical throughput
(c) Efficiency peak
(d) Design capacity

31. The system that has the highest equipment flexibility is:
(a) Job shop
(b) Batch system
(c) Repetitive production
(d) Continuous processing

32. Volume of output is very low in case of:
(a) Job shop
(b) Batch system
(c) Repetitive production
(d) Continuous processing

33. Unit cost is very low in case of:
(a) Job shop
(b) Batch system
(c) Repetitive production
(d) Continuous processing

34. Variety of job is very high in case of:
(a) Job shop
(b) Batch system
(c) Repetitive production
(d) Continuous processing

35. An operations strategy for process selection should recognize that:
(a) process selection is easily changed
(b) engineering is solely responsible for choosing what process is best
(c) flexibility may not the best choice
(d) managers need not have technical skills

36. Product profiling is linked to:
(a) Process capabilities
(b) Pricing strategies
(c) Large order sizes
(d) Environmental scanning

37. Which one of the following is not generally regarded as an advantage of product layouts?
(a) Labor costs are low per unit.
(b) The system is fairly flexible to changes in volume of output.
(c) unit costs may be lower than with other processes
(d) There is high utilization of labor and equipment

38. Which of the following layout uses standardized processing operations to achieve smooth, rapid, high-volume flow
(a) Product Layout
(b) Process Layout
(c) Fixed Position Layout
(d) Combination Layout

39. Which of the following Layout that can handle varied processing requirements?
(a) Product Layout
(b) Process Layout
(c) Fixed Position Layout
(d) Combination Layout

40. In which layout the product or project remains stationary, and workers, materials, and equipment are moved as needed
(a) Product Layout
(b) Process Layout
(c) Fixed Position Layout
(d) Combination Layout

41. Which one of the following is not generally regarded as an advantage of product layouts?
(a) In-process inventory costs can be high
(b) Highly susceptible to shutdowns
(c) Needs preventive maintenance
(d) None of the above

42. In a product layout, the process of deciding how to assign tasks to work stations is referred to as:
(a) process balancing
(b) task allocation
(c) line balancing
(d) scheduling

43. If a line is balanced with 90 percent efficiency, the "balance delay" would be:
(a) 10 percent
(b) 90 percent
(c) 100 percent
(d) None of the above

44. Job Analysis can be done using
(a) Flow Process Chart
(b) Worker-Machine Chart
(c) Both of the above
(d) None of the above

45. Chart used to examine the overall sequence of an operation by focusing on movements of the operator or flow of materials is
(a) Flow Process Chart
(b) Worker-Machine Chart
(c) Pareto Chart
(d) None of the above

46. Chart used to determine portions of a work cycle during which an operator and equipment are busy or idle
(a) Flow Process Chart
(b) Worker-Machine Chart
(c) Pareto Chart
(d) None of the above

47. The symbols of a flow process chart stand for:
(a) Inspection, delay, transportation, storage, operation
(b) Transportation loaded, search, grasp, store, perform operation
(c) Release, transport, store, grasp, search
(d) Input, processing and output

48. Job design does not focus on:
(a) What will be done in the job
(b) How the job will be done
(c) Why will be the job done
(d) Where the job will be done

49. Giving a worker a larger portion of the total task by horizontal loading is known as:
(a) Job Enlargement
(b) Job Rotation
(c) Job Enrichment
(d) None of the above

50. Workers periodically exchange jobs
(a) Job Enlargement
(b) Job Rotation
(c) Job Enrichment
(d) None of the above

51. Increasing responsibility for planning and coordination tasks, by vertical loading
(a) Job Enlargement
(b) Job Rotation
(c) Job Enrichment
(d) None of the above

52. The systematic study of the human motions used to perform an operation is known as
(a) Method Analysis
(b) Motion Study
(c) Work Sampling
(d) None of the above

53. Analysis of therbligs is most closely related to:
(a) Methods analysis
(b) Simo chart
(c) Learning curve effect
(d) Motion study

54. Among the following, all are advantages of specialization except:
(a) Simplifies training
(b) Difficult to motivate quality
(c) High productivity
(d) Relatively low wages

55. Which one of the following statements concerning work sampling is not true?
(a) Observations are made at random intervals
(b) A detailed description of activities and times is obtained
(c) Multiple observations are necessary
(d) There is little or no disruption of work

56. The need for methods analysis may arise from any of these sources except:
(a) changes in seasonal demand
(b) changes in product design
(c) new products or services
(d) changes in tools and equipment

57. Stop watch time studies require determination of
(a) Allowance factors
(b) The number of cycles to observe
(c) Standard times
(d) Work measurement

58. Manufacturing firms focus more on _______ for selecting a location.
(a) Cost
(b) Revenue
(c) Both of the above
(d) None of the above

59. Service/Retail firms focus more on _______ for selecting a location
(a) Cost
(b) Revenue
(c) Both of the above
(d) None of the above

60. Which one of the following is more a community factor than a site factor?
(a) Rail access (b) Zoning
(c) Tax rates (d) Room for expansion

61. Which step should occur first in making location decisions?
(a) Breakeven analysis
(b) Identifying factors that are important, such as location of markets
(c) Identifying criteria that will be used to evaluate alternatives
(d) Evaluate each of the specific alternatives

62. Communities offer financial and other incentives in or to ____ new businesses.
(a) tax
(b) Attract
(c) Capitalize
(d) Control

63. In location planning, environmental regulations, availability of utilities, and direct and indirect taxes are:
(a) regional factors
(b) community factors
(c) site-related factors
(d) none of the above

64. Which of the following is used as a technique for evaluating location alternatives
(a) Locational Cost-Profit-Volume Analysis
(b) Factor Rating
(c) Centre of Gravity Method
(d) All of the above

65. Which tool organizes data into logical categories?
(a) Relationship diagrams
(b) Pareto charts
(c) Affinity diagrams
(d) Process charts

66. Which name is associated with a cause-and-effect diagram?
(a) Deming (b) Crosby
(c) Baldrige (d) Ishikawa

67. Which of the following may not be considered as a determinant of quality?
(a) Ease of use
(b) Design
(c) Conformance to design
(d) Focus groups

68. Warranty service, processing of complaints, and costs of litigation are examples of:
(a) internal failure costs
(b) external failure costs
(c) appraisal costs
(d) prevention costs

69. Which one of these is the conceptual basis for continuous improvement?
(a) Benchmarking
(b) Cause-and-effect diagram
(c) Pareto analysis
(d) The plan-do-check-act cycle

70. Pareto is most closely associated with
(a) 80/20
(b) 50/50
(c) Brainstorming
(d) Cause and effect diagrams

71. Giving workers responsibility for quality improvements and authority to make changes is known as:
(a) Continuous improvement
(b) Delegation
(c) Employee empowerment
(d) Employee involvement

72. A cause of variation in a process probably wouldn't be due to:
(a) People
(b) Standards
(c) Methods
(d) Environment

73. What kind of variation is undeterminable?
(a) Assignable
(b) Controllable
(c) Random
(d) Theoretical

74. What type of control chart would be used to monitor the number of defects in the output of a process?
(a) x-bar chart (b) p chart
(c) c chart (d) R chart

75. A c-chart is best suited for this type of data:
(a) Count (b) Attribute
(c) Measurement (d) Proportion defective

76. Which of the following quality control sample statistics indicates a quality characteristic that is an attribute?
(a) Mean
(b) Variance
(c) Standard deviation
(d) Proportion

77. A point that is outside of the lower control limit on an R-chart:
(a) Requires that the lower control limit be raised
(b) Should be ignored because it signifies better than average quality
(c) Should be investigated because an assignable cause of variation might be present
(d) Is impossible since the lower limit is always zero

78. Which are the elements of an acceptance-sampling plan?
(a) The lot size, the sample size, and the lot tolerance percent defective
(b) The lot size, the sample size, and the acceptance number
(c) The sample size, the acceptance number, and the lot tolerance percent defective
(d) The sample size, the acceptance number, and the average outgoing quality limit

79. In order to effectively plan, estimates of each of the following items must be available to planners except:
(a) Inventory lot sizes
(b) Demand for each period
(c) Capacity for each period
(d) Costs (regular time, overtime, subcontracting, backorders, etc.)

80. Master schedules are the result of:
(a) Backorders
(b) Subcontracting
(c) Disaggregation
(d) Mission statements

81. Which one of the following is not a basic option for altering capacity?
(a) Hire and fire workers
(b) Use overtime
(c) Use subcontracting
(d) Use backorders

82. Which one of these is not an output of a master production schedule?
(a) Production requirements
(b) Projected inventory
(c) Customer orders
(d) Available-to-promise inventory

83. Aggregate planners try to bring a balance between
(a) make vs buy
(b) capacity and inventories
(c) fixed and variable costs
(d) capacity and demand

84. An option for altering the availability of capacity is:
(a) use of overtime or slack time
(b) pricing
(c) promotion
(d) backorders

85. The principal input files and programs that are incorporated into a computerized MRP system are:
(a) Inventory record, master schedule, BOM, MRP computer program
(b) Inventory record, master file, part-period balancing program
(c) Capacity requirements planning file, master schedule, inventory record
(d) MRP computer program, master schedule, capacity requirements planning file

86. Which one of these would not be a possible lot size that would be identified by MRP?
(a) EOQ
(b) Gross requirements
(c) Net requirements
(d) An amount less than net requirements

87. A list of all parts and materials needed to assemble one unit of a product is called:
(a) A master schedule
(b) A Kanban
(c) An inventory record file
(d) A bill of materials

88. The real time aspect of ERP makes it a valuable:
(a) Strategic Planning tool
(b) Telecommuting system
(c) Downloading system
(d) Forecasting system

89. Which of the following results is not a benefit of JIT?
(a) Flexibility
(b) Lower Inventories
(c) Reduced lead times
(d) Lower total setup costs

90. In the JIT philosophy, the ideal lot size is:
(a) The economic order quantity
(b) The economic run size
(c) One unit
(d) The capacity of the standard container

91. A basic requirement for operating with low inventories in JIT systems is:
(a) Adequate space is available to store inventory
(b) Inventory investment must be saved
(c) Major problems must have been solved
(d) Safety stock must be sufficient to cover all possible stockouts

92. Inventory management requires two questions to be answered: How many and when JIT answers those two questions with?
(a) Exact quantities, when orders are received
(b) Economic order quantities, reorder point
(c) Economic run sizes, reorder point
(d) Economic lot sizes, lead time

93. The degree of preventive maintenance that is desirable is influenced by all of the following except:
(a) The cost of equipment failure (breakdown)
(b) Salvage value of equipment.
(c) The degree of technology involved
(d) The cost of preventive maintenance

94. Cost of equipment breakdown does not include:
(a) The cost of idle workers
(b) Possible damage to other equipment
(c) Increased fixed costs
(d) Possible injury of operators

95. Ideally, preventive maintenance will be scheduled:
(a) On a periodic basis
(b) At the start of each workday
(c) Just prior to a breakdown or failure
(d) None of the above

96. Which one of the following is not an appropriate job-shop scheduling rule?
(a) Earliest due date
(b) Assignment method
(c) First come first served
(d) Johnson's rule

97. Which one of the following is an optimizing technique?
(a) Shortest processing time
(b) Assignment method
(c) First come first served
(d) Critical Ratio

98. Which scheduling rule usually minimizes lateness?
(a) SPT (b) FCFS
(c) Critical Ratio (d) EDD

99. Which statement is correct?
(a) Vertical loading involves infinite loading
(b) Horizontal loading involves infinite loading
(c) Vertical loading involves finite loading
(d) Finite loading is a subset of infinite loading

100. Finite loading, as compared to infinite loading, assumes that:
(a) There are only a finite number of jobs to be done
(b) Work centers can be loaded without considering their capacity
(c) A work centers' capacity (limitations) must be taken into account
(d) Work centers' capacity should be ignored as a constraint

101. In a task assignment situation, in how many different ways can four jobs be assigned to four machines?
(a) 4
(b) 8
(c) 16
(d) 24

102. A major advantage of Gantt charts in project management is that they:
(a) Monitor progress over time
(b) Identify the critical path
(c) Allow for probabilistic estimates
(d) Reveal all activity relationships

103. Using the computing algorithm, activity slack can be computed as:
(a) LS – ES
(b) LF – EF
(c) Both of the above
(d) None of the above

104. An important assumption in PERT calculations is:
(a) Activity times are normally distributed
(b) Activity times are independent
(c) Alternate activities are available for every element
(d) Early start times must equal early complete times

105. Concerning PERT and CPM:
(a) These are two different names for essentially the same technique
(b) CPM tends to focus more on the critical path while PERT computes the minimum slack times
(c) PERT gives a clearer indication of activity slack, while CPM gives a better picture of path slack
(d) Because of its recent discovery, much less is known about CPM than PERT

106. The set of interrelated management activities which are involved in manufacturing certain products is called as _____
(a) Production management
(b) Quality management
(c) Operation management
(d) Strategic management

107. The set of interrelated management activities which are involved in providing services are called as_____
(a) Production management
(b) Quality management
(c) Operations management
(d) Strategic management

108. Construction of flats is a type of _____
(a) Product (b) Project
(c) Service (d) Output

109. Operational decisions are taken at _____ level of management.
(a) Top (b) Middle
(c) Bottom (d) both a and c

110. Input is converted into output is called as_____ process.
(a) Conversion (b) Transformation
(c) both a and b (d) none

111. Services are comes under _____
(a) Input (b) Output
(c) Product (d) Project

112. _____ uses facilities, equipment and operating methods to produce goods that satisfy consumers.
(a) Production system
(b) Operations system
(c) Continuous system
(d) Intermittent system

113. Medical facilities is an example of _____
(a) Product (b) Project
(c) Service (d) Output

114. _____ is the conversion process in which successive units of output undergo the same sequence of operations, using specialized equipment usually positioned along a production line.
(a) Flow shop (b) Job shop
(c) Project (d) Continuous

115. _____flow shop will produce the same type of output.
(a) Intermittent (b) Continuous
(c) Batch (d) Project

116. _____flow shop will produce the products in which the process is interrupted to set it up to handle different specifications of the same basic design.
(a) Intermittent (b) Continuous
(c) Batch (d) Project

117. Construction of bridge is an example of _____ type of manufacturing system.
(a) Project (b) Job shop
(c) Flow shop (d) Continuous

118. _____ is a conversion process in which units of different types of products follow through different sequences through different shops.
(a) Batch (b) Project
(c) Job shop (d) Flow shop

119. _____ is a conversion process intermediate variety of products with intermediate volumes.
(a) Batch (b) Project
(c) Job shop (d) Flow shop

120. Complex scheduling is involved in _____ production process.
(a) Batch (b) Project
(c) Job shop (d) Flow shop

121. _____ refers to the process of creating a complex one kind-of product or service.
(a) Batch (b) Project
(c) Job shop (d) Flow shop

122. The range of possible operations is narrower in _____ manufacturing process.
(a) Batch (b) Project
(c) Job shop (d) Flow shop

123. Auto repair is an example of _____ production process.
(a) Batch (b) Project
(c) Job shop (d) Flow shop

124. _____ is the relationship between the output and input of a business system.
(a) Production
(b) Productivity
(c) Operations
(d) Transformation

125. Auto factory is an example of _____ production process.
(a) Batch (b) Project
(c) Job shop (d) Flow shop

126. The efficient and effective generation and development of ideas through a process that leads to new products.
(a) Product planning
(b) Process layout
(c) Product design
(d) Process planning

127. The creation of new product is _____
(a) Process design
(b) Product development
(c) Process development
(d) Process design

128. The designing of equipment required for the production of a product according to a pre-determined process.
(a) Process design
(b) Product development
(c) Process development
(d) Process design

129. The process of converting an invention into useful product which can satisfy human wants either directly or indirectly is _____
(a) Manufacture
(b) Design
(c) Innovation
(d) None

130. _____ is a floor plan of the physical facilities which are used in production.
(a) Process layout
(b) Product layout
(c) Group layout
(d) Plant layout

131. In _____ layout, similar machines and services are located together.
(a) Process layout
(b) product layout
(c) group layout
(d) plant layout

132. In _____ layout, when machines and auxiliary services according to the processing sequence of the product.
(a) Process layout
(b) Product layout
(c) Group layout
(d) Fixed position layout

133. The combination of the product layout and process layout is _____
(a) Process layout
(b) Product layout
(c) Group layout
(d) Fixed position layout

134. In _____ layout, the physical characteristics of the product dictate as to which type of machines and men are to be brought to the product.
(a) Process layout
(b) Product layout
(c) Group layout
(d) Fixed position layout

135. In which layout, breakdown of one machine in a product line may cause stoppages of machines in the downstream of the line?
(a) Process layout
(b) product layout
(c) group layout
(d) fixed position layout

136. The primary goal of _____ is to achieve zero inventories within an organization.
(a) Stock inventory
(b) Kanban system
(c) Just-In-Time
(d) Kaizen system

137. Operations management includes all of these activities except: _____
(a) Scheduled work
(b) secure financial resources
(c) maintain quality
(d) manage inventories

138. Which of the following is not included in "5Ps" of operations management?
(a) Product (b) Plant
(c) Process (d) Price

139. The major difference between goods and services is _____
(a) It is difficult to inventory resources
(b) Services lack the profitability that goods enjoy
(c) Goods production usually requires more labour than services
(d) Goods are always customized, services are always standardized

140. Technology choices don't have a major effect on _____
(a) Cost
(b) Employee turnover
(c) Productivity
(d) Quality

141. Which is the correct hierarchy of operations management decisions?
(a) Operational, strategic, tactical
(b) Goals, mission, productivity
(c) Strategic, operational, tactical
(d) Strategic, tactical, operational

142. Which of the following has least likely source of input for developing an operations strategy?
(a) Planning personnel
(b) Suppliers
(c) Top management
(d) Production personnel

143. Quality, supply chains, inventory and capacity are strategic decisions generally made by __
(a) Marketing
(b) Human resources
(c) Project managers
(d) Operations

144. Labour productivity cannot be measured in terms of _____
(a) Units of Labour per hour
(b) Units of output per shift
(c) Value-added per Labour hour
(d) Units of output per rupee input

145. Efficiency in capacity terms, is the ratio of
(a) Actual output to effective capacity
(b) Effective capacity to actual output
(c) Design capacity to effective capacity
(d) Design capacity to actual capacity

146. A bottlenecked operations will result in _____
(a) Unbalanced system
(b) Labor unrest
(c) Increased capacity
(d) Job shop efficiency

147. Ideally, preventive maintenance will be scheduled _____
(a) On a periodic basis
(b) At the start of each workday
(c) Just prior to a breakdown or failure
(d) None of the above

148. Cost of equipment breakdown doesn't include _____
(a) The cost of idle workers
(b) Increased fixed cost
(c) Possible damage to other equipment
(d) Possible injury of operations

149. Service/retail firms focus more on _____ for selecting a location.
(a) Cost
(b) Revenue
(c) both of the above
(d) none

150. A cause of variation in a process probably wouldn't be due to _____
(a) People
(b) Standards
(c) Methods
(d) Environment

151. The symbols of a flow process chart stands for _____
(a) Inspection, delay, transportation, storage, operation
(b) Transportation loaded, search, grasp, store, perform operation
(c) Release, transport, store, grasp, search
(d) Input, processing and output

152. When organization has high volume of production, flow will be _____
(a) Streamlined (b) Jumbled
(c) Customized (d) Standardized

153. When organization has high number of varieties, flow will be _____
(a) Streamlined
(b) Jumbled
(c) Customized
(d) Standardized

154. Inventory which cannot be carried to future periods is ____
(a) Stock
(b) Project
(c) Service
(d) Product

155. _____aims to have fewer choices of components and products that will be produced so that there will be more focus on the production activities.
(a) Simplification
(b) New product development
(c) Standardization
(d) None of the above

156. The quality of service can be described by its _____
(a) Reliability
(b) Responsiveness
(c) Time
(d) All the above

157. Intention of designers to include or exclude features in a product or service is _____
(a) Quality of product
(b) Quality of service
(c) Quality of design
(d) Quality of process

158. The degree to which goods or services confirm to the intent of the designers is _____
(a) Quality of assurance
(b) Quality of reliability
(c) Quality of convenience
(d) Quality of conformance

159. Cost of activities designed to ensure quality or uncover defects is _____
(a) Appraisal cost
(b) Prevention cost
(c) Failure cost
(d) Setup cost

160. Cost caused by defective parts or products or by faculty services are _____
(a) Appraisal cost
(b) Prevention cost
(c) Failure cost
(d) Setup cost

161. Cost related to reducing the potential for quality problems are _____
(a) Appraisal cost (b) Prevention cost
(c) Failure cost (d) Setup cost

162. A philosophy that involves everyone in an organization in a continual effort to improve quality and achieve customer satisfaction is _____
(a) Production management
(b) Total quality management
(c) Strategic management
(d) Operations management

163. A business process for improving quality, reducing costs, and increasing customer satisfaction is _____
(a) TQM
(b) Quality at the service
(c) Quality at the source
(d) Six Sigma

164. The philosophy of making each worker responsible for the quality of his or her work is ___
(a) TQM
(b) Quality at the service
(c) Quality at the source
(d) Six Sigma

165. DMAIC: ____
(a) Define, Monitor, Analyse, Invention & Control
(b) Describe, Monitor, Analyse, Improve & Control
(c) Define, Measure, Analyse, Invention & Control
(d) Define, Measure, Analyse, Improve & Control

166. A diagram of the step in a process is ____
(a) Check sheet (b) Flow sheet
(c) Histogram (d) Pareto chart

167. A tool for recording and organizing data to identify a problem is ____
(a) Check sheet (b) Flow sheet
(c) Histogram (d) Pareto chart

168. A chart of an empirical frequency of distribution
(a) Check sheet (b) flow sheet
(c) Histogram (d) Pareto chart

169. A diagram that arranges categories from highest to lowest frequency of occurrence
(a) Control chart
(b) Cause & Effect diagram
(c) Scatter diagram
(d) Pareto chart

170. A statistical chart of time- occurred values of a sample statistic
(a) Control chart
(b) cause & effect diagram
(c) scatter diagram
(d) Pareto chart

171. A graph that shows the degree and direction of relationship between two variables
(a) Control chart
(b) Cause & Effect diagram
(c) Scatter diagram
(d) Pareto chart

172. A diagram used to organize a search for the cause of a problem known as ____
(a) Control chart
(b) Cause & Effect diagram
(c) Scatter diagram
(d) Pareto chart

173. Cause and effect diagram is also known as ____
(a) Fishbone Diagram
(b) Ishikawa Diagram
(c) both a & b
(d) none

174. Process of measuring performance against the best in the same or another industry is ____
(a) Performance appraisal
(b) Bench marking
(c) Evaluation
(d) Monitoring

175. Cost and effect diagram with action cards is ____
(a) DMAIC
(b) Poke Yoke diagram
(c) CEDAC
(d) Run chart

176. Tools for tracking results over a period of time is ____
(a) DMAIC
(b) Poke Yoke diagram
(c) CEDAC
(d) Run chart

177. A process that evaluates output relative to a standard and takes corrective action is ____
(a) Quality management
(b) Quality control
(c) Return on Quality
(d) Monitoring

178. A process is generally considered to be in control of ____
(a) There are no sample points outside the control limits.
(b) Points appear to be randomly distributed around the centre line
(c) Equal number of points occurs above and below the centre line
(d) All of these

179. p- chart uses ____
(a) Amount of dispersion in a sample
(b) Process average of a sample
(c) Portion defective in a sample
(d) Number of defects in a sample

180. Range chart uses ____
(a) Amount of dispersion in a sample
(b) Process average of a sample
(c) Portion defective in a sample
(d) Number of defects in a sample

181. Attribute chart require _____ sample sizes whereas variable charts requires ____ samples.
(a) Smaller, larger
(b) Larger, smaller
(c) Smaller, smaller
(d) Larger, larger

182. _____ is design specifications reflecting customer requirements for a product.
(a) Control charts
(b) Tolerances
(c) Process capability
(d) Process variability

183. The main elements associated with process capability ____
(a) Process variability
(b) Process Centre
(c) Design specifications
(d) all of these

184. A product characteristic that can be measured ____
(a) Variable (b) attribute
(c) c- chart (d) p- chart

185. The probability of rejecting a lot that has an AQL _____
(a) Producer's risk
(b) Consumer's risk
(c) α
(d) both a & c

186. An acceptable portion of defects in a lot to the consumer _____
(a) LTPD (b) AQL
(c) Producer risk (d) Consumer risk

187. The probability of accepting a lot in which the fraction of defective items exceeds LTPD __
(a) Consumer risk
(b) Producer risk
(c) β
(d) both a & c

188. _____ system relies on pre-determined schedule.
(a) Push (b) Pull
(c) Kanban (d) Kaizen

189. A system of continuous improvement ____
(a) Push (b) Pull
(c) Kanban (d) Kaizen

190. A card that corresponds to standard quality of production ____
(a) Kaizen (b) Kanban
(c) Push (d) None

191. Just in Time concept includes _____
(a) Zero Lead Time
(b) Zero Failure
(c) Zero Inventory
(d) All of these

192. Master Sheet is prepared in –
(a) Intermittent Manufacturing System
(b) Continuous Manufacturing System
(c) Mass production System
(d) Both b & c

193. The administrative costs that are made within the organization is called as
(a) Set up cost
(b) Holding cost
(c) Ordering cost
(d) Appraisal cost

194. The administrative cost that are made outside the organization is called as
(a) Set up cost
(b) Holding cost
(c) Ordering cost
(d) Appraisal cost

195. Process of changing and improving the design of a product or system –
(a) Value Engineering
(b) Re-engineering
(c) Reverse Engineering
(d) Quality engineering

196. Investigating the design of components from a cost-value perspective –
(a) Value engineering
(b) Re-engineering
(c) Reverse Engineering
(d) Quality engineering

197. Process of deciding the path and sequence of operations to be performed on the job from one machine to another ____
(a) Routing (b) Scheduling
(c) Loading (d) Transporting

198. Type I error is also called ___
(a) Producer's risk
(b) Consumer's risk
(c) α
(d) both a & c

199. Dealing with the problems or breakdowns when they occur ____
(a) Breakdown Maintenance
(b) Preventive Maintenance
(c) Total Maintenance
(d) predictive maintenance

200. Preventive Maintenance is ____
(a) Reactive Approach
(b) Proactive Approach
(c) Active approach
(d) Passive approach

201. Quality tool includes ____
(a) Pareto analysis
(b) Scatter Diagram
(c) Check Sheets
(d) All

202. The upper limit on the percentage of defects that a consumer is willing to accept ____
(a) AQL (b) LTPD
(c) AQO (d) EOQ

203. The movement of materials, services, cash and information in supply chain____
(a) Value Chain
(b) Logistics
(c) Supply Chain Management
(d) Value Engineering

204. The order size that minimizes total annual cost ____
(a) Economic Order Quantity
(b) Threshold Order Quantity
(c) Inventory Order
(d) Acceptable Quality Level

205. Control charts used for attributes ____
(a) p & c – chart
(b) r and x- chart
(c) only p- chart
(d) none

ANSWER KEY (Production & Operations Management)

1. (b) **2.** (d) **3.** (a) **4.** (b) **5.** (d) **6.** (b) **7.** (d) **8.** (b) **9.** (b) **10.** (d)
11. (b) **12.** (a) **13.** (a) **14.** (c) **15.** (a) **16.** (c) **17.** (d) **18.** (a) **19.** (a) **20.** (b)
21. (d) **22.** (c) **23.** (b) **24.** (c) **25.** (d) **26.** (b) **27.** (a) **28.** (a) **29.** (a) **30.** (d)
31. (a) **32.** (a) **33.** (d) **34.** (a) **35.** (c) **36.** (a) **37.** (b) **38.** (a) **39.** (b) **40.** (c)
41. (a) **42.** (c) **43.** (a) **44.** (c) **45.** (a) **46.** (b) **47.** (a) **48.** (c) **49.** (a) **50.** (b)
51. (c) **52.** (b) **53.** (d) **54.** (b) **55.** (b) **56.** (a) **57.** (b) **58.** (a) **59.** (b) **60.** (c)
61. (b) **62.** (b) **63.** (b) **64.** (d) **65.** (c) **66.** (d) **67.** (d) **68.** (b) **69.** (d) **70.** (a)
71. (c) **72.** (b) **73.** (c) **74.** (c) **75.** (a) **76.** (d) **77.** (c) **78.** (b) **79.** (a) **80.** (c)
81. (d) **82.** (c) **83.** (d) **84.** (a) **85.** (a) **86.** (d) **87.** (d) **88.** (a) **89.** (d) **90.** (c)
91. (c) **92.** (a) **93.** (b) **94.** (c) **95.** (a) **96.** (b) **97.** (b) **98.** (d) **99.** (a) **100.** (c)

101. (d) **102.** (a) **103.** (c) **104.** (b) **105.** (a) **106.** (a) **107.** (b) **108.** (a) **109.** (c) **110.** (c)
111 (b) **112.** (a) **113.** (c) **114.** (a) **115.** (b) **116.** (a) **117.** (a) **118.** (c) **119.** (a) **120.** (c)
121.. (b) **122.** (b) **123.** (c) **124.** (b) **125.** (d) **126.** (c) **127.** (b) **128.** (a) **129.** (a) **130.** (d)
131. (a) **132.** (b) **133.** (c) **134.** (d) **135.** (b) **136.** (c) **137.** (b) **138.** (d) **139.** (a) **140.** (b)
141. (d) **142.** (b) **143.** (d) **144.** (d) **145.** (a) **146.** (a) **147.** (a) **148.** (b) **149.** (b) **150.** (b)
151. (a) **152.** (a) **153.** (b) **154.** (c) **155.** (a) **156.** (d) **157.** (c) **158.** (d) **159.** (a) **160.** (c)
161. (b) **162.** (b) **163.** (d) **164.** (c) **165.** (d) **166.** (b) **167.** (a) **168.** (c) **169.** (d) **170.** (a)
171. (c) **172.** (b) **173.** (c) **174.** (b) **175.** (c) **176.** (d) **177.** (b) **178.** (d) **179.** (c) **180.** (a)
181. (b) **182.** (b) **183.** (d) **184.** (a) **185.** (d) **186.** (b) **187.** (d) **188.** (a) **189.** (d) **190.** (b)
191. (d) **192.** (d) **193.** (a) **194.** (c) **195.** (b) **196.** (a) **197.** (a) **198.** (d) **199.** (a) **200.** (b)
201. (d) **202.** (b) **203.** (b) **204.** (a) **205.** (a)

8. Operations Research

1. Operations Research (OR) is a very powerful tool for
(a) Research (b) Decision - Making
(c) Operations (d) None of the above

2. The innovative science of Operations Research was discovered during
(a) Civil War
(b) World War I
(c) World War II
(d) Industrial Revolution

3. Which technique is used to find a solution for optimizing a given objective, such as profit maximization or cost minimization under certain constraints?
(a) Linear Programming
(b) Waiting Line
(c) Both A and B
(d) None of the above

4. The first step in formulating a linear programming problem is
(a) Identify any upper or lower bounds on the decision variables
(b) State the constraints as linear combinations of the decision variables
(c) Identify the decision variables
(d) State the objective function as a linear combination of the decision variables

5. The reason for studying the graphical method of solving LP problems is
(a) Lines are easy to draw on paper
(b) To develop an understanding of the linear programming strategy
(c) It is faster than computerized methods
(d) It provides better solutions than computerized methods

6. Which one of the following steps is often the most crucial in the decision-making process?
(a) Identifying the problem
(b) Specifying objectives
(c) Developing alternatives
(d) Selecting the best alternative

7. Which one of the following won't generally be given as a reason for poor decisions?
(a) Unforeseeable circumstances
(b) Bounded rationality
(c) Mistakes in the decision process
(d) All of the above

8. In ________ models, everything is defined and the results are certain.
(a) Deterministic Models
(b) Probabilistic Models
(c) Both A and B
(d) None of the above

9. In ____________ models there is risk and uncertainty.
(a) Deterministic Models
(b) Probabilistic Models
(c) Both A and B
(d) None of the above

10. Which of the following statements is NOT true?
(a) A feasible solution satisfies all constraints
(b) An optimal solution satisfies all constraints
(c) An infeasible solution violates all constraints
(d) A feasible solution point does not have to lie on the boundary of the feasible region

11. What do you mean by Slack variable?
    (a) Is the difference between the left and right sides of a constraint
    (b) Is the amount by which the left side of a ≤ constraint is smaller than the right side
    (c) Is the amount by which the left side of a ≥ constraint is larger than the right side
    (d) Exists for each variable in a linear programming problem
12. The improvement in the value of the objective function per unit increase in a right-hand side is the
    (a) sensitivity value
    (b) dual price
    (c) constraint coefficient
    (d) slack value
13. The iso-profit line is
    (a) a method for solving a linear programming problem
    (b) a line with the same profit at all points
    (c) parallel to all other iso-profit lines in the same problem
    (d) all of the above
14. The corner point solution method requires
    (a) Finding the value of the objective function at the origin
    (b) Moving the iso-profit line to the highest level that still touches some part of the feasible region
    (c) Moving the iso-profit line to the lowest level that still touches some part of the feasible region
    (d) Finding the coordinates at each corner of the feasible solution space
15. The region which satisfies all the constraints in a graphical linear programming is called the
    (a) optimal solution space
    (b) feasible solution space
    (c) profit maximization space
    (d) region of optimality
16. A constraint that does not affect the feasible region is a
    (a) Non-negativity constraint
    (b) Redundant constraint
    (c) Standard constraint
    (d) Slack constraints
17. Which of the following is not a requirement of a linear programming problem?
    (a) An objective function, expressed in terms of linear equations
    (b) Constraint equations, expressed as linear equations
    (c) Alternative courses of action
    (d) For each decision variable, there must be one constraint or resource limit
18. In graphical approach to Linear programming problem, when there is no common point in the feasible area for the constraints of a problem then ________ exists.
    (a) Infeasibility
    (b) Multiple optimal solution
    (c) Unbounded solution
    (d) Degeneracy
19. When the objective function is parallel to a constraint, which is binding and which forms an edge or boundary on the feasible region
    (a) Infeasibility
    (b) Multiple optimal solution
    (c) Unbounded solution
    (d) Degeneracy
20. Graphical method can be applied to solve a LPP when there are only ________ variable.
    (a) One
    (b) More than One
    (c) Two
    (d) Three
21. A minimization problem can be converted into a maximization problem by changing the sign of coefficients in the ________.
    (a) Constraints
    (b) Objective Functions
    (c) Both A and B
    (d) None of the above

22. If in a LPP , the solution of a variable can be made infinitely large without violating the constraints the solution is:
(a) Infeasible (b) Unbounded
(c) Alternative (d) None of the above

23. Pre-requisites for applying simplex method for solving a linear programming problem is/are
(a) Each of the decision variables be non-negative
(b) R.H.S. of each constraints should be non-negative
(c) Both of the above
(d) None of the above

24. In maximization cases, ________ are assigned to the artificial variables as their coefficients in the objective function
(a) +m (b) –m
(c) 0 (d) None of the above

25. In simplex method, ________ variables are added in the case of '=' type of problem
(a) Slack Variable
(b) Surplus Variable
(c) Artificial Variable
(d) None of the above

26. In simplex method, if there is tie between a decision variable and a slack (or surplus) variable, ________ should be selected
(a) Slack variable
(b) Surplus variable
(c) Decision variable
(d) None of the above

27. Every LPP is associated with another LPP is called
(a) Primal
(b) Dual
(c) Non linear programming
(d) None of the above

28. If there are more than one optimal solution for the decision variable, the solution is
(a) Infeasible
(b) Unbounded
(c) Multiple Optimal
(d) None of the above

29. Dual of the dual is
(a) Primal
(b) Dual
(c) Alternative
(d) None of the above

30. Linear programming has been successfully applied in
(a) Agricultural
(b) Industrial applications
(c) Both of the above
(d) None of the above

31. LP model is based on the assumptions of
(a) Proportionality
(b) Additivity
(c) Certainty
(d) All of the above

32. Simple linear programming problem with ______ decision variables can be easily solved by the graphical method
(a) One
(b) Four
(c) Three
(d) Two

33. solution to a LPP which satisfies the non- negativity restrictions of the LPP is called its
(a) Unbounded solution
(b) Optimal solution
(c) Feasible solution
(d) Both A and B

34. Any feasible solution which optimizes (minimizes or maximizes) the objective function of the LPP is called its
(a) Optimal solution
(b) Non-basic variables
(c) Solution
(d) Basic feasible solution

35. In case of Simplex method of solving linear programming problem, if any Non-basic variable has zero Zj – Cj value in the final simplex table having optimal solution, then it has
(a) Multiple Optimal Solution
(b) Infeasible Solution
(c) Unbounded solution
(d) Degenerate solution

36. In case of Simplex method of solving linear programming problem, if the artificial variable does not leave the basis in two phase method or remains in the final table in Big M method, then it has
(a) Multiple Optimal Solution
(b) Infeasible Solution
(c) Unbounded solution
(d) Degenerate solution

37. In case of Simplex method of solving linear programming problem, if there is no non-negative replacement ratio, then it is has
(a) Multiple Optimal Solution
(b) Infeasible Solution
(c) Unbounded solution
(d) Degenerate solution

38. _________ is a method to solve a given LPP involving some artificial variable
(a) Graphical method
(b) Method of penalties
(c) Two-phase simplex method
(d) None of the above

39. Which variables are fictitious and cannot have any physical meaning?
(a) Optimal variable
(b) Decision variable
(c) Artificial variable
(d) None of the above

40. An assumption that implies that finite numbers of choices are available to a decision – maker and the decision variables do not assume negative values is known as
(a) Certainty
(b) Continuity
(c) Finite choices
(d) None of the above

41. A basic solution which also satisfies the condition in which all basic variables are non -negative is called
(a) Basic feasible solution
(b) Feasible solution
(c) Optimal solution
(d) None of the above

42. In Linear Programming, all the constraints are expressed as equations and the right hand side of each constraint and all variables are non-negative is called
(a) Canonical variable
(b) Canonical form
(c) Canonical solution
(d) Both A and B

43. An objective function is maximized when it is a _________ function
(a) Passive (b) Profit
(c) Cost (d) None of the above

44. In LPP, If the primal problem has an optimal solution, then the dual has
(a) Infeasible solution
(b) Unbounded solution
(c) Multiple optimal solution
(d) An optimal solution

45. In sensitivity analysis, a zero shadow price (or dual value) for a resource ordinarily means that
(a) The resource is scarce
(b) The resource constraint was redundant
(c) The resource has not been used up
(d) Something is wrong with the problem formulation

46. A shadow price (or dual value) reflects which of the following in a maximization problem?
(a) The marginal gain in the objective realized by subtracting one unit of a resource
(b) The market price that must be paid to obtain additional resources
(c) The increase in profit that would accompany one added unit of a scarce resource
(d) The reduction in cost that would accompany a one unit decrease in the resource

47. Which objective function is parallel to the line $4x + 2y = 200$?
(a) $2x + 4y$ (b) $2x + y$
(c) $2x + 2y$ (d) $2x - y$

48. Which of the following is not an assumption of linear programming models?
(a) Divisibility (b) Certainty
(c) Negativity (d) Linearity

49. Corner point coordinates are substituted into the objective function in the approach called
(a) Corner constraint analysis
(b) Multiple regression
(c) Enumeration
(d) Graphical regression

50. Constraints in linear programming Indicates
(a) Minimum available resource
(b) Maximum resource requirements
(c) Both of the above
(d) None of the above

51. The logical approach, from beginning to end, for assembling a linear programming model begins with
(a) identifying the decision variables
(b) identifying the objective function
(c) specifying the objective function parameters
(d) identifying the constraint coefficients

52. Which of the following is a component of mathematical models?
(a) Decision Variables
(b) Result Variables
(c) Uncontrollable Variables
(d) All of the above

53. Which of the following is an essential condition in a situation for linear programming to be useful?
(a) Nonlinear constraints
(b) Bottlenecks in the objective function
(c) Homogeneity
(d) Uncertainty

54. Which of the following is a typical application of linear programming in operations management?
(a) Cycle counting analysis
(b) Cost allocation studies
(c) Plant location studies
(d) Product design decisions

55. The maximum number of constraints allowed in a linear programming is
(a) Less than 5
(b) Less than 512
(c) Less than 1,024
(d) Unlimited

56. Exclusion of ____________ does not affect the optimal solution to a LPP.
(a) Basic Variable
(b) Non Basic Variable
(c) Redundant Variable
(d) All of the above

57. A solution in LPP is said to be degenerate if some basic variable has a solution value equal to
(a) Non-Negative
(b) Positive
(c) Zero
(d) All of the above

58. Slack variable is used in case of ________ type of constraints in Linear Programming Problem
(a) $\leq$
(b) $\geq$
(c) Both of the above
(d) None of the above

59. Surplus variable is used in case of ____________ type of constraints in a Linear Programming Problem
(a) $\leq$
(b) $\geq$
(c) Both of the above
(d) None of the above

60. Artificial variable is used in case of _____ type of constraints in Linear Programming problem.
(a) $\leq$ (b) $\geq$
(c) $=$ (d) Both b) and c)

61. Iso-profit/Iso-cost line runs parallel to the _______ in graphical approach to Linear Programming Problem.
(a) Constraint line
(b) Objective line
(c) None of the above
(d) Both of the above

62. For a 4 variable and 5 constraint primal problem, the dual would be a ___ variable and _____ constraint problem.
(a) 5, 4 (b) 4, 5
(c) 4, 4 (d) 5, 5

63. A transportation model has 6 rows and 5 columns. What is the minimum required number of occupied cells?
(a) 10 (b) 11
(c) 12 (d) 30

64. The cost differences in the Vogel's approximation method indicate the ________ for not using the respective least cost route.
(a) Bonus (b) Penalty
(c) Cost (d) Profit

65. Method for testing the optimality of a solution to a transportation problem is
(a) Stepping Stone Method
(b) Modified Distribution Method
(c) Both the above
(d) None of the above

66. In a transportation model, an alternate optimal solution exists if
(a) The values of all of the unoccupied cells are positive
(b) The value of at least one unoccupied cell is zero
(c) The values of all of the unoccupied cells are negative
(d) An alternative optimal solution never exists

67. In transportation model analysis the "intuitive method" is used to obtain:
(a) An initial feasible solution
(b) An initial optimal solution
(c) An improvement to the initial solution
(d) An improvement to the previous solution

68. VAM stands for:
(a) Vogeal's Approximation Method
(b) Vogel's Approximate Method
(c) Vangel's Approximation Method
(d) Vogel's Approximation Method

69. Once the initial basic feasible solution has been computed , what is the next step in the problem?
(a) VAM
(b) Modified distribution method
(c) Optimality test
(d) None of the above

70. One can find the initial basic feasible solution by using ____________
(a) VAM
(b) MODI
(c) Optimality test
(d) None of the above

71. Which of the following is used in order to determine the optimal solution of a transportation problem?
(a) LPP
(b) VAM
(c) MODI Method
(d) None of the above

72. A given transportation model said to be unbalanced, if the total supply is not equal to the total________.
(a) Optimization
(b) Demand
(c) Cost
(d) None of the above

73. If the total supply is less than the total demand, a dummy source (row) is included in the cost matrix with ________.
(a) Dummy Demand
(b) Dummy Supply
(c) Zero Cost
(d) Both A and B

74. For maximization in Transportation Problem , the objective is to maximize the total _______
(a) Solution
(b) Profit Matrix
(c) Profit
(d) None of the above

75. In applying Vogel's approximation method to a profit maximization problem, row and column penalties are determined by:
(a) Finding the largest unit cost in each row or column
(b) Finding the smallest unit cost in each row or column
(c) Finding the difference between the two lowest unit costs in each row and column
(d) Finding the difference between the two highest unit costs in each row and column

76. The northwest corner rule requires that we start allocating units to shipping routes in the:
(a) Lower right corner of the table
(b) Upper right corner of the table
(c) highest costly cell of the table
(d) Upper left-hand corner of the table

77. In a transportation problem, when the number of occupied routes is less than the number of rows plus the number of columns -1, we say that the solution is
(a) Unbalanced (b) Infeasible
(c) Optimal (d) Degenerate

78. The only restriction applied on the initial solution of a transportation problem is that:
(a) There must be nonzero quantities in a majority of the boxes
(b) All constraints must be satisfied
(c) Demand must be equal to supply
(d) None of the above

79. The purpose of the stepping-stone method is to
(a) Identify the relevant costs in a transportation problem
(b) Assisting in moving from an initial feasible solution to the optimal solution
(c) Develop the initial solution to the transportation problem
(d) Determine whether a given solution is feasible or not

80. The purpose of a dummy source or dummy destination in a transportation problem is to
(a) Obtain a balance between total supply and total demand
(b) Provide a means of representing a dummy problem
(c) Make certain that the total cost does not exceed some specified figure
(d) Prevent the solution from becoming degenerate

81. Which of the following is a method for improving an initial solution in a transportation problem?
(a) stepping-stone
(b) northwest-corner
(c) intuitive lowest-cost
(d) southeast-corner rule

82. In a transportation problem with total demand equal to 1200 and total supply equal to 900, we should add a ____________ with a quantity equal to __________ to convert it to a balanced problem.
(a) Dummy supply, 300
(b) Dummy supply, 2100
(c) Dummy demand, 300
(d) Dummy demand, 2100

83. An m source, n destination transportation, when written as a transshipment problem would have
(a) m sources, n destinations
(b) m sources, m destinations
(c) n sources, m destinations
(d) m+n sources, m+n destinations

84. In transportation problem a basic feasible solution is degenerate if no. of allocations is
(a) $=m+n-1$ (b) $<m+n-1$
(c) $>m+n-1$ (d) None of the above

85. In the linear programming formulation of the transshipment problem, demand at the destination points are required to be satisfied from shipment from
(a) Origins
(b) Intermediate points
(c) Either origins or intermediate points
(d) Origins or external sources

86. A transshipment problem has 3 origins, 2 intermediate points and 4 destinations. The number of decision variables in the linear programming formulation of this problem will be
(a) 24 (b) 9
(c) 48 (d) 14

87. For every prohibited assignment the given cost element is replaced by __________ value.
(a) Very high (b) Very small
(c) Zero (d) Any value

88. An efficient method for solving Assignment Problem is
(a) Hungarian method
(b) Simples method
(c) Transportation method
(d) All of the above

89. An Assignment Problem can be
(a) Only balanced problem
(b) Either Balanced or Unbalanced
(c) Only Unbalanced
(d) None of the above

90. The operations Research technique which helps in minimizing total waiting and service costs is
(a) Queuing Theory
(b) Decision Theory
(c) Both A and B
(d) None of the above

91. A three-step service with one server at each step would be an example of a
(a) Multiple-phase, multiple-channel system
(b) Single-phase, multiple-channel system
(c) Multiple-phase, single-channel system
(d) Solo server, trio channel

92. Waiting lines are due to
(a) Multiple-channels
(b) Single servers
(c) Arrival and Service Variability
(d) Infinite source

93. A system with an arrival rate of two per hour, a service time of 20 minutes, and one server would have a utilization of
(a) 10% (b) 33%
(c) 40% (d) 67%

94. For an infinite-source system with an arrival rate of 6 per hour (Poisson) and service time of 6 minutes per customer (exponential), the average number being served is
(a) 0.10 (b) 0.60
(c) 1.0 (d) none of these

95. The term "queue discipline" refers to
(a) The willingness of customers to wait in line for service
(b) Having multiple waiting lines without customers switching from line to line
(c) The order in which customers are processed
(d) The use of barriers to prevent lane changing

96. When the cost incurred by all customers is not the same, an appropriate queuing model is:
(a) single channel, single phase
(b) single channel, multiple phase
(c) multiple channel, single priority
(d) multiple channel, multiple priority

97. A queuing system has four crews with three members each. The number of "servers" is
(a) 3 (b) 4
(c) 7 (d) 12

98. Which of the following are the three major components of a queuing system?
(a) The source population, how customers exit the system and the queuing discipline
(b) The number of servers, the service speed and the waiting line
(c) The source population, how the customer exits the system and the servicing system
(d) The service speed, the queue discipline and the waiting line

99. Which of the following is an example of a finite population in a queuing system?
(a) People waiting to place their order at a fast food restaurant
(b) The departmental faculty in line at the copier
(c) People waiting in line at an ATM
(d) Patients seeking help in a hospital emergency room

100. Buying food at a large food store with multiple checkout counters features which type of queuing system?
(a) Single channel, single phase
(b) Single channel, multiphase
(c) Multichannel, single phase
(d) Multichannel, multiphase

101. For an infinite queuing situation, if the arrival rate for loading trucks is 5 trucks per hour, what is the mean time between arrivals?
(a) 5 hours
(b) 2.5 hours
(c) 0.2 hours
(d) None of the above

102. Which of the following is an advantage of simulation?
(a) Ability to provide optimal solutions
(b) Ability to conduct experiments on system behavior
(c) Multiple solutions because of randomizing
(d) Approximate answers are more realistic than exact answers

103. Which theory concerns making sound decisions under conditions of certainty, risk and uncertainty
(a) Game Theory
(b) Network Analysis
(c) Decision Theory
(d) None of the above

104. A table showing the present values for each alternative in every possible state of nature is called a ________ table
(a) Decision Environment
(b) Decision Risk
(c) Payoff
(d) Present Value

105. If the minimum expected regret is computed, it indicates to a decision-maker the:
(a) expected value of perfect information
(b) expected payoff under certainty
(c) minimum guaranteed profitability
(d) expected payoff under risk

106. In decision making under uncertainty, Maximax is essentially a ________ approach
(a) Optimistic (b) Pessimistic
(c) Deterministic (d) Probabilistic

107. In decision making under uncertainty, maximin is ____________ approach
(a) Optimistic
(b) Pessimistic
(c) Deterministic
(d) Probabilistic

108. In decision making under uncertainty, which approach chooses the alternative with the best of the worst possible payoff?
(a) Maximin
(b) Maximax
(c) Laplace
(d) Minimax Regret Criterion

109. In decision making under uncertainty, which approach chooses the alternative with the best possible payoff?
(a) Maximin
(b) Maximax
(c) Laplace
(d) Minimax Regret Criterion

110. In decision making under uncertainty, which approach chooses the alternative with the best average payoff of any of the alternatives?
(a) Maximin
(b) Maximax
(c) Laplace
(d) Minimax Regret Criterion

111. In decision making under uncertainty, which approach chooses the alternative that has the least of the worst regrets?
(a) Maximin
(b) Maximax
(c) Laplace
(d) Minimax Regret Criterion

112. In Game theory, if the pay-off of the maximin strategy matches with the pay-off of the minimax strategy, then ________ exists.
(a) Saddle point (b) Pay-off
(c) Mixed game (d) None of the above

113. In game theory, the maximizing player uses ________ strategy
(a) Maximin (b) Minimax
(c) Maximax (d) All of the above

114. In game theory, the minimizing player uses ________ strategy
(a) Maximin (b) Minimax
(c) Maximax (d) All of the above

115. A Mixed strategy game can be solved using
(a) Algebraic Method
(b) Calculus
(c) Linear Programming
(d) All of the above

116. Key concept under which technique are network of events and activities, resource allocation, time and cost considerations, network paths and critical paths?
(a) Game Theory
(b) Network Analysis
(c) Decision Theory
(d) None of the above

117. Which technique is used to imitate an operation prior to actual performance?
(a) Simulation
(b) Integrated Production Models
(c) Inventory Control
(d) Game Theory

118. ________ are the representation of reality
(a) Models
(b) Phases
(c) Both A and B
(d) None of the above

119. Simulation is a ________ technique
(a) Deterministic
(b) Optimizing
(c) Descriptive
(d) Probabilistic technique

120. Probabilistic models are also known as
(a) Deterministic Models
(b) Stochastic Models
(c) Dynamic Models
(d) Static Models

121. ________ models assumes that the values of the variables do not change with time during a particular period
(a) Static Models
(b) Dynamic Models
(c) Both A and B
(d) None of the above

122. ________ models considers time as one of the important variable
(a) Static Models
(b) Dynamic Models
(c) Both A and B
(d) None of the above

123. In the basic EOQ model, if annual demand doubles, the effect on the EOQ is
(a) It doubles
(b) It is four times its previous amount
(c) It is about 70 percent of its previous amount
(d) It increases by about 40 percent

124. What items do students typically buy on an ROP basis?
(a) textbooks
(b) wedding gifts
(c) Petrol for their bikes
(d) computer software

125. Firms keep supplies of inventory for which of the following reasons?
(a) To maintain dependence of operations
(b) To provide a feeling of security for the workforce
(c) To meet variation in product demand
(d) To hedge against wage increases

126. Which of the following is not a reason to carry inventory?
(a) To provide a safeguard for variation in raw material delivery time
(b) To maintain independence of operations
(c) To meet variation in product demand
(d) To keep the stock out of the hands of competitors

127. In making any decision that affects inventory size, which of the following costs do not need to be considered?
(a) Holding costs (b) Ordering costs
(c) Fixed costs (d) Shortage costs

128. Which of the following is an assumption of the basic fixed-order quantity inventory model?
(a) Lead times are averaged
(b) Ordering costs are variable
(c) Price per unit of product is constant
(d) Back orders are allowed

129. Assuming no safety stock, what is the re-order point (R) given an average daily demand of 50 units, a lead time of 10 days and 625 units on hand?
(a) 550 (b) 500
(c) 715 (d) 450
(e) 475

130. In a reorder point model, the daily demand rate is normally distributed with a mean of 50 units and a standard deviation of 5 units. The lead time is 9 days, Q = 500 units, and the lead time service level (LTSL) is 90%. What are approximate values for Safety Stock and Reorder Point?
(a) 19, 69 (b) 4, 69
(c) 6, 469 (d) 19, 469

131. If both the lead time and daily demand are constant (no variation), the ROP is equal to:
(a) Safety stock
(b) Expected demand minus safety stock
(c) Expected demand plus safety stock
(d) Daily demand times lead time

132. Which one of the following is not a requirement for effective inventory management?
(a) A system to keep track of inventory on hand
(b) A classification system for inventory items
(c) Reasonable estimates of holding and shortage costs
(d) Using an EOQ model for determining reorder point

133. Which one of the following would be considered a potential cost of inventory?
(a) Shortage
(b) Carrying
(c) Holding
(d) Overtime

134. In a two-bin inventory system, the amount contained in the second bin is equal to the
(a) ROP
(b) EOQ
(c) Amount in the first bin
(d) Safety stock

135. The Basic EOQ model ignores
(a) Purchasing costs
(b) Ordering costs
(c) Holding costs
(d) All of the above

136. In the basic EOQ model, if annual demand doubles, the effect on the EOQ is:
(a) It doubles
(b) It is half its previous amount
(c) It is about 70 percent of its previous amount
(d) It increases by about 40 percent

137. If annual demand is 12,000 units, the ordering cost is `6 per order and the holding cost is `2.50 per unit per year, which of the following is the optimal order quantity using the fixed-order quantity model?
(a) 576
(b) 240
(c) 120.4
(d) 60.56

138. If it takes a supplier 25 days to deliver an order once it has been placed and the standard deviation of daily demand is 20, which of the following is the standard deviation of usage during lead time?
(a) 50
(b) 100
(c) 400
(d) 1,000

139. A company wants to determine its reorder point (R). Demand is variable and they want to build a safety stock into R. If the average daily demand is 12, the lead time is 5 days, the desired "z" value is 1.96, and the standard deviation of usage during lead time is 3, which of the following is the desired value of R?
(a) About 6
(b) About 16
(c) About 61
(d) About 66

140. The Pareto principle is best applied to which of the following inventory systems?
(a) EOQ
(b) Fixed-time period
(c) ABC classification
(d) Fixed-order quantity

141. Using the ABC classification system for inventory, which of the following is a true statement?
(a) The "C" items are of moderate annual usage value
(b) The "A" items are of low annual usage value
(c) The "A" items are of high annual usage value
(d) Inexpensive and low usage items are classified as "C" no matter how critical

142. If annual demand is 8,000 units, the ordering cost is `20 per order and the holding cost is `12.50 per unit per year, what is the optimal order quantity using the fixed-order quantity inventory model?
(a) 320
(b) 160
(c) 80
(d) 40

143. Which of the following is not one of the categories of manufacturing inventory?
(a) Finished products
(b) Component parts
(c) Just-in-time
(d) Supplies

144. When developing inventory cost models, which of the following is not included as costs to place an order?
(a) Phone calls
(b) Taxes
(c) Calculating quantity to order
(d) Postage

145. Which of the following is usually included as an inventory holding cost?
(a) Breakage
(b) Typing up an order
(c) Quantity discounts
(d) Annualized cost of materials

146. From the following statements relating to classical EOQ mode, mark the incorrect one:
(a) Total Ordering Cost = Total Holding Cost
(b) Total Relevant Cost = $\sqrt{(2DAh)}$
(c) If annual demand doubles with all other parameters remaining constant, the EOQ is double
(d) The lead-time is constant

147. Which of the following is fixed-order quantity inventory model?
(a) Economic order quantity model
(b) The ABC model
(c) Periodic replenishment model
(d) Cycle counting model

148. Which of the following is fixed-time period inventory model?
(a) The least cost method
(b) The Q model
(c) Periodic system model
(d) Just-in-time model

149. Which of the following is a perpetual system for inventory management?
(a) Fixed-time period
(b) Fixed-order quantity
(c) P model
(d) First-in-first-out

150. The time-period occurring between successive procurement actions is known as
(a) Procurement cycle
(b) Inventory cycle
(c) Procurement frequency
(d) None of the above

151. The penalty incurred for being unable to meet a demand as it occurs is known as
(a) Shortage cost
(b) Penalty cost
(c) Storage cost
(d) None of the above

152. The elapsed time between the initiation of an order and the receipt of replenishment stock is known as
(a) Lead-time
(b) Procurement time
(c) Order time
(d) None of the above

153. Among all models, the EOQ, with similar inputs, under classical model is
(a) Minimum
(b) Maximum
(c) Both of the above
(d) None of the above

154. In general, when back-ordering cost is small, orsders are made
(a) Less often but of higher quantity
(b) Frequently but of lower quantity
(c) Less often with low quantity
(d) Frequently with higher quantity

155. When back-ordering cost becomes infinite, the EOQ formula for the planned shortages model becomes
(a) Same as that of the classical model
(b) Infinite
(c) Zero
(d) None of the above

156. Which of the following is not an assumption underlying classical EOQ model?
(a) The annual demand is constant and uniform.
(b) Shortages are not permitted at all.
(c) The lead-time is zero.
(d) The holding cost per unit per year is constant.

157. For a 25% increase in order quantity (under classical EOQ model) the total relevant cost would
(a) Decrease by 2.5%
(b) Decrease by 0.25%
(c) Increase by 2.5%
(d) Increase by 1.025%

158. Safety stock should be less for ____________ category of items under ABC system of Inventory classification.
(a) A item
(b) B items
(c) C items
(d) All the above

159. Safety stock is required at the time of
(a) Variable lead time
(b) Variable demand
(c) Both of the above
(d) None of the above

160. If ordering cost/order and holding cost/unit are both doubled, EOQ will become
(a) Double
(b) Increase Four times
(c) Becomes half
(d) No change

161. Which of the following is not true?
(a) In deterministic inventory models, Re-order level = Demand during lead time (DDLT)
(b) In cases where demand is probabilistic, ROL = Expected DDLT + Safety Stock
(c) Greater the variation in demand and/or lead-time, greater the amount of safety stock needed
(d) Safety stock is the total stock kept to meet the demand during lead-time

162. _________ approach of inventory classification is based on the value of stocks on hand
(a) VED analysis
(b) XYZ analysis
(c) FSN analysis
(d) HML analysis

163. ________ approach of inventory classification is based on the annual usage value
(a) VED analysis (b) XYZ analysis
(c) FSN analysis (d) ABC analysis

164. __________ approach of inventory classification is based on the unit value of stock
(a) VED analysis
(b) XYZ analysis
(c) FSN analysis
(d) HML analysis

165. __________ approach of inventory classification is based on the consumption figure of the items
(a) VED analysis (b) XYZ analysis
(c) FSN analysis (d) HML analysis

ANSWER KEY (Operations Research)

1. (b) **2.** (d) **3.** (a) **4.** (c) **5.** (b) **6.** (a) **7.** (d) **8.** (a) **9.** (b) **10** (c)
11. (b) **12.** (b) **13.** (d) **14.** (d) **15.** (b) **16.** (b) **17.** (d) **18.** (a) **19.** (b) **20.** (c)
21. (b) **22.** (b) **23.** (c) **24.** (a) **25.** (c) **26.** (c) **27.** (b) **28.** (c) **29.** (a) **30** (c)
31. (d) **32.** (d) **33.** (c) **34.** (a) **35.** (a) **36.** (b) **37.** (c) **38.** (c) **39.** (c) **40.** (c)
41. (a) **42.** (b) **43.** (b) **44.** (d) **45.** (c) **46.** (c) **47.** (b) **48.** (c) **49.** (c) **50.** (c)
51. (a) **52.** (d) **53.** (c) **54.** (c) **55.** (d) **56.** (c) **57.** (c) **58.** (a) **59.** (b) **60.** (b)
61. (a) **62.** (a) **63.** (a) **64.** (b) **65.** (c) **66.** (b) **67.** (a) **68.** (d) **69.** (c) **70.** (a)
71. (c) **72.** (b) **73.** (c) **74.** (c) **75.** (d) **76.** (d) **77.** (d) **78.** (b) **79.** (b) **80.** (a)
81. (a) **82.** (a) **83.** (d) **84.** (b) **85.** (c) **86.** (b) **87.** (a) **88.** (a) **89.** (b) **90.** (a)
91. (c) **92.** (c) **93.** (d) **94.** (b) **95.** (c) **96.** (d) **97.** (b) **98.** (c) **99.** (b) **100.** (c)
101. (c) **102.** (b) **103.** (c) **104.** (c) **105.** (a) **106.** (a) **107.** (b) **108.** (a) **109.** (b) **110.** (c)
111. (d) **112.** (a) **113.** (a) **114.** (b) **115.** (d) **116.** (b) **117.** (a) **118.** (a) **119.** (c) **120.** (b)
121. (a) **122.** (b) **123.** (d) **124.** (c) **125.** (c) **126.** (d) **127.** (c) **128.** (c) **129.** (b) **130.** (d)
131. (d) **132.** (d) **133.** (d) **134.** (a) **135.** (a) **136.** (d) **137.** (b) **138.** (b) **139.** (d) **140.** (c)
141. (c) **142.** (b) **143.** (c) **144.** (b) **145.** (a) **146.** (c) **147.** (a) **148.** (c) **149.** (b) **150.** (b)
151. (a) **152.** (a) **153.** (a) **154.** (a) **155.** (a) **156.** (c) **157.** (c) **158.** (a) **159.** (c) **160.** (d)
161. (d) **162.** (b) **163.** (d) **164.** (d) **165.** (c)

9. Managerial Economics and Farm Business Management

1. Micro-economic theory is also called as
(a) Income theory (b) Price theory
(c) Both (d) None

2. Macro-economic theory is also called as
(a) Income, Employment and Growth theory
(b) Price theory
(c) Both
(d) None

3. Economic Analysis is a
(a) Positive Science
(b) Normative Science
(c) Both
(d) None

4. The theory of Income and Employment is given by
(a) Robbins (b) Marshall
(c) Lord Keynes (d) Pigou

5. Demand for product implies
(a) Desires to acquire it
(b) Willingness to pay for it
(c) Ability to pay for it
(d) All of the above

6. In market, demand varies with price of a commodity
(a) Directly
(b) Indirectly
(c) Zig Zac way
(d) Not sufficient condition to answer

7. When the price of a commodity falls, the demand for its substitute product will
(a) Increase
(b) Decrease
(c) Both of the above
(d) Can't say

8. If two commodities are complementary, then a rise in the price of one commodity will induce
(a) An upward shift in demand curve
(b) A rise in the price of the other commodity
(c) No shift in the demand for the other commodity
(d) A backward shift in demand for the other commodity

9. *ceteris paribus* means
(a) Perfect competition
(b) Tastes remain unchanged
(c) Other things being equal
(d) State of equilibrium

10. Cardinal measurement of utility is also called as
(a) Indifference curve analysis
(b) Budget line
(c) Marginal utility analysis
(d) Isoquant analysis

11. Marginal utility reflects the change in total utility with
(a) Change in price of the product
(b) Change in cost of the product
(c) Change in quantity consumed
(d) Change in population of a country

12. The consumer adjusts his purchase till
(a) $Mu_{money} > Mu_{consumption}$
(b) $Mu_{money} < MU_{consumption}$
(c) $Mu_{money} = MU_{consumption}$
(d) None

13. Consumer's willingness to pay minus (-) actual payment is called as
(a) Producer's surplus
(b) Supplier's surplus
(c) Consumer's surplus
(d) Utility cost

14. Law of Equi-Marginal utility is also called as
(a) Law of substitution
(b) Law of Law of indifference
(c) Law of Economy of expenditure
(d) All of the above

15. The law of 'Diminishing Marginal Utility' is stated by
(a) Adam Smith (b) Alfred Marshall
(c) J R Hicks (d) Cannon

16. 'Law of Equi-Marginal Utility' is stated by
(a) Adam Smith
(b) Alfred Marshall
(c) J R Hicks
(d) Cannon

17. If the price of a product increases then its demand will also increases. This behavior is seen in
(a) Giffen good (b) Veblen good
(c) Both (d) None

18. In case of giffen goods, the demand curve is
(a) Negative slope
(b) Positive Slope
(c) Parallel to X- Axis
(d) Parallel to Y-Axis

19. In the case of Veblen goods, demand curve will slope
(a) Upward (b) Downward
(c) Horizontal (d) Vertical

20. What are the unique features of Giffen goods?
(a) For which demand increases as price increases
(b) Which are in short supply
(c) Which have high income elasticity of demand
(d) Which give rise to a decrease in demand

21. In case of Giffen goods, the demand curve is
(a) Concave to origin
(b) Convex to origin
(c) Parallel to X-Axis
(d) Parallel to Y-Axis

22. The doctrine 'conspicuous consumption' was given by
(a) Alfred Marshall
(b) Robert Giffen
(c) Thorstein Veblen
(d) J M Keynes

23. Giffen goods are those goods
(a) For which demand increases as price increases
(b) Which are in short supply
(c) Which have high elasticity of demand
(d) Which gives rise to a Cob-Web situation

24. Non Price Quantity relationship is found in
(a) Contraction of demand
(b) Extension of demand
(c) Increase in demand
(d) All of the above

25. Price Quantity relationship is found in
(a) Decrease in demand
(b) Extension of demand
(c) Increase in demand
(d) None

26. A right ward shift in supply curve indicates
(a) A decrease in supply
(b) An increase in quantity demanded
(c) An increase in supply
(d) Law of variable proportion

27. If both supply and demand increase by the same proportion then
(a) Quantity remains constant
(b) Price remains constant
(c) Price increases
(d) Quantity increases

28. If both supply and demand decrease by the same proportion then
(a) Quantity remains constant
(b) Price remains constant
(c) Price decreases
(d) Quantity decreases

29. If demand decreases and supply increases in the same proportion then
(a) Quantity remains constant
(b) Price remains constant
(c) Price decreases
(d) Quantity decreases

30. If demand increases and supply decreases in the same proportion then
(a) Quantity remains constant
(b) Price remains constant
(c) Price increases
(d) Quantity increases

31. When the demand for a commodity is decreased as other consumers are also buying the same is called as
(a) Bandwagon effect
(b) Veblen effect
(c) Snob effect
(d) Ratchet effect

32. When the demand for a commodity is increased as other consumers are also buying the same is called as
(a) Bandwagon effect
(b) Veblen effect
(c) Snob effect
(d) Ratchet effect

33. Degree of responsiveness or sensitiveness of demand for a commodity with change in parameter is called as
(a) Increase in demand
(b) Contraction of demand
(c) Elasticity of demand
(d) Demand Forecasting

34. The method that is mostly followed to measure the elasticity of demand for a commodity
(a) Total expenditure method
(b) Geometrical method
(c) Proportional method
(d) Discounting method

35. In measuring elasticity of demand at a particular point on the demand curve, the formula that is mostly used is
(a) Elasticity = Lower segment x Upper segment
(b) Elasticity = Lower segment/Upper segment
(c) Elasticity = Lower segment + Upper segment
(d) Elasticity = Lower segment - Upper segment

36. When there is unit elasticity of demand, as the price is varied then the total revenue will
(a) Increases (b) Decreases
(c) Remain constant (d) None

37. When there is inelastic demand, then the increase in price will
(a) Increase the revenue
(b) Decrease the revenue
(c) Revenue remain unchanged
(d) There is no relationship

38. When there is elastic demand, then the increase in price will
(a) Increase the revenue
(b) Decrease the revenue
(c) Revenue remain unchanged
(d) There is no relationship

39. If the demand curve is a rectangular hyperbola, elasticity is
(a) 1 (b) 0
(c) Infinity (d) Less than one

40. When the price is perfectly inelastic, the demand curve is
(a) Parallel to X Axis
(b) Parallel to Y Axis
(c) Parallel to Z Axis
(d) Pass through the origin

41. When the price is perfectly elastic, the demand curve is
(a) Parallel to X Axis
(b) Parallel to Y Axis
(c) Parallel to Z Axis
(d) Pass through the origin

42. The price elasticity of demand, in general ranges from
(a) -1 to 1
(b) 0 to 1
(c) 0 to ∞
(d) -1 to ∞

43. At the middle point of the demand line, the price elasticity is
(a) 0
(b) 1
(c) -1
(d) None

44. When Average Price is P, elasticity is e and Marginal revenue is MR then the correct relationship between them is
(a) M_R= P (1 + e) (b) M_R = P (1-e)
(c) M_R = P/e (d) M_R = P(1-1/e)

45. Quadratic cost function is written as
(a) Tc = a + bQ
(b) Tc = a + bQ + cQ
(c) Tc = a + bQ + CQ^2
(d) Tc = a + b + c

46. Degree of responsiveness of demand for a commodity to the changes in its substitutes/complementary goods is called as
(a) Income elasticity
(b) Complementary elasticity
(c) Complex elasticity
(d) Cross elasticity

47. Cross elasticity of substitute products are
(a) Positive (b) Negative
(c) Neutral (d) Zero

48. If cross elasticity in response to the price of substitutes is greater than 1, it is advisable to
(a) Decrease the price
(b) Increase the price
(c) Can't say
(d) More information is required

49. The cross demand curve for substitutes is
(a) Convex (b) Concave
(c) Parabola (d) Hyperbola

50. The cross demand curve for complements
(a) Convex (b) Concave
(c) Parabola (d) Hyperbola

51. Income elasticity for an essential good is
(a) Less than 1
(b) Greater than 1
(c) Equal to 1
(d) zero

52. Price elasticity of demand for luxury goods will be
(a) Relatively elastic
(b) Relatively inelastic
(c) Infinitely inelastic
(d) Infinitely elastic

53. Income elasticity for on luxurious good is
(a) Less than 1 (b) Greater than 1
(c) Equal to 1 (d) zero

54. If elasticity of price expectations is greater than 1, then
(a) Seller will be sell more in future with higher price
(b) Seller will be sell more in future with lower price
(c) Seller will sell everything in present price
(d) Seller will not decide any thing

55. Total consumer expenditure on a good falls as its price falls, this indicates that its price elasticity is
(a) One
(b) Less than one
(c) More than One
(d) None of the above

56. If demand is Q, price is P and demand function is Q= 100-4P, then price elasticity ep is equal to
(a) 0.5 (b) 1
(c) -0.67 (d) 0.67

57. Which is not the short term demand forecasting method?
(a) Complete enumeration
(b) Delphi method
(c) Econometrics method
(d) Expert Opinion method

58. In demand forecasting important method used in trend projection is
(a) Scientific method
(b) Delphi method
(c) Box-Jenkin method
(d) Logistics method

59. The input that is in-elastic in short run is called as
(a) Labour
(b) Land
(c) Power
(d) Raw material

60. In a very long run situation, it is supposed that
(a) Land for production will change
(b) Labour quantity will change
(c) Technology for production will change
(d) Raw material requirement will change

61. That is not assumed in a production process
(a) Perfect divisibility of both inputs and out puts
(b) Maximum number of substitution of one factor with other
(c) Inelastic supply of fixed factors in the short run
(d) Labour and Capital are two factors of production process

62. The Cobb Dauglass Production function is often used as
(a) $Q = AK^aL^b$
(b) $Q = AK^aL^{1-a}$
(c) Logq = alogK + logC
(d) $Q = Ab^a\text{-}L^a$

63. Short run laws of production is otherwise called as
(a) Law of returns to a variable input
(b) Law of diminishing return
(c) Both
(d) None

64. The change in total product with unit change in input is called as
(a) Average Product
(b) Increment Product
(c) Marginal Product
(d) Productivity

65. Rational stage of production function is
(a) Stage I (b) Stage II
(c) Stage III (d) None

66. In short run, a farmer should produce something where
(a) Marginal Productivity is highest
(b) Average Productivity is highest
(c) Total productivity is highest
(d) It has subjective criteria

67. When the total productivity is its highest point then the marginal Productivity is?
(a) Less than Total Productivity
(b) Equal to Total Productivity
(c) More than Total Productivity
(d) Zero

68. The stage I in production function is that point where
(a) AP>MP (b) AP=MP
(c) AP<MP (d) None

69. Irrational zone of production function is
(a) Stage I (b) Stage II
(c) Stage III (d) None

70. The curve that represents the locus of points representing various combinations of two inputs-capital and labour yielding the same output is called as
(a) Production Indifference Curve
(b) Equal Product Curve
(c) Isoquant Curve
(d) All of the above

71. In the law of variable proportions, phase 'two' indicates
(a) Increasing return
(b) Constant return
(c) Diminishing return
(d) Negative return

72. That is not true about Iso-quant curve is
(a) It is convex to origin
(b) Upper Isoquant represent higher level of out put
(c) Isoquants cannot intersect or be tangent to each other
(d) Isoquants have a positive slope

73. The behavior of output in response to a proportional change in input in the long run is called as
(a) Law of variable proportion
(b) Law of return of scale
(c) Law of returns to a variable input
(d) Law of diminishing return

74. A firm produces output using labour and capital. If prices of both inputs double, then the total minimum cost of producing any given level of output is
(a) Necessarily doubled
(b) Necessarily less than doubled
(c) Necessarily more than doubled
(d) Doubled provided the firm has constant returns to scale technology

75. Law of Increasing return to a scale comes due to
(a) Condition of techno-managerial Indivisibility
(b) Higher degree of specialization
(c) Matter of dimensional relation
(d) All of the above

76. The least cost combination of various inputs when represented through locus of points is called as
(a) Budget Line
(b) Isocost Lines
(c) Both of the above
(d) None of the above

77. To make maximum productivity, Isoquant line should be
(a) Parallel to Budget line
(b) Perpendicular to budget line
(c) Crossing the budget line
(d) None of the above

78. For least cost or optimal input combination
(a) MRP of inputs is greater than their price ratio
(b) MRP of inputs is smaller than their price ratio
(c) MRP of inputs is equal to their price ratio
(d) None

79. When marginal productivity is Mp, Average Productivity is Ap and Elasticity of production is Ep then
(a) Ep = Mp x Ap
(b) Ep = Mp/Ap
(c) Ep = Ap/Mp
(d) Ep = Mp + Ap

80. Profit will be maximum when
(a) Total cost =Total revenue
(b) Average cost = Average revenue
(c) Marginal Cost = Marginal revenue
(d) All of the above

81. A line or curve connecting the least cost combination of inputs for all output levels is known as
(a) Budget Line
(b) Iso-cost line
(c) Isoclines
(d) None of the above

82. The isocline which is considered to be the most appropriate over a particular period is called as
(a) Ridge line
(b) Iso-cost line
(c) Expansion Path
(d) Boundary Line

83. At any particular time, number of expansion path will be formed is
(a) 0
(b) 1
(c) 2
(d) Infinity

84. Points of maximum output from each input, given a fixed amount of the other input is called as
(a) Expansion Path
(b) Budget Line
(c) Isoclines
(d) Ridge Line

85. Marginal rate of substitution between supplementary product is
(a) Positive
(b) Negative
(c) Zero
(d) None of the above

86. Marginal rate of substitution between complementary product is
(a) Positive (b) Negative
(c) Zero (d) None of the above

87. Marginal rate of substitution between competitive product is
(a) Positive (b) Negative
(c) Zero (d) None of the above

88. The curve that represents all possible combinations of two products that could be produced with given amounts of inputs is called as
(a) Iso-resource curve
(b) Opportunity curve
(c) Production possibility curve
(d) All of the above

89. When two products are competitive and they substitute with each other at increasing rate then then production possible curve is
(a) Pass through the origin
(b) Convex to the origin
(c) Concave to the origin
(d) None of the above

90. When two products are competitive and they substitute with each other at decreasing rate then then production possible curve is
(a) Pass through the origin
(b) Convex to the origin
(c) Concave to the origin
(d) None of the above

91. If the products are substituted at increasing rate, then a farmer should go for
(a) Specialization in farming
(b) Diversification in farming
(c) Both
(d) None

92. Find the wrong the statement in the following options
(a) In order to maximize net revenue, variable costs are the relevant costs to be considered
(b) In short run farmer has the objective to minimize the loss
(c) In the long run farmer has the objective to maximize the revenue
(d) All are correct

93. If the substitution ratio is greater than the price ratio, one can reduce the cost by
(a) Using more of replaced resources
(b) Using more of added resources
(c) Try to make both added and replaced resource to be equal
(d) There is no such standing principle

94. Those are not the joints products
(a) Hide and beef
(b) Wheat and straw
(c) Cow milk and buffalo milk
(d) Cattle and manure

95. The change in overall costs that results from particular decision is made is called as
(a) Implicit cost
(b) Marginal cost
(c) Future Cost
(d) Incremental cost

96. Incremental cost is also called as
(a) Avoidable cost
(b) Escapable cost
(c) Differential cost
(d) All of the above

97. The cost that is not affected or altered by a change in the level or nature of business activity is called as
(a) Prime cost
(b) Sunk cost
(c) Historical cost
(d) Prime cost

98. If the price of bronze metal is Rs 15 in 1974 and it is Rs 20 in 2014, then the historical cost and replacement cost respectively
(a) Rs 18, Rs 15
(b) Rs 15, Rs 18
(c) Both are Rs 18
(d) Both are Rs 15

99. The cost that firms' forgoes in order to avail itself of the returns from the best use of the resources is called as
(a) Implicit cost
(b) Explicit cost
(c) Opportunity Cost
(d) Direct cost

100. When any plant is permanently closed down, some costs are to be incurred for disposing the fixed assets. It is called as
(a) Shut down cost
(b) Out of pocket cost
(c) Book Cost
(d) Abandonment cost

101. Variable cost is also called as
(a) Overhead cost
(b) Prime cost
(c) Both
(d) None

102. The term opportunity cost refers to
(a) Variable cost
(b) Short run cost
(c) Cost related to an optimum level of production
(d) Cost of one product in terms of production of other foregone

103. Short run fixed cost is
(a) Parallel to X axis
(b) Parallel to Y axis
(c) Pass through the origin
(d) None of the above

104. In short run ,marginal cost depends on value of
(a) Average cost
(b) Total variable cost
(c) Total fixed cost
(d) Average variable cost

105. The cost that is approaching both axes asymptotically is
(a) Total variable cost
(b) Average cost
(c) Average Fixed cost
(d) Marginal cost

106. When Average Variable Cost (AVC) attains minimum then
(a) MC > AVC
(b) MC = AVC
(c) MC < AVC
(d) None

107. When Average Total Cost is at its minimum, then
(a) AFC = ATC (b) MC = ATC
(c) AVC = ATC (d) TFC = ATC

108. The statement that is true is
(a) Both Long run and short run marginal cost is U-shaped
(b) Short run Average Variable cost is U-shaped
(c) Long run Average Cost is U-shaped
(d) All are true

109. The relationship between Marginal cost (MC), Average Cost (AC) and Cost elasticity (EC) is
(a) $E_C = M_C X A_C$
(b) $E_C = M_C / A_C$
(c) $E_C = M_C X 2A_C$
(d) $E_C = M_C / 2A_C$

110. The cost curve that is also known as envelope curve is
(a) Short run Average variable Cost curve
(b) Short run Average cost curve
(c) Long run Average cost curve
(d) Marginal cost curve

111. At the breakeven point
(a) TC > TR (b) TC < TR
(c) TC = TR (d) 2TC = TR

112. Margin of safety is
(a) Above the breakeven point
(b) Below the breakeven point
(c) Equals to breakeven point
(d) Not related to breakeven point

113. Profit is maximized when
(a) MC < MR (b) MC = MR
(c) MC > MR (d) MC ≤ MR

114. Market with one buyer and one seller is called
(a) Monopsony (b) Monopoly
(c) Duopoly (d) Bilateral Monopoly

115. The situation where market consists of only two buyers, then it is called as
(a) Monopsony (b) Duopsony
(c) Monopoly (d) Duopoly

116. The feature that is not seen in perfect market condition
(a) Large number of buyer and seller
(b) Heterogeneous products
(c) Perfectly elastic demand curve
(d) No government intervention

117. A perfectly competitive firm is
(a) Price maker
(b) Price taker
(c) Both
(d) None

118. In a very short run period, a perfectly competitive firm has a supply curve parallel to
(a) X- Axis
(b) Y-Axis
(c) Pass through the origin
(d) None of the above

119. The demand of a perfectly competitive firm parallel to
(a) X- Axis
(b) Y-Axis
(c) Pass through the origin
(d) None of the above

120. Shut down point is that point where price is
(a) Less than variable cost
(b) Equal to average cost
(c) Equal to variable cost
(d) None of the above

121. In the short run, profit maximizing firm produces additional units of production, as long as
(a) Elasticity of demand is infinite
(b) Total revenue is increasing
(c) Price covers at least average fixed cost
(d) Additional revenue per unit exceeds additional cost per unit

122. In the long run, the perfectly competitive firm will enjoy
(a) Subnormal Profit
(b) Normal Profit
(c) Super Normal Profit
(d) All

123. In the long run, the Long run Average Cost is equal to
(a) Average Revenue
(b) Marginal Revenue
(c) Short run Average Cost
(d) All of the above

124. What leads to rationing and black marketing under perfect competition?
(a) Price support
(b) Decrease in supply
(c) Excess demand
(d) Price control

125. What leads to black marketing under perfect competition?
(a) Price support
(b) Decrease in supply
(c) Excess demand
(d) Price control

126. The cross elasticity demand for a monopoly product is
(a) Negative (b) Zero
(c) Both (d) None

127. Monopolist can decide
(a) Quantity
(b) Price
(c) Both
(d) Either price or quantity

128. The statement that is not true for monopoly market is
(a) The entire market is under control of a single firm
(b) There is a clear cut difference between monopoly firm with its industry
(c) Monopoly firm is a price maker
(d) There is no substitute or close substitute for monopoly product

129. The relation between the slope of Marginal revenue and Average revenue in monopoly condition is
(a) $(\text{Marginal revenue})_{slope} = (\text{Average revenue})_{slope}$
(b) $(\text{Marginal revenue})_{slope} = 2\,(\text{Average revenue})_{slope}$
(c) $2(\text{Marginal revenue})_{slope} = (\text{Average revenue})_{slope}$
(d) There is no relationship

130. The relationship between short run marginal cost(SMC) and the long run marginal cost(LMC) for a monopoly firm is
(a) SMC=LMC (b) SMC>LMC
(c) SMC<LMC (d) None of the above

131. Under monopoly market structure, the degree of freedom in pricing decision is
(a) Very low (b) Quite good
(c) Very high (d) Zero

132. Which of the following is the condition of equilibrium for the monopolist?
(a) MR= MC
(b) AR = MC = Price
(c) MC = AR
(d) AC = AR

133. The demand curve facing the pure monopolist is
(a) Positively sloped
(b) Horizontal
(c) Negatively sloped
(d) None of the above

134. A monopolist always charge a price which is
(a) Greater than marginal cost
(b) Equal to marginal cost
(c) Equal to average cost
(d) Equal to average variable cost

135. Practice of buying goods from a cheaper market and selling them in a costlier market is known as
(a) Price discrimination
(b) Speculation
(c) Arbitrage
(d) Forward Marketing

136. Output under price discrimination will
(a) Generally be larger than under simple monopoly
(b) Be indeterminate because, we cannot know what prices can be charged
(c) Usually be the same as under perfect competition
(d) Be produced at higher average cost than under simple monopoly

137. The worst form of price discrimination is
(a) First degree form
(b) Second degree form
(c) Third degree form
(d) All of the above

138. Form of price discrimination in which seller fix price at highest level and then gradually he creases is
(a) First degree form
(b) Second degree form
(c) Third degree form
(d) All of the above

139. Form of price discrimination, when the seller divides consumer in groups and then discriminates on the basis of consumer surplus is
(a) First degree form
(b) Second degree form
(c) Third degree form
(d) None of the above

140. The degree of price discrimination in which seller sets different prices in different market having demand curves with different elasticity is
(a) First degree
(b) Second degree
(c) Third degree
(d) None of the above

141. Monopolistic competition is a market structure where we find
(a) Differentiated products
(b) Large number of buyers and sellers
(c) The products are close, not perfect substitutes
(d) All of the above

142. The model of price and output determination under monopolistic competition was developed by
(a) Augustin Cournot
(b) Paul Sweezy
(c) Edward H. Chamberlin
(d) None of the above

143. Demand curve under monopolistic competition is
(a) Downward sloping and flat
(b) Downward sloping and steep
(c) Parallel to X- Axis
(d) Parallel to Y- Axis

144. Action, reaction and counteraction phenomena is seen among the competitors of
(a) Perfect Market Condition
(b) Monopoly Market Condition
(c) Monopolistic Market Condition
(d) Oligopolistic Market Condition

145. According to Cournot's duopoly Model, equilibrium comes when both the sellers' having
(a) 1/2 of the market share
(b) 1/3 of the market share
(c) 1/4 of the market share
(d) 1/5 of the market share

146. According to Cournot's duopoly Model, if the number of sellers are five and they are competing in three different markets, then the share of each seller is
(a) 1/2 (b) 1/3
(c) 1/4 (d) 2/5

147. The concept of kinked demand curve or price rigidity model was stated by
(a) Augustin Cournot
(b) Paul Sweezy
(c) Edward H. Chamberlin
(d) None of the above

148. The kinked demand curve hypothesis predicts
(a) Stable costs for the firm
(b) More advertising for the firm
(c) Sticky prices in the market
(d) None of the above

149. According to price rigidity model, when a firm will increases its price, then
(a) Other will follow the same pattern
(b) Others will not follow
(c) Others will decrease their price
(d) Others first increase or then decrease their price

150. The barometric firm having
(a) Lowest cost than its competitor
(b) Dominant market demand than its competitor
(c) Better acumen to forecast the prevailing market conditions than its competitor
(d) All of the above

151. OPEC is an example of the type of producer's organization known as a
(a) Marketing Board
(b) Producer's Cooperative
(c) Trust
(d) Cartel

152. An economy has no economic transactions with the rest of the world is called as
(a) Close economy
(b) Open economy
(c) Mixed economy
(d) Under developed economy

153. Market value of all final goods and services plus income earned locally by foreigners minus income earned abroad by the nationals is called as
(a) Gross National Product
(b) Gross Domestic Product
(c) Net National Product
(d) Net National Income at Factor Cost

154. Market value of all final goods and services minus income earned locally by foreigners plus income earned abroad by the nationals is called as
(a) Gross National Product
(b) Gross Domestic Product
(c) Net National Product
(d) Net National Income at Factor Cost

155. Gross National Product minus Depreciation is
(a) Gross National Product
(b) Gross Domestic Product
(c) Net National Product
(d) Net National Income at Factor Cost

156. National Income is
(a) NDP at Factor cost
(b) NNP at factor cost
(c) NNP at market price
(d) GNP at factor cost

157. Transfer Pricing is referred to as
(a) Price charged to a customer
(b) Price charged to different firms
(c) Price charged under intrafirm transfer of product
(d) Interfirm transfer price

158. What are the methods are used to measure the national income?
(a) Expenditure method
(b) Income method
(c) Output method
(d) All of the above

159. In India, which method is used to measure National Income in service sector?
(a) Expenditure method
(b) Income method
(c) Output method
(d) Value added method

160. The periodic boom and slumps in economic activity is called as
(a) Manufacturing Cycle
(b) Production Cycle
(c) Business Cycle
(d) Operation Cycle

161. When the growth rate falls below the steady growth line, the economy enters into?
(a) Expansion
(b) Recession
(c) Depression
(d) Recovery

162. 'Labour markets become the seller's market' in which phases of business cycle
(a) In the initial phase of prosperity
(b) In the beginning phase of recovery
(c) In the starting phase of recession
(d) In the later phase of prosperity

163. The impact of recession is first visualized in
(a) Automobile sector
(b) Household Appliance Sector
(c) Real Estate Sector
(d) Service Sector

164. The phase of reversal process in business cycle begins in
(a) Raw material Industry
(b) Banking Sector
(c) Labour Market
(d) Consumer side

165. Profit motive and Private dominated business enterprise are the root philosophy of
(a) Socialistic economy
(b) Capitalistic economy
(c) Mixed economy
(d) All of the above

166. When public sector has a measure chunk of control in their economic development then it is called as
(a) Socialistic economy
(b) Capitalistic economy
(c) Mixed economy
(d) All of the above

167. The type of socialism in which it has been believed that business organization, factories and productive resources should be handed over to the labour organization is
(a) Guild Socialism
(b) Syndicalism
(c) Communism
(d) Anarchism

168. 'Strike is a measure weapon for economic warfare' is the philosophy of
(a) Guild Socialism
(b) Syndicalism
(c) Communism
(d) Anarchism

169. The Book, 'General Theory of Employment, Interest and Money' is written by
(a) J B Say
(b) J M Keynes
(c) Adam Smith
(d) Alfred Marshall

170. 'Supply creates its own demand' is stated by
(a) J B Say (b) J M Keynes
(c) James Mill (d) Pigou

171. Frictional unemployment occur when one changes himself from one job to another. This was the statement of one of the classical theorists of employment. He was
(a) J B Say (b) James Mill
(c) Pigou (d) David Ricardo

172. According to 'Say's law' of employment
(a) The government always interfere in the decision making
(b) The size of the market is unlimited
(c) Savings are equal to investment
(d) There is always a hindrance in the free flow of the money

173. 'Aggregate effective demand is the tool for creation of employment in the economy', was the concept of
(a) J B Say
(b) J M Keynes
(c) Adam Smith
(d) Alfred Marshall

174. According to Keynes' theory of employment, the true and the valid statement is
(a) In the short run, propensity to consume is relatively stable
(b) Investment depends on the marginal efficiency of capital and the rate of interest
(c) Total income depends on the volume of total employment
(d) All are true

175. If the marginal propensity to consume is 3/4 then the value of the multiplier is
(a) 1
(b) 2
(c) 3
(d) 4

176. An individual is not merely influenced by current income, but also by the standard of living he has enjoyed in the past' stated by
(a) Arthur C Pigou
(b) Prof. Duesenberry
(c) J M Keynes
(d) Milton Friedman

177. 'consumption goes up as income increases, but does not fall in the same proportion', this phenomena is called as
(a) Demonstration effect
(b) Ratchet effect
(c) Pigou effect
(d) Duesenberry hypothesis

178. 'Poor people imitate the living style of the rich', this pattern of consumption is called as
(a) Demonstration effect
(b) Ratchet effect
(c) Pigou Effect
(d) Duesenberry hypothesis

179. Demonstration effect is otherwise called as
(a) Ratchet effect
(b) Snob effect
(c) Bandwagon effect
(d) Pigou effect

180. Demonstration effect of consumption is given by
(a) Arthur C Pigou
(b) Prof. Duesenberry
(c) J M Keynes
(d) Milton Friedman

181. Autonomous investment is
(a) Income-elastic
(b) Income-inelastic
(c) Both
(d) None

182. Inducement for investment directly depends on
(a) Interest rate
(b) Marginal efficiency of capital
(c) Both
(d) None

183. The shape of autonomous investment curve is
(a) Parallel to X-Axis
(b) Parallel to Y-Axis
(c) Pass through the origin
(d) None of the above

184. The shape of Induced investment curve is
(a) Parallel to X-Axis
(b) Parallel to Y-Axis
(c) Pass through the origin
(d) Slopes upward from left to right

185. Autonomous investment which is made by the government to revive the economy from the phase of depression is called as
(a) Reflation
(b) Regressive Tax
(c) Pump priming
(d) None of the above

186. For luxury good, Engel curve bents towards
(a) X- Axis (b) Y-Axis
(c) Both (d) None

187. For necessity good, Engel curve bents towards
(a) X- Axis (b) Y-Axis
(c) Both (d) None

188. In Keynesian analysis, the marginal propensity to consume (MPC) lies in between
(a) -1 to 1 (b) 0 to 1
(c) -1 to 0 (d) >1

189. Psychological law of consumption was propounded by
(a) Pigou (b) J M Keynes
(c) J B Say (d) Ricardo

190. The concept of investment multiplier was first developed by
(a) J B Say (b) J M Keynes
(c) Pigou (d) Ricardo

191. International demonstration effect is stated by
(a) J m Keynes (b) J B say
(c) J R Hicks (d) Nurkse

192. A sustained rise of less than 3 per cent in prices per annum is called as
(a) Creeping inflation
(b) Walking inflation
(c) Running inflation
(d) Galloping inflation

193. When the rise in price falls in the range of 3 to 6%, it is called as
(a) Creeping inflation
(b) Walking inflation
(c) Running inflation
(d) Galloping inflation

194. When the sustained price rise is about 10 per cent per annum is called as
(a) Creeping inflation
(b) Walking inflation
(c) Running inflation
(d) Galloping inflation

195. When the sustained price rise is more than 16 per cent per annum is called as
(a) Creeping inflation
(b) Walking inflation
(c) Running inflation
(d) Galloping inflation

196. When the prices of few commodities show an upward trained, it is called as
(a) Comprehensive inflation
(b) Sporadic inflation
(c) Both
(d) None

197. Moderate degree of controlled inflation is called as
(a) Hyper inflation (b) Deflation
(c) Reflation (d) Stagflation

198. The decline in the rate of inflation is called as
(a) Deflation (b) Reflation
(c) Disinflation (d) Stagflation

199. When the general level of price is falling is called as
(a) Deflation (b) Reflation
(c) Disinflation (d) Stagflation

200. The inflation corresponding to stagnation in development is called as
(a) Deflation (b) Reflation
(c) Disinflation (d) Stagflation

201. The unemployment that arises due to lack of adjustment between demand and supply of labour in a growing economy is called as
(a) Structural unemployment
(b) Frictional unemployment
(c) Cyclical Unemployment
(d) Seasonal Unemployment

202. Find the wrong matching
(a) Long term unemployment- Marxian unemployment
(b) Cyclical Unemployment – Keynesian unemployment
(c) Structural Unemployment - Marxian unemployment
(d) All are correct

203. The cultivation of crops in areas where rainfall is less than 750mm is called as
(a) Dry land farming
(b) Dry farming
(c) Rainfed farming
(d) None of the above

204. The cultivation of crops in areas where rainfall is more than 750 mm is called as
(a) Dry land farming
(b) Dry farming
(c) Rainfed farming
(d) None of the above

205. When the annual rainfall is more than 1150 mm, the farming that is carried on is
(a) Dry land farming
(b) Dry farming
(c) Rainfed farming
(d) None of the above

206. Types of cooperative farming in which both ownership and operation is indivizual, is
(a) Cooperative better farming
(b) Cooperative joint farming
(c) Cooperative tenant farming
(d) Cooperative collective farming

207. Types of cooperative farming in which ownership is indivisual but operation is collective, is
(a) Cooperative better farming
(b) Cooperative joint farming
(c) Cooperative tenant farming
(d) Cooperative collective farming

208. Types of cooperative farming in which ownership is collective but operation is indivisual, is
(a) Cooperative better farming
(b) Cooperative joint farming
(c) Cooperative tenant farming
(d) Cooperative collective farming

209. Types of cooperative farming in which both ownership and operation is collective, is
(a) Cooperative better farming
(b) Cooperative joint farming
(c) Cooperative tenant farming
(d) Cooperative collective farming

210. If the prices and/or incomes of the enterprises have correlation nearer to 1, then it is recommended to go for
(a) Diversification in farming
(b) Specialization in farming
(c) Both
(d) None

211. When wage is paid for completion of a specific task is called as
(a) Cash wage
(b) Kind wage
(c) Piece wage
(d) Task wage

212. Which is not a type of hired labour?
(a) Casual labour
(b) Attached labour
(c) Family labour
(d) Contract labour

213. Type of method for valuation of asset that are primarily meant for sale is
(a) Net Selling Price Method
(b) Cost less depreciation Method
(c) Market Price Method
(d) Cost Method

214. Land value can be evaluated through
(a) Cost less depreciation Method
(b) Market Price Method
(c) Income capitalization method
(d) Cost Method

215. Current value of the farm produced inputs like seed, FYM can be evaluated through
(a) Net Selling Price Method
(b) Cost less depreciation Method
(c) Market Price Method
(d) Cost Method

216. An average farmer require following records except
(a) Land use record
(b) Feed record
(c) Water harvesting record
(d) Log book

217. The land holding capacity of marginal farmer is
(a) Below 1 hectare
(b) Between 1 to 2 hectare
(c) Between 2 to 4 hectare
(d) Between 4 to 10 hectare

218. The land holding capacity of small farmer is
(a) Below 1 hectare
(b) Between 1 to 2 hectare
(c) Between 2 to 4 hectare
(d) Between 4 to 10 hectare

219. The land holding capacity of medium farmer is
(a) Below 1 hectare
(b) Between 1 to 2 hectare
(c) Between 2 to 4 hectare
(d) Between 4 to 10 hectare

220. The land holding capacity of large farmer is
(a) Below 1 hectare
(b) Between 1 to 2 hectare
(c) Between 4 to 10 hectare
(d) 10 hectare and above

221. According to Say's law
(a) Supply creates its own demand
(b) Demand creates its own supply
(c) Oversupply and constant demand causes excess
(d) Undersupply and more demand causes shortage

222. Stagflation refers to:
(a) co-existanceof recession with high inflation
(b) High unemployment
(c) Both (a) and (b)
(d) Stagnant prices

223. Who explained the rate of growth of income that must take place if the steady growth of the economy is to be achieved?
(a) Harrod and Domar
(b) Mahalonobis
(c) Kaldor and Robinson
(d) Solow

224. Laffer curve explains the relationship between
(a) Government revenue and rate of tax
(b) Government revenue and expenditure
(c) Government revenue and employment
(d) Government revenue and cost

225. Macro economics is concerned with the analysis of
(a) The behaviour of the economic system in totality
(b) Analysis of small individual units
(c) Resource allocation among various products
(d) Income distribution among factors

226. The subject matter of macroeconomics is to explain
(a) What determines the level of total economic activity
(b) What causes the general price level to rise
(c) Factors which determine the national income in long run
(d) All of the above

227. According to the rational expectations theory
(a) Involuntary unemployment can not prevail
(b) Involuntary employment can prevail
(c) Involuntary unemployment can prevail
(d) Involuntary employment cannot prevail

228. An attempt to save more by all people cause their income or standard of living to decline is called
(a) Paradox of thrift
(b) Wage employment paradox
(c) Paradox of consumption
(d) Consumption employment paradox

229. A decline in aggregate demand causes
(a) National output and income to fall and unemployment to decrease
(b) National output to increase, income and unemployment to decrease
(c) National output and income to fall and unemployment to increase
(d) National output to decrease, income and unemployment to increase

230. Keynes' theory of multiplier states that
(a) Income and employment fall more than the original rise in saving
(b) Income and employment fall less than the original rise in saving
(c) Income and employment remain static
(d) Income and employment are not related to saving in any way

231. The Wage Employment Paradox refers
(a) Proves that what is true for the individual may not be true for the society as a whole
(b) Proves that a cut in wages leads to increased unemployment in society as a whole
(c) Both (a) and (b)
(d) Proves that what is true for the individual may be true for the society as a whole

232. Macro economics helps in understanding the problems of
(a) Economic growth
(b) Business Cycles
(c) Monetary policies
(d) All of these

233. National income of a country is defined as:
(a) Total market value of all goods and services produced in the economy in a year
(b) Total market value of all goods produced in the economy in a year
(c) Total market value of all services in the economy in a year
(d) Total market value of all intermediate goods produced in the economy in a year

234. Final goods are:
(a) Those goods which are purchased for final use and not for resale or further processing
(b) Those goods which are processed and finally turned into a product
(c) Those goods which are paid for after final bargain
(d) Those goods which can be resold for final consumption

235. The sale of intermediate goods is excluded from gross national product
(a) True
(b) False

236. The sale of intermediate goods is excluded from gross national product because
(a) Its inclusion will involve double counting
(b) Its inclusion will exaggerate estimate of gross national product
(c) Both (a) and (b)
(d) Neither (a) nor (b)

237. Gross national product has the following components
(a) Value of final goods and services produced in a year
(b) Gross Private Investment
(c) Value of output of General Government
(d) Net Exports and Net factor income from abroad
(e) All of the above

238. Value of new capital goods produced and addition to the inventories of goods such as raw materials, unfinished goods and consumer goods produced but not sold during a year is called
(a) Gross capital investment
(b) Gross private investment
(c) Gross national investment
(d) Gross capital budget

239. The sum of factor incomes such as wages and salaries (i.e. compensation of employees), rent, interest and profits generated within the domestic territory of a country in a year is called
(a) Domestic factor income
(b) Domestic income
(c) Domestic product
(d) Domestic mean

240. Net factor income from abroad consists of
(a) Net compensation of employees
(b) Net income from property
(c) Net retained earnings of the resident companies working in foreign countries
(d) All of the above

241. C+I+G+Xn where Xn=(X-M), defines
(a) Gross Domestic Product
(b) Gross National Product
(c) Net National Product
(d) Net Domestic Product

242. Net factor income from abroad + Gross Private Investment + Net Exports (Xn) + Government purchases + Consumption Expenditure C.
(a) Gross Domestic Product
(b) Gross National Product
(c) Net National Product
(d) Net Domestic Product

243. (Net Private Investment+Net Exports+Grovernment Expenditure) less depreciation defines
(a) Gross Domestic Product
(b) Gross National Product
(c) Net National Product
(d) Net Domestic Product at market prices

244. Net Income or Net Income at Factor cost is defined as
(a) Net income at market Prices- indirect taxes + subsidies
(b) Net income at market prices + Indirect taxes-subsidies
(c) Net income at market prices- indirect taxes- subsidies
(d) Net income at market prices + indirect taxes + subsidies

245. Relationship among National income, Net National Product and Net Indirect Taxes could be established as
(a) National income = Net National Product- Net Indirect taxes
(b) National income = Net National Product+ Net Indirect taxes
(c) National income = 2(Net National Product)- Net indirect taxes
(d) National income = (Net National Product)-2(Net indirect taxes)

246. Which of these about the disposable income is correct
(a) Disposable income = Personal income-personal taxes
(b) Disposable income = Consumption + Saving
(c) Disposable income = Personal income + personal taxes
(d) Both (a) and (b)

247. The National income can be measured by
(a) Value added method
(b) Income method
(c) Expenditure method
(d) All of the above

248. Income method approaches national income from
(a) Production side
(b) Distribution side
(c) Expenditure side
(d) None of the above

249. The relation between national income (GNP) and Net Economic Welfare (NEW) can be represented as follows
(a) Real GNP - Depreciation + Value of leisure + Value of non market activities - Environment pollution - Regrettable costs = Net Economic Welfare
(b) Real GNP - Depreciation - Value of leisure + Value of non market activities - Environment pollution - Regrettable costs = Net Economic Welfare
(c) GNP - Depreciation + Value of leisure - Value of non market activities - Environment pollution- Regrettable costs = Net Economic Welfare
(d) Real GNP + Depreciation + Value of leisure + Value of non market activities - Environment pollution - Regrettable costs = Net Economic Welfare

250. In circular flow of income in a two sector economy
(a) Resources flow from household to business firms
(b) Money flows from business to household firms
(c) Goods and services flow from households to business firms
(d) All of the above

251. In two sector simple economy with neither Government, nor Foreign trade
(a) Investment is identically equal to saving
(b) Investment is not identically equal to saving
(c) Investment is more than saving
(d) Investment is less than saving

252. Which of the following is true?
(a) Government borrowing increases the demand for credit, which causes rate of interest to rise
(b) Government borrowing increases the demand for credit, which causes rate of interest to fall
(c) Government borrowing decreases the demand for credit, which causes rate of interest to rise
(d) Government borrowing decreases the demand for credit, which causes rate of interest to fall

253. In a three sector economy with Government sector
(a) G-T=S-I (b) G+T= S+I
(c) G-T=S+I (d) G+T= S-I

254. In a three sector economy, if the Government Expenditure (G) is greater than the tax revenue (T), i.e. G>T, the government will have
(a) Excess budget
(b) Deficit budget
(c) Revenue budget
(d) Revenue Excess

255. In a four sector economy
(a) C+I+G+Xn=C+S+T
(b) C+I+G = C+S+T+Xn
(c) C+I+T+S= G-Xn+C
(d) C+I+T+G+Xn= S+C

256. Keynesian theory of income and employment is a
(a) Short run theory
(b) Long run theory
(c) Very long run theory
(d) Very short run theory

257. Keynesian theory of income and employment is a short run theory because
(a) Keynes assumes that the amount of capital; the size of the population, labour force, technology , etc, do not change
(b) Keynes assumes that the amount of capital; the size of the population, labour force, technology , etc, changes
(c) None of the above

258. With the expansion in money supply, nominal wage rate and price level rise, but real wage rate, level of employment and output remain constant. This shows
(a) Neutrality of money
(b) Rigidity of money
(c) Flexibility of money
(d) Velocity of money

259. Tax multiplier is
(a) -MPC/1-MPC (b) MPC/1-MPC
(c) MPC/1+MPC (d) -MPC/1+MPC

260. Marginal Propensity to consume is
(a) Change in consumption/change in income
(b) Change in consumption/change in saving
(c) Change in consumption/change in investment
(d) Change in consumption/ change in tx

261. Average propensity to save is
(a) The proportion of disposable income that is consumed
(b) The proportion of disposable income that is saved
(c) The proportion of disposable income that is invested
(d) The proportion of disposable income that is taxed

262. Mathematically, Average propensity to save is
(a) S/T (b) S/I
(c) S/Y (d) S/G

263. Sum of marginal Propensity to Consume and Marginal Propensity to Save is
(a) Equal to 1
(b) Less than 1
(c) More than 1
(d) Not defined

264. According to Keynes, factors which influence the consumption function are
(a) Changes in real price effect
(b) Fiscal policy and rate of interest
(c) Stock of wealth
(d) All of these

265. According to Keynes, factors which influence the consumption function are
(a) Credit conditions and consumer indebtedness
(b) Income distribution
(c) Windfall gains and losses
(d) Change in expectations
(e) All of the above

266. Keynes' consumption function is
(a) C=a-by
(b) C=a/by
(c) C=a+by
(d) C=a*by

267. Duesebberry's relative income hypothesis suggests that as income increases, consumption function curve shifts above so that average propensity to consume
(a) increases
(b) decreases
(c) remains constant
(d) changes infinite

268. The relative income hypothesis suggests that individuals or households try to imitate or copy the consumption levels of their neighbors or other families in a particular community. This is called
(a) Demonstration or Duesenberry effect
(b) Paradox of thrift
(c) Trickle down effect
(d) Keynes' effect

269. Ratchet effect suggests
(a) When income of individual or households increases, their consumption expenditure falls much
(b) When income of individual or households fall, their consumption expenditure falls much
(c) When income or individual or households increase, their consumption expenditure does not fall much
(d) When income or individual or households fall, their consumption expenditure does not fall much

270. Life cycle theory of consumption is given by
(a) Franco Modigliani
(b) Franco Modigliani and a. Ando
(c) a.Ando
(d) J. M. Keynes

271. According to the life cycle theory
(a) The consumption in any period is not only the function of current income of that period but also of the whole lifetime expected income
(b) The consumption in any period is the function of current income of that period and not of the whole lifetime expected income
(c) The consumption in any period is the function of current income of that period
(d) The consumption in any period is not the function of current income of that period but of the whole lifetime expected income

272. The essence of multiplier is
(a) The total increase in income is manifold the original increase in investment
(b) The total increase in output is manifold the original increase in investment
(c) The total increase in employment is manifold the original increase in investment
(d) All of the above

273. The value of the multiplier is
(a) Greater than one but less than infinity
(b) Less than one but less than infinity
(c) Equal to one but less than infinity
(d) None of the above

274. Keynes multiplier is also known as
(a) Instantaneous multiplier
(b) Dynamic multiplier
(c) Static multiplier
(d) Investment multiplier

275. A state of affairs when in a country, there are large number of able bodied persons of working age who are willing to work but cannot find work at the current wage levels is called
(a) Unemployment
(b) Underemployment
(c) Dragged employment
(d) Frictional employment

276. People, who are only partially employed or are engaged in inferior jobs, though they can do better jobs, are not adequately employed It is called a state of
(a) Unemployment
(b) Underemployment
(c) Dragged employment
(d) Frictional employment

277. The main kinds of unemployment are
(a) Frictional Unemployment
(b) Structural Unemployment
(c) Cyclical Unemployment
(d) All of the above

278. Some people remain unemployment for a relatively short period of time before they get a new job It is called
(a) Frictional Unemployment
(b) Structural Unemployment
(c) Cyclical Unemployment
(d) Disguised Unemployment

279. The feature that the unemployed workers lack skills required by the expanding industries shows
(a) Frictional Unemployment
(b) Structural Unemployment
(c) Cyclical Unemployment
(d) Disguised Unemployment

280. The unemployment due to deficiency of effective demand is called
(a) Frictional Unemployment
(b) Structural Unemployment
(c) Cyclical Unemployment
(d) Disguised Unemployment

281. Full employment does not prevail, if
(a) The number of unemployed is greater than frictional and structural unemployment
(b) The number of unemployed is lesser than frictional and structural unemployment
(c) The number of unemployed is equal to frictional and structural unemployment
(d) The number of unemployed is greater than structural employment only

282. Various causes which account for unemployment and underemployment are
(a) Lack of the stock of physical capital
(b) Use of capital intensive techniques
(c) Inequitable distribution land
(d) All of the above

283. Various causes which account for unemployment and underemployment are
(a) Rigid protective labour legislation
(b) Neglect of the role of agriculture in employment generation
(c) Lack of infrastructure
(d) All of the above

284. Which of the following is the function of a country's central bank?
(a) It acts as note issuing agency
(b) It acts as a banker to the state
(c) It controls credit
(d) All of the above

285. Which of the following is the function of a country's central bank?
(a) It acts as a banker's bank
(b) It acts as the lender of last resort
(c) It manages exchange rate
(d) All of the above

286. Quantitative or General methods of credit control comprise
(a) Changing in bank rate
(b) Open market operations
(c) Changing the cash reserve ratio
(d) All of the above

287. Qualitative or selective methods of credit control include
(a) Changes in minimum margin for lending by banks against the stock of goods kept or against other type of securities
(b) The fixation of minimum discriminatory rates of interest chargeable on credit for particular purpose
(c) Prohibition of the discounting of bills of exchange involving sale of sensitive commodities
(d) All of the above

288. What are the objectives of monetary policy?
(a) Ensuring price stability, i.e. containing inflation
(b) To encourage economic growth
(c) To ensure stability of currency exchange rate
(d) All of the above

289. Real rate of interest is
(a) Nominal rate of interest - rate of inflation
(b) Nominal rate of interest + rate of inflation
(c) Nominal rate of interest/ rate of inflation
(d) Nominal rate of interest * rate of inflation

290. Inflation is
(a) A persistent rise in the general price level
(b) A once for all rise in the general price level
(c) A persistent rise in the demand
(d) A persistent rise in the supply

291. According to Keynes, inflation arises when
(a) The full employment does not exist
(b) Aggregate demand exceeds aggregate supply at full employment level of output
(c) Aggregate supply exceeds aggregate demand at full employment level of output
(d) Aggregate demand equals aggregate supply at full employment level of output

292. Depending on the specific causes, types of inflation have been distinguished
(a) Demand pull inflation
(b) Cost-push inflation
(c) Structural inflation
(d) All of the above

293. The booklet ," How to pay for the war", was written by
(a) a. c. Pigou (b) Adam Smith
(c) J.M. Keynes (d) Amartya Sen

294. In Hyper-inflation
(a) Wages and price chase each other at a very quick speed
(b) Wages and price chase each other at moderate speed
(c) Wages and price chase each other at slow speed
(d) Wages and prices do not chase each other

295. When aggregate demand for all purposes-consumption, investment, and government expenditure exceeds the supply of goods at current prices, there is a rise in prices. It is termed as
(a) Cost push inflation
(b) Demand pull inflation
(c) Structural inflation
(d) Hyperinflation

296. The average number of times per year that the money stock is used in making payment for final goods and services is called
(a) Inflation
(b) Velocity of money
(c) Nominal income
(d) Disposable income

297. According to the constant monetary rule
(a) Money supply should not be increased each year at the same annual rate as the potential growth of real GNP
(b) Money supply should be increased each year but not at the same annual rate as the potential growth of real GNP
(c) Money supply should be increased each year at the same annual rate as the potential growth of real GNP
(d) Money supply should be increased each year at an increased annual rate as the potential growth of real GNP

298. Autonomous increases in cost which generate cost push inflation are
(a) Wage push inflation
(b) Profit push inflation
(c) Increases in prices of raw materials, especially energy inputs such as rise in crude oil prices
(d) All of the above

299. The theory of structural inflation was put forward by
(a) G. Myrdal and P. Streetan
(b) G. Myrdal
(c) P. Streetan
(d) F. Nixon

300. Inflation affect economic growth by
(a) Inflation serves as a disincentive to save
(b) Inflation lead to unproductive form of investment
(c) Inflations accentuates the problem of poverty
(d) It affects balance of payments
(e) All of the above

301. Advantages of falling prices (Deflation) are
(a) Increase in real wages and income of workers
(b) A check on speculative activity in the economy
(c) Rising value of money
(d) Favorably affect the balance of payment
(e) All of the above

302. Falling prices involve several dangers. They are
(a) Bleak profit expectation of entrepreneurial class which adversely affects investment
(b) Declined investment causes recession or depression
(c) It may lead to unemployment and decline in National income
(d) All of the above

303. The inverse relationship between rate of unemployment and rate of inflation is explained by
(a) IS-LM curve
(b) Giriffth curve
(c) Phillips curve
(d) Consumption curve

304. The explanation of occurrence of a higher rate of inflation simultaneously with a higher rate of unemployment was provided by
(a) Milton Friedman
(b) a. W. Phillips
(c) a. c. Pigou
(d) Adam Smith

305. The alternative periods of expansion and contraction in economic activity is called
(a) Business cycles
(b) Business expansion
(c) Growth
(d) Recession

306. The four phases of business cycle are
(a) Expansion (Boom, Upswing and Prosperity)
(b) Peak (Upper turning point)
(c) Contraction (Downswing, Recession or Depression)
(d) Trough (Lower turning point)
(e) All of the above

307. The characteristic of expansion is
(a) Increased output and employment
(b) Decreased output and employment
(c) Constant output and employment
(d) None of the above

308. The characteristic of contraction and depression in economy is
(a) Fall in GNP and low level of employment
(b) Excess capacity with industries
(c) Fall in prices
(d) All of the above

309. The characteristic of trough and revival is
(a) Increased investment
(b) Increased consumption
(c) Both (a) and (b)
(d) Neither (a) nor (b)

310. Following are the features of a business cycle
(a) Business cycles occur periodically
(b) Business cycles are synchronic
(c) Consumption of non durable goods does not vary much during different phases
(d) Profits fluctuate more than any other type of income
(e) All if the above

311. Which of the following is not a feature of business cycle?
(a) Business cycles are international in character
(b) Depression an expansion affects inventories of goods
(c) Fluctuations occur at the level of production only
(d) Business cycles occur periodically

312. The sun spot theory of business cycle says
(a) Weather cycles cause fluctuations in agricultural output which in turn cause instability in economy
(b) Sun spots cause weather cycles
(c) Both (a) and (b)
(d) Neither (a) nor (b)

313. Sun-spot theory was given by
(a) Jevons
(b) R.G. Hawtrey
(c) J. a.Hobson
(d) F.a. Hayck

314. Hawtrey's theory of business cycle still holds valid because
(a) It shows how changes in money supply affects economic activity through changes in price level and rate of interest
(b) It shows how changes in money supply affects economic activity through changes in employment
(c) It shows supply creates its own demand
(d) It explains how a free enterprise economy could enter a long run economic slowdown

315. Under consumption theory of business cycle was propounded by
(a) Sismodi and Hobson
(b) Karl Marx
(c) R.G. Hawtrey
(d) F.a. Hayck

316. Karl Marx is known as the philosopher of
(a) Capitalism
(b) Socialism
(c) Business cycles
(d) None of the above

317. Over Investment theory of business cycle was propounded by
(a) Sismodi and Hobson
(b) Karl Marx
(c) R.G. Hawtrey
(d) Knur Wicksell

318. A systematic record of economic transactions of the residents of a country with the rest of the world during a given period of time is known as
(a) Balance of trade
(b) Balance of payments
(c) Balance of money
(d) Balance of economic activity

319. Balance of trade refers to the
(a) Difference of value of imports and exports of commodities only
(b) Difference of economic transactions of all kind- invisible and visible
(c) Difference of supply and demand
(d) None of the above

320. An example of visible and invisible items, to record for balance of payments is
(a) Crude oil and gold, respectively
(b) Crude oil and shipping, respectively
(c) Crude oil and Iron, respectively
(d) Tourism and insurance, respectively

321. Cyclical disequilibrium in Balance of Payments is caused by
(a) The fluctuations in the economic activity or what are known as trade cycles
(b) The technological changes
(c) The long run and deep seated changes in an economy
(d) When a change in demand or supply of exports alters a previously existing equilibrium

322. Technological disequilibrium in Balance of Payments is caused by
(a) The fluctuations in the economic activity or what are known as trade cycles
(b) The technological changes
(c) The long run and deep seated changes in an economy
(d) When a change in demand or supply of exports alters a previously existing equilibrium

323. Secular or Long run disequilibrium in Balance of Payments is caused by
(a) The fluctuations in the economic activity or what are known as trade cycles
(b) The technological changes
(c) The long run and deep seated changes in an economy
(d) When a change in demand or supply of exports alters a previously existing equilibrium

324. Technological disequilibrium in Balance of Payments is caused by
(a) The fluctuations in the economic activity or what are known as trade cycles
(b) The technological changes
(c) The long run and deep seated changes in an economy
(d) When a change in demand or supply of exports alters a previously existing equilibrium

325. The "Flow of Expenditures" approach is calculated by adding
(a) Consumption, investment, government spending, and net exports
(b) Real GDP plus Nominal GDP
(c) Wages, rents, interest, and profits

326. If the government wants to use monetary policy to fight inflation, it will
(a) Decrease the money supply
(b) Increase the money supply
(c) Decrease government spending
(d) Increase government spending

327. The Monetarists believe that inflation is caused by
(a) Printing too much money
(b) Printing too little money
(c) Counterfeiting money

328. If the quantity of money increases in the quantity theory of money equation, what must increase?
(a) Quantity
(b) Velocity
(c) Price

329. Suppose people spend 60 cents of every dollar of their disposable income and save 40 cents. What is the MPC?
(a) .60 (b) .20
(c) .40 (d) .80

330. The longer the term of a loan
(a) The higher the interest rate
(b) The lower the interest rate
(c) There is no difference in the interest rate

331. The money multiplier equals:
(a) 1/RR
(b) 1/MPS
(c) 1/RR – MPC

332. Which type of gap is monetary policy likely to be most effective at closing, according to the Keynesians?
(a) Inflationary gap
(b) Recessionary gap
(c) Stagflationary gap

333. Which type of unemployment requires the most targeted policies?
(a) Cyclical
(b) Frictional
(c) Persistent
(d) Structural

334. Cost-push inflation is caused by
(a) Too much money chasing too few goods
(b) Supply-side shocks like drought and oil price hikes
(c) A rise in the savings rate

335. The key policy implication of the Phillips Curve theory is that
(a) You can use expansionary policies to reduce unemployment and the only price will be a bit more inflation
(b) Keynesian fiscal policy will not be effective at stimulating the economy
(c) Keynesian monetary policy will not be effective at stimulating the economy

336. According to Keynesian economists, the traditional solution to an inflationary spiral is to
(a) Impose wage and price controls until inflation dissipates
(b) Increase the money supply
(c) Cut taxes

337. If the inflation rate was 5% last year, rational expectations would predict that inflation next year will be
(a) Cannot be determined from the information
(b) 10%
(c) 5%

338. Which of the following are elements of the GDP forecasting equation?
(a) Consumption
(b) Investment
(c) Government spending
(d) Net exports
(e) All of the above

339. What contributed to the slow growth of the United States during the 2000s?
(a) Wars in Iraq and Afghanistan
(b) Collapse of a housing bubble
(c) China's entry into the World Trade Organization
(d) All of the above

340. John Maynard Keynes believed that:
(a) Unemployment would be solved through government expenditures
(b) Unemployment would be solved through adjustments in wages and prices
(c) Unemployment would be solved by tax increases

341. The "real balance" or "wealth effect" means that as prices fall:
(a) Purchasing power rises
(b) Purchasing power falls
(c) Purchasing power stays the same

342. The Foreign Trade multiplier is equal to
(a) The reciprocal of marginal propensity to save (s) plus marginal propensity to import (m)
(b) The marginal propensity to save (s) plus marginal propensity to import (m)
(c) The reciprocal of marginal propensity to consume (b) plus marginal propensity to import (m)
(d) The reciprocal of marginal propensity to consume (b) plus marginal propensity to import (m)

343. Floating exchange rate is in which
(a) The value of the currency is allowed to adjust freely or to float as determined by demand for and supply of foreign exchange
(b) The value of the currency is fixed by the government
(c) The value of the currency is fixed by the central bank
(d) The value of the currency has nothing to do with the foreign exchange

344. Fixed exchange rate is in which
(a) The value of the currency is allowed to adjust freely or to float as determined by demand for and supply of foreign exchange
(b) The value of the currency is fixed by the government
(c) The value of the currency is fixed by the central bank
(d) The value of the currency has nothing to do with the foreign exchange

345. If the value of an Indian rupee in terms of US dollars changes from Rs 70.10 to Rs 69. 50, it is called
(a) Depreciation (b) Appreciation
(c) Stagnation (d) None of these

346. If the value of an Indian rupee in terms of US dollars changes from Rs 69.50 to Rs 71.10, it is called
(a) Depreciation
(b) Appreciation
(c) Stagnation
(d) None of these

347. The possible factors that could lead to a rise in demand for dollars by Indians are
(a) The Indian individuals, firms or Government who import goods form the USA into India
(b) The Indians travelling and studying in the USA would require dollars to meet their travelling and education expenses
(c) The Indian firms who wish to invest in building factories, sale facilities, shops in the USA
(d) All of the above

348. If a country's Consumer Price Index was 180 last year and 160 this year, which of the following happened
(a) The quantity of the goods produced has decreased
(b) The inflation rate increased
(c) The price level has decreased
(d) The unemployment has decreased

349. In the phase of expansion and prosperity
(a) Gap between potential and actual GNP is zero
(b) Level of production is the maximum production level
(c) Both (a) and (b)
(d) Neither (a) nor (b)

350. The industries that are hit hard during depression are
(a) Capital goods
(b) Durable consumer goods
(c) Both (a) and (b)
(d) Neither (a) nor (b)

351. Q1, 2018-19 growth rate of India's GDP is
(a) 8.2%
(b) 8.1%
(c) 8.3%
(d) 8.4%

352. India is the world's ....largest economy by nominal GDP and .... largest by purchasing power parity
(a) Second, fifth
(b) Third, second
(c) Fifth, fourth
(d) Sixth, third

ANSWER KEY (Managerial Economics and Farm Business Management)

1. (b) **2.** (a) **3.** (c) **4.** (c) **5.** (d) **6.** (d) **7.** (b) **8.** (a) **9.** (c) **10.** (c)
11. (c) **12.** (c) **13.** (c) **14.** (d) **15.** (b) **16.** (b) **17.** (c) **18.** (b) **19.** (a) **20.** (a)
21. (a) **22.** (c) **23.** (d) **24.** (c) **25.** (b) **26.** (c) **27.** (d) **28.** (d) **29.** (c) **30.** (c)
31. (c) **32.** (a) **33.** (c) **34.** (c) **35.** (b) **36.** (c) **37.** (a) **38.** (b) **39.** (a) **40.** (b)
41. (a) **42.** (c) **43.** (b) **44.** (d) **45.** (c) **46.** (d) **47.** (a) **48.** (a) **49.** (b) **50.** (a)
51. (a) **52.** (a) **53.** (b) **54.** (a) **55.** (b) **56.** (c) **57.** (c) **58.** (c) **59.** (b) **60.** (c)
61. (b) **62.** (a) **63.** (c) **64.** (c) **65.** (b) **66.** (b) **67.** (d) **68.** (b) **69.** (c) **70.** (d)
71. (b) **72.** (d) **73.** (b) **74.** (d) **75.** (d) **76.** (c) **77.** (b) **78.** (c) **79.** (b) **80.** (c)
81. (c) **82.** (c) **83.** (c) **84.** (d) **85.** (c) **86.** (a) **87.** (b) **88.** (d) **89.** (c) **90.** (b)
91. (b) **92.** (d) **93.** (b) **94.** (c) **95.** (d) **96.** (d) **97.** (b) **98.** (b) **99.** (c) **100.** (d)
101. (b) **102.** (d) **103.** (a) **104.** (b) **105.** (c) **106.** (b) **107.** (b) **108.** (d) **109.** (b) **110.** (c)
111. (c) **112.** (a) **113.** (b) **114.** (d) **115.** (b) **116.** (b) **117.** (b) **118.** (b) **119.** (a) **120.** (c)
121. (d) **122.** (b) **123.** (d) **124.** (b) **125.** (d) **126.** (c) **127.** (c) **128.** (b) **129.** (b) **130.** (b)
131. (b) **132.** (a) **133.** (b) **134.** (b) **135.** (c) **136.** (b) **137.** (a) **138.** (a) **139.** (b) **140.** (c)
141. (d) **142.** (c) **143.** (a) **144.** (d) **145.** (b) **146.** (a) **147.** (b) **148.** (c) **149.** (b) **150.** (c)
151. (d) **152.** (a) **153.** (b) **154.** (a) **155.** (c) **156.** (b) **157.** (c) **158.** (a) **159.** (c) **160.** (c)
161. (c) **162.** (d) **163.** (c) **164.** (c) **165.** (b) **166.** (a) **167.** (a) **168.** (b) **169.** (b) **170.** (a)
171. (c) **172.** (c) **173.** (b) **174.** (d) **175.** (d) **176.** (b) **177.** (b) **178.** (a) **179.** (c) **180.** (b)
181. (b) **182.** (c) **183.** (a) **184.** (d) **185.** (c) **186.** (a) **187.** (b) **188.** (b) **189.** (b) **190.** (b)
191. (d) **192.** (a) **193.** (b) **194.** (c) **195.** (d) **196.** (b) **197.** (c) **198.** (c) **199.** (a) **200.** (d)
201. (b) **202.** (d) **203.** (b) **204.** (a) **205.** (c) **206.** (a) **207.** (b) **208.** (c) **209.** (d) **210.** (b)
211. (d) **212.** (c) **213.** (a) **214.** (c) **215.** (d) **216.** (c) **217.** (a) **218.** (b) **219.** (d) **220.** (d)
221. (a) **222.** (c) **223.** (a) **224.** (a) **225.** (a) **226.** (d) **227.** (a) **228.** (a) **229.** (c) **230.** (a)
231. (c) **232.** (d) **233.** (a) **234.** (a) **235.** (a) **236.** (c) **237.** (e) **238.** (b) **239.** (a) **240.** (d)
241. (a) **242.** (b) **243.** (d) **244.** (a) **245.** (a) **246.** (d) **247.** (d) **248.** (b) **249.** (a) **250.** (d)
251. (a) **252.** (a) **253.** (a) **254.** (b) **255.** (a) **256.** (a) **257.** (a) **258.** (a) **259.** (a) **260.** (a)
261. (b) **262.** (c) **263.** (a) **264.** (d) **265.** (e) **266.** (c) **267.** (c) **268.** (d) **269.** (d) **270.** (b)
271. (d) **272.** (d) **273.** (a) **274.** (a) **275.** (a) **276.** (b) **277.** (d) **278.** (a) **279.** (b) **280.** (c)
281. (a) **282.** (d) **283.** (d) **284.** (d) **285.** (d) **286.** (d) **287.** (d) **288.** (d) **289.** (a) **290.** (a)
291. (b) **292.** (d) **293.** (c) **294.** (a) **295.** (b) **296.** (b) **297.** (c) **298.** (d) **299.** (a) **300.** (e)
301. (e) **302.** (d) **303.** (c) **304.** (a) **305.** (a) **306.** (e) **307.** (a) **308.** (d) **309.** (c) **310.** (e)

311. (c) **312.** (c) **313.** (a) **314.** (a) **315.** (a) **316.** (b) **317.** (d) **318.** (b) **319.** (a) **320.** (b)
321. (a) **322.** (b) **323.** (c) **324.** (d) **325.** (a) **326.** (a) **327.** (a) **328.** (c) **329.** (a) **330.** (a)
331. (a) **332.** (a) **333.** (d) **334.** (b) **335.** (a) **336.** (a) **337.** (a) **338.** (e) **339.** (d) **340.** (a)
341. (a) **342.** (a) **343.** (a) **344.** (b) **345.** (b) **346.** (a) **347.** (d) **348.** (c) **349.** (c) **350.** (c)
351. (a) **352.** (d)

10. Agribusiness Policy

1. Agreement on Textile and clothing was set out as a transitional process on .... for the ultimate removal of quotas by WTO by 01 Jan 2007
   (a) 01 Jan 1995
   (b) 01 Jan1996
   (c) 15 Jan 1995
   (d) 15 Jan 1996

2. The first ever set of multilateral, legally enforceable rules covering international trade in services is the........
   (a) General Agreement on Trade in services
   (b) General Agreement on Tariffs and Trade
   (c) Intellectual Property Rights
   (d) None of the above

3. Trade Related Investment Measures (TRIMS) apply for
   (a) Measures that affect trade in goods.
   (b) Discouraging measures that limit a company's trade.
   (c) Both of the above
   (d) None of the above.

4. General Agreement on Trade in Services is applicable to
   (a) Cross border supply of services
   (b) Trade across borders
   (c) Individuals traveling from own country to supply services in another
   (d) All of the above

5. As a part of WTO guidelines, Agreement on Agriculture (AOA) doesn't consider
   (a) Direct payments to farmers.
   (b) Indirect assistance and support to farmers by the Government
   (c) Domestic policies which directly effect on production and trade have to be cut back.
   (d) Least developed countries do not need to make any cuts.

6. .... is the only international organization dealing with the rules of trade between organizations.
   (a) GATT
   (b) WTO
   (c) IMF
   (d) World Bank

7. Quantitative restrictions refer to limit set by countries to curb
   (a) Imports
   (b) Exports
   (c) Imports & exports
   (d) None of the above

8. A Most Favored nation status doesn't necessarily refers to
   (a) Trade with favorite nation
   (b) Same and equal economic treatment
   (c) Non-discriminatory treatment
   (d) Same tariff rates applicable

9. The world trade organization was formed in the year __________ with GATT as it basis
   (a) 1993
   (b) 1994
   (c) 1995
   (d) 1996

10. Professional culture refers to
(a) shared belief of top managers about how they should manage about themselves and their employees
(b) Shared code of ethics and other commonalities shared by members of a given group
(c) Both (a) and (c)
(d) None of the above

11. Specific cultural dimensions that does not have a significant impact on cross national business interactions is
(a) Geography
(b) Monochromatic vs polychromatic tune
(c) Communication
(d) Religious beliefs

12. Theory of Comparative advantage demonstrates that mutually advantageous trade cannot occur when one trading partner has an absolute advantage
(a) False
(b) True
(c) Not absolutely true
(d) None of the above

13. Product life cycle theory states that, an exporter country of a particular product can become an importer country during the product cycle of a product
(a) False
(b) True
(c) Somewhat true
(d) Somewhat false

14. ------------ theory states that, lack of resources often helps countries to become competitive
(a) Competitive theory
(b) Theory of International Trade
(c) Theory of Mercantilism
(d) Product life cycle theory

15. Theory of Mercantilism propagates
(a) Encourage exports
(b) Encourage imports
(c) Discourage imports
(d) Encourage exports and discourage imports, both

16. Identify a factor that doesn't play an important role in attracting FDI
(a) Laws, rules and regulations
(b) Administrative procedures and efficiency
(c) Infrastructure related factors
(d) Language

17. The country that attract the largest FDI inflow is
(a) India (b) China
(c) USA (d) Brazil

18. B2B sector specific policy of Indian Government is
(a) FDI up to 100% subject to condition that the foreign company will invest 26 % in favor of Indian Public within 5 yrs
(b) FDI up to 100%
(c) FDI up to 50%
(d) None of the above

19. The term economies of scale refer to the reductions in cost achieved by producing a large volume of the product is
(a) False
(b) True
(c) Somewhat true
(d) Somewhat false

20. Contractual agreements as global market entry involve transfer of trademarks or human skills is
(a) False
(b) True
(c) Somewhat true
(d) Somewhat false

21. In 90's the global management perception was based on
(a) Standardization v/s adaptation
(b) Local responsiveness
(c) Global integration v/s Local Responsiveness
(d) of the above

22. CISF pricing includes the following
(a) Commission is paid to the agent involved in the transaction
(b) Cost of the goods exported
(c) Freight & Insurance charges for the goods exported
(d) All of the above

23. The basic objective of Export Promotion Council is to promote and develop the Exports of the
(a) High valued products of country
(b) Only attractive products and services of the country
(c) Only services and industry products of the country.
(d) Overall exports of the country

24. Ms. Christine Lagarde is the Managing Director of
(a) World Bank (b) IMF
(c) WTO (d) IDA

25. Countries experiencing balance of trade difficulties can avail overdraft from
(a) IDA (b) IMF
(c) WTO (d) World Bank

26. Increasing inequalities in developing countries is explained by the curve
(a) Lorenz curve
(b) IS-LM curves
(c) Gini Coefficient
(d) Kuznet's Curve

27. Kyoto Convention deals with procedures of
(a) Insurance (b) Aviation
(c) Defence (d) Environment

28. A Bill of Lading is issued by
(a) Carrier (b) Consignee
(c) Shipper (d) Indenting Agent

29. VAT is a/an
(a) income tax
(b) customs duty
(c) sales tax
(d) travel tax

30. Hague Protocol and Warsaw Conventions deal with
(a) air cargo
(b) marine cargo
(c) post parcel
(d) None of the above

31. EuroDollars are held
(a) Inside Europe
(b) Outside Europe
(c) In the USA
(d) Out of the USA

32. The main characteristic of international trade is that
(a) Goods must travel within the country
(b) Goods must be scrutinized before transport anywhere
(c) Taxes must be paid on goods crossing international boundaries
(d) Goods and/or services must cross international boundaries

33. Which of the following exceptions in the WTO rules does not permit a country to use a tariff or quota that is more restrictive than its bound tariff?
(a) Anti-dumping
(b) Countervailing duties
(c) Preferential trade agreements
(d) Safeguard measures

34. The "principal supplier" and "principal demander" are the
(a) Countries that negotiate tariff reductions on a particular product in a trade round
(b) Sources of the MFN and National Treatment cornerstones of the GATT
(c) Members of a dispute settlement panel that argue for and against the defendant
(d) Points at which supply and demand curves cross the axes

35. World Intellectual Property Rights (WIPO) deals primarily in
(a) Labor rights
(b) Tariffs and other trade barriers
(c) Patents, copyrights, Geographical indicators, and trademarks
(d) Trade and other economic transactions

36. The upper limit on the tariff that a country can levy on a particular good, according to its commitments under the GATT and WTO, is known as
(a) Advanced tariff
(b) Bound tariff
(c) Basic tariff
(d Education cess

37. The tariff is
(a) A tax on specific home-produced products;
(b) A general sales tax;
(c) A tax on imported goods
(d) A tax on holiday-makers

38. Which of the following is NOT a direct non-tariff barrier?
(a) A quota
(b) Exchange control
(c) An import deposit scheme
(d) None of the above

39. A system that limits citizens' ability to conduct transactions requiring the use of foreign currencies is known as
(a) A quota (b) Exchange control
(c) An embargo (d) A tariff

40. When the overall trade balance is zero
(a) Merchandise exports equal merchandise imports
(b) The service trade balance must be zero
(c) Exports equal imports
(d) All of the above

41. Over a given period of time if exports are greater than imports, the result is
(a) A trade war (b) A trade deficit
(c) An embargo. (d) A trade surplus

42. Increased opportunity to trade increases production by
(a) Protecting countries from competition
(b) Improving efficiency through specialization
(c) Improving efficiency through diversification
(d) None of the above

43. Which of the following is an argument for free trade?
(a) Lower consumption possibilities.
(b) The need to protect goods and services essential to national security.
(c) Greater efficiency through specialization.
(d) Aiding infant industries

44. If a country is completely self-reliant in producing goods for its own consumption needs, then
(a) It consumes more than it can with trade.
(b) Its consumption possibilities equal its production possibilities.
(c) It promotes specialization.
(d) None of the above

45. In the absence of trade, a country's consumption possibilities are
(a) Limited to its domestic production possibilities.
(b) Less than its trade balance.
(c) More than its terms of trade.
(d) Greater than with trade.

46. It is not likely that a country will specialize completely in one good even if it has a lower opportunity cost because
(a) Opportunity costs increase as less of a good is produced
(b) Opportunity costs increase as more of a good is produced.
(c) Opportunity costs decrease as more of a good is produced
(d) Opportunity costs decrease as less of a good is produced

47. A country can specialize in a good for which it has a low opportunity cost, if
(a) There is a significant demand in the country for that good.
(b) The country also has an absolute advantage in the production of that good.
(c) The production-possibilities curve is a straight line.
(d) None of the above

48. Two countries will have no incentive to trade in two goods, if
(a) The opportunity cost for both countries in producing two goods is same.
(b) One country has an absolute advantage in the production of both goods
(c) One country has a comparative advantage in the production of both goods
(d) None of the above

49. Consumption possibilities, during a given time period, refer to the
(a) The amount of good and services a country can consume at given prices
(b) The amount of goods and services a country can consume of imported products
(c) Alternative combinations of goods and services that a country can consume.
(d) Maximum amount that a country can consume if imports exceed exports

50. The benefits from international trade include
(a) Increased world output of goods and services.
(b) Optimum resource utilization to produce goods and services.
(c) Higher quality of goods and services produced.
(d) All of the above

51. Specialization increases the production of goods and services because
(a) Resources are used more efficiently.
(b) Increased supply decreases the prices and generates higher demand
(c) Incomes increase due to increased production that further leads to increased investment
(d) All of the above

52. Comparative advantage in production is achieved by:
(a) Subsidizing the price of an exported good.
(b) Producing a good with minimum resources available
(c) Having better terms of trade with other countries
(d) Having a lower opportunity cost of producing a good relative other countries

53. Comparative advantage refers to
(a) The ability of a country to produce more of specialized product
(b) The ability of a country to sell a certain good at higher cost than its cost of produ ction in other country
(c) The ability of a country to sell a certain good at lower cost than its cost of production in other country
(d) The ability of a country to produce a specific good at a lower opportunity cost than other countries

54. The expansion of world output as a result of trade is mainly due to the effects of
(a) Trade barriers.
(b) Improved terms of trade.
(c) Comparative advantage.
(d) Absolute advantage.

55. To say that a country has a comparative advantage in the production of milk is to say that
(a) It can produce milk with fewer resources than any other country can.
(b) Its opportunity cost of producing milk is greater than any other countries.
(c) Its opportunity cost of producing milk is lower than any other countries.
(d) The relative price of milk is higher in that country than in any other

56. Suppose the production of 100 tons of copper in India requires the same amount of resources as the production of 10 ton of Bauxite. In China, 100 tons of copper requires the same amount of resources as 20 tons of Bauxite. Implicitly that
(a) India has the comparative advantage in producing copper.
(b) China has an absolute advantage in producing Bauxite.
(c) India has an absolute advantage in producing copper.
(d) All of the above

57. Suppose 2 cars take the same resources as producing 2 mini trucks in India. But in Sri Lanka 2 cars take the same resources as 1 mini truck. In the production of cars we can conclude
(a) Sri Lanka has the absolute advantage.
(b) India has the absolute advantage.
(c) India has the comparative advantage.
(d) Sri Lanka has the comparative advantage

58. In China suppose 2 hours of labor produces either 4 bicycles or 3 water purifiers , but in Japan 2 hours produces either 3 bicycles or 2 water purifiers. Implicitly, Japan has
(a) Both an absolute and a comparative advantage in bicycles
(b) An absolute but not necessarily a comparative advantage in bicycles.
(c) A comparative but not necessarily an absolute advantage in bicycles.
(d) Neither a comparative nor an absolute advantage in bicycles

59. Absolute advantage refers to
(a) The ability of a country to produce a certain good with fewer resources than other countries
(b) Monopoly of one country in the production and sales of a certain good.
(c) The ability of a country to produce a specific good at a lower opportunity cost than its trading partners.
(d) All of the above

60. If the world trade would have been based on worker's productivity rather than comparative advantage:
(a) Production will have no impact.
(b) International trade would decrease.
(c) Trade will have no impact
(d) International trade would increase

61. Terms of trade refers to
(a) The opportunity costs incurred in trade.
(b) The rate at which goods are exchanged.
(c) Both of the above.
(d) None of the above.

62. To ensure that the trade is mutually beneficial, the terms of trade between two countries should always
(a) Between their respective opportunities costs in production.
(b) Be in favor of country with higher output
(c) Be in favor of country with lower comparative advanatge
(d) None of the above

63. If the terms of trade between any two countries is somewhat between their respective opportunity costs, then the effects of trade will include
(a) Higher specialization by both countries
(b) Consumption outside the production-possibilities curve for both countries.
(c) Less specialization by both countries.
(d) All of the above

64. A country will not trade unless
(a) It has an absolute advantage.
(b) The terms of trade are superior to domestic opportunities.
(c) Its balance of trade is in a surplus position.
(d) All of the above

65. Assume that country A can produce a unit of good X with 1 labor hour and a unit of good Y with 3 labor hours. Country B can produce a unit of good X with 2 labor hours and a unit of good Y with 6 labor hours. Labor is the only resource used. What would be the terms of trade for 1 unit of good X
(a) Between 2 and 3 units of Y
(b) Between 1/3 and 1/4 unit of Y
(c) Between 3 and 6 units of Y......
(d) Between 2 and 6 units of Y

66. With regard to international trade, the market:
(a) Provides a profit incentive to producers who have absolute advantage in production of goods and services.
(b) Provides a profit incentive to producers who have comparative advantage in production of goods and services.
(c) Determines the terms of trade.....
(d) All of the above

67. Suppose that Sri Lanka has a comparative advantage in coffee and India has a comparative advantage in tea. Which of the following groups would be worse off if these countries specialize and trade?
(a) Sri Lankan coffee producers.
(b) Indian coffee producers.
(c) Indian tea producers.
(d) All groups are better off when specialization and trade take place

68. Which group is most likely to be adversely affected by the importation of iron?
(a) Consumers of domestic automobiles which use iron as an input.
(b) Domestic iron workers.
(c) Domestic automobile workers.
(d) All of the above

69. Which of the following groups will not want free trade?
(a) Workers producing goods for export
(b) Producers in import-competing industries
(c) None of the above
(d) All of the above

70. increased imports will............, if trade restrictions are eliminated
(a) Will cause the supplies to sell below cost of production
(b) Increase the demand
(c) Shift the allocation of resources away from import-competing industries.
(d) Prices will crash

71. The elimination of import restrictions will
(a) Alter the mix of output from export industries toward domestic industries.
(b) Redistribute income from import-competing industries to export industries.
(c) Alter the mix of output from export industries to import-competing industries.
(d) Redistribute income from domestic to foreign producers

72. All the gains from specialization and trade minus all the losses would be
(a) Zero
(b) Positive
(c) Negative
(d) Depends

73. ....is more desirable from consumer's viewpoint
(a) Quantitative restrictions
(b) Tariffs on imported goods.
(c) Free trade.
(d) No trade.

74. International trade
(a) Lowers prices to consumers.
(b) Alters the mix of domestic production.
(c) Redistributes income from import-competing industries to export industries.
(d) All of the above.

75. Under the situation, when trade is mutually beneficial for two countries, increasing trade
(a) Increases the welfare of producers that compete with importers.
(b) Makes those countries less interdependent.
(c) Leads to increased output in export industries.
(d) All of the above.

76. A number of trade restrictions
(a) Increase a trade deficit in the short run.
(b) Alter a nation's production possibilities.
(c) Reduce total consumption possibilities.
(d) All of the above.

77. Dumping is defined as
(a) Foreign producers selling goods in India at prices lower than their marginal cost of production
(b) Foreign producers selling goods in India at prices lower than the Indian average cost of production
(c) Foreign producers selling goods in the India at prices lower than those prevailing in their own countries or below its cost of production
(d) None of the above

78. A country adopting to the policy of protectionism, means
(a) It is protecting infant industries on its land
(b) Greater specialization
(c) Protection of goods
(d) Protection of services

79. The infant-industry argument can be justified because
(a) A new industry may never develop in a protected, noncompetitive environment
(b) A new industry may be difficult to start in the face of existing foreign competition
(c) The government may not be able politically to end protectionism even when protectionism is no longer justified
(d) Government may choose to protect industries which place heavy import burdens on the economy

80. Which of the following is a barrier to trade?
(a) Tariffs (b) Embargoes
(c) Quotas (d) All of the above

81. An embargo is
(a) A prohibition on exports or imports
(b) A tax
(c) A quota
(d) A limit on imports

82. A country imposing tariffs means it aims at
(a) Increased quantities of imports
(b) Higher prices for the import-competing goods
(c) Lower prices for domestic production
(d) Less expensive exports

83. Tariffs reduce the volume of imports by
(a) Setting maximum allowable import limits
(b) Placing severe quality restrictions on imported goods
(c) Making them more expensive
(d) Reducing prices of domestically produced goods

84. Tariffs on imports cause
(a) Higher prices of the imported item
(b) Higher prices for the import-competing items
(c) Gains by workers of import-competing industries
(d) All of the above

85. When tariffs are imposed, the losers include
(a) Domestic consumers and the domestic government
(b) Foreign consumers and domestic producers of import-competing goods
(c) Domestic consumers and domestic producers of import-competing goods
(d) Domestic consumers and foreign producers

86. A country may achieve..................by imposing tariffs on imported goods
(a) Greater profitability of import-competing firms
(b) Higher exports
(c) Higher efficiency in domestic production
(d) Higher production possibilities

87. If tariff is imposed on imports, the equilibrium price and quantity in a market will
(a) Go up
(b) Equilibrium price should increase, and equilibrium quantity should decrease
(c) Equilibrium price should decreases, and equilibrium quantity should increase
(d) Can't say

88. A quota is
(a) A prohibition against trading a good
(b) A limit on the quantity of a good that may be imported
(c) A tax imposed on imported goods
(d) An elimination of trade to nurture an infant industry

89. Quotas pose a greater threat than tariffs because
(a) Quotas allow imports but only at a higher price
(b) Tariffs do not reduce the quantity sold and quotas do
(c) Tariffs are voluntary and quotas are not
(d) Quotas preclude additional imports at any price

90. Which of the following is not a non-tariff barrier?
(a) A quota on milk powder
(b) A voluntary export restraint on cars
(c) A tax equal to 25% of value on crude oil
(d) A protectionist policy

91. When quotas are eliminated, losers include
(a) There are no losers when free international trade is established
(b) Domestic producers
(c) Foreign producers
(d) Domestic consumers

92. The General Agreement on Tariffs and Trade (GATT)
(a) Encourages trade restrictions
(b) Established the European Common Market
(c) Committed GATT members to a reduction of protectionism
(d) Has now eliminated all nontariff barriers to trade

93. A principal objective of GATT is to
(a) Protect domestic producers
(b) Reduce barriers to trade
(c) Increase employment and income
(d) Increase imports

94. The objective of the World Trade Organization (WTO) is to
(a) Protect the local industry in any country
(b) Enforce the rules of free trade
(c) Protect producers from foreign competition
(d) Ensure there are no losers as a result of free trade

95. If exports are being excluded unfairly from a market, the World Trade Organization (WTO) may authorize
(a) Quotas.
(b) Retaliatory comparative advantages
(c) Retaliatory tariffs
(d) Other non-tariff barriers

96. The BRIC nations are
(a) Brazil, Russia, India, and China
(b) Bulgaria, Russia, Istanbul and Catan
(c) Bolivia, Russia, Ireland, and Cyprus
(d) Bahrain, Russia, Iberia, and Ceylon

97. Protectionist policies include all except
(a) non-tariff barriers
(b) Tariffs
(c) martial law
(d) buy local" policies

98. Free Trade area is an arrangement in which countries give each other .... by eliminating tariffs and other barriers on goods.
(a) Discriminatory treatment
(b) Treatment based on competitive advantage
(c) Preferential treatment
(d) Treatment based on absolute advantage

99. ....is the most advanced form of a trade bloc.
(a) free trade area
(b) Confederation
(c) customs union
(d) common market

100. Regional trading blocs such as the EU and NAFTA were established to ___ trade between the member countries.
(a) Increase (b) Liberalize
(c) Smoothen (d) Stagnate

101. The concept of fair trade is _____.
(a) An approach to global trade aiming to ensure that producers in developing countries receive a fair price for their products .
(b) Synonymous with free trade.
(c) Requires a market clearing or equilibrium price set by market forces.
(d) A dynamic of the law of comparative advantage in which countries produce with their strengths.

102. The PESTLE model includes all of the following except ______.
(a) Production
(b) Political
(c) Economic
(d) Environmental

103. High tech products are produced by research-intensive industries using _____.
(a) the standardized technology available
(b) the most advanced technology available
(c) the widely available technology
(d) the adaptive technology available

104. Post globalization, the term global economy is more referred to and that means that the world economies have now become more ____.
(a) Detached (b) Interconnected
(c) liberalized (d) United

105. By definition, the Green Economy must be:
(a) environmentally sustainable
(b) socially just
(c) locally rooted
(d) All of the above

106. Which of these is not typically analyzed as part of the economic environment?
(a) exchange rates
(b) legal systems
(c) balance of payments
(d) taxation

107. Which of the following is not a part of cultural environment?
(a) national income
(b) national character
(c) time orientation
(d) space orientation

108. Trademark covers all of the following except that
(a) Trademarks can be renewed forever as long as they are used
(b) Trademark protection requirements vary by country
(c) Trademarks protect the words, names, color or symbols that distinguish goods
(d) Trademarks always cover similarly named goods and services

109. A patent law means all of the following except that
(a) It provides protection giving the exclusive right to make, use or sell a concept or invention for its duration.
(b) Patents can be renewed forever as long as they are used.
(c) Patent agreements seek to simplify the process.
(d) Patent agreements seek to grant patent rights in all of the member nations that are party to the agreement.

110. Which is the correct about abut copyright?
(a) Copyright can extend past the death of the author.
(b) Copyright dies with the author of the work.
(c) Copyright covers inventions and brand names.
(d) Copyright law is uniform worldwide.

111. Trade secrets may be protected by
(a) Revocation of the professional licenses of disclosing employees.
(b) Formal "trade secret" registration that retains the "trade secret" status.
(c) Nondisclosure agreements and injunctions.
(d) None of the above

112. Public Distribution System(PDS) is operated under the responsibility of the
(a) Central Government
(b) State Government
(c) Both of the above
(d) None of the above

113. Which of the following agricultural commodity of India gives largest in terms of export value?
(a) Spices
(b) Basmati Rice
(c) Meat and Meat preparations
(d) Cotton

114. Arrange the following in the descending order of the largest agricultural imports of India
(a) Cashewnuts , B. Wood and Wood Products , C. Edible oils
(a) A-B-C (b) A-C-B
(c) B-A-C (d) C-B-A

115. Public Distribution System (PDS) means distribution of essential commodities to a large number of people through a network of Fair Price Shops on a recurring basis. The Commodities distributed under PDS are: (1) Wheat (2) Rice (3) Sugar (4) Pulses (5) Kerosene Select the correct option:
(a) 1 ,2 and 4 (b) 1, 2, 3 and 4
(c) 1, 2, 3 and 5 (d) All of the above

116. The central nodal agency for implementing the price support operations for commercial crops is
(a) NAFED
(b) NABARD
(c) TRIFED
(d) FCI

117. International trade brings the benefit of
(a) Economic growth for all countries.
(b) More self-sufficiency.
(c) Greater equality between countries.
(d) Specialization and the resulting economies of scale.

118. A significant portion of actual world trade patterns results from
(a) The industrial policies of governments.
(b) The different tastes and preferences of people in different countries.
(c) Different sizes of the countries.
(d) Different factor endowments between countries.

119. David Ricardo gave the theory of
(a) Absolute advantage.
(b) Mutual advantage.
(c) Comparative advantage.
(d Multilateral advantage.

120. A country has a comparative advantage in the production of a product if the good's________cost is different from the good's ________ cost in another country.
(a) Cost of production, opportunity
(b) Cost of production, cost of production
(c) opportunity, opportunity
(d) Opportunity, cost of production

121. Terms of trade for a country are the ratio of __________ to __________.
(a) its currency, other currencies
(b) its opportunity costs, world opportunity costs
(c) value of exports, value of imports
(d) export prices, import prices

122. One of the main advantages of trade, economists suggest, is
(a) Lower tariffs.
(b) Development of tourism.
(c) Competition with foreign suppliers.
(d) Technological change.

123. The term tariff, as used in international trade, refers to
(a) A government payment to encourage exports.
(b) The price of goods when they leave the producing country.
(c) A limit on the quantity of a good that can be imported into a country.
(d) A tax on imports.

124. If India places a limit on the amount of milk powder that could be imported into India in a particular period will be called as imposition of
(a) Dumping.
(b) An export subsidy.
(c) A quota.
(d) A tariff.

125. Dumping occurs when Indian goods are sold for ________ abroad than ______.
(a) less, their Indian cost of production
(b) less, their Indian price
(c) more, their Indian cost of production
(d) more, their Indian price

126. Tariffs are used to protect "Infant Industries", these industries are those in which
(a) Are not large enough to achieve economies of scale.
(b) Employ many young or untrained workers.
(c) Are competing with well-established overseas firms.
(d) Use a new technology.

127. If a group of countries abolish trade barriers between them and set the same tariffs on goods coming in from other countries they are a
(a) Common market.
(b) Free trade area.
(c) Federation.
(d) Customs union.

128. The member countries trade with each other grows faster in free trade due to
(a) Trade creation.
(b) Tradechanneling.
(c) Trade diversion.
(d) None of the above

129. The size of the trade bloc has the effect of improving countries'
(a) Balance of payments.
(b) Absolute advantage
(c) Terms of trade.
(d) Comparative advantage.

130. Developing countries have:
(a) A lower infant mortality rate.
(b) A smaller percentage of the urban labour force
(c) AHigher literacy rate.
(d) A low investment on food.

131. Import substitution policy is administered by several countries when
(a) The prices of the imported were falling.
(b) Terms of trade were improving.
(c) The prices of the goods exported goods were falling.
(d) Terms of trade were deteriorating.

132. For development to occur, it is necessary for an economy to
(a) Stress the importation of agricultural products and the export of manufactured goods.
(b) Promote agricultural development
(c) Promote industrial development.
(d) Adopt strategy that promotes both agricultural and industrial development.

133. ....led to increase in agricultural output in developing countries is
(a) Land reforms
(b) Abolition of intermediaries in value chain
(c) Abolition of zamindari
(d) All of the above

134. Land reform could increase agricultural output because
(a) Farmer was no more exploited by landlords.
(b) Farmer was not required to pay rent on land/ cultivate land on lease
(c) Farmer was the owner of the land and so was able to take his own decisions related to production
(d) All of the above

135. Gross National Product comprises
(a) Current account balance of a country
(b) Consumption and Investment and net exports.
(c) Consumption and net exports and government expenditure.
(d) Consumptionand savings.

136. Balance of payments
(a) Provides detailed information concerning the demand and supply of a country's currency.
(b) Can be used to evaluate the performance of a country in international economic competition.
(c) None of the above
(d) Both a and b

137. If a country is grappling with a major balance-of-payment difficulty, it may not be able to expand imports from the outside world. Instead, the country may be tempted to
(a) Impose measures to restrict imports.
(b) Impose measures to discourage capital outflows.
(c) Both of the above
(d) None of the above

138. If India exports more than it imports, then
(a) The supply of dollars is likely to exceed the demand in the foreign exchange market, ceteris paribus.
(b) One can infer that the rupee would be likely to appreciate against othe currencies.
(c) Both of the above
(d) None of the above

139. The balance of payments covers
(a) Exports and imports.
(b) only cross-border investments
(c) both of the above
(d) none of the above

140. A country having surplus balance of payment means it has
(a) Expanded imports
(b) not posed restriction on foreign exchange
(c) Both of the above
(d) None of the above

141. The current account includes
(a) All purchases and of stocks, bonds, debentures, etc
(b) The export and import of goods and services.
(c) All purchases of gold and debt incurred
(d) All of the above

142. The current account deficit implies that a country
(a) Saved less and invested more.
(b) Produced more and consumed less.
(c) Consumed more and produced less
(d) Saved more and invested less

143. The factors of production are
(a) land, labor, capital, and entrepreneurial ability.
(b) interest, wages and dividends.
(c) Both a & b
(d) Neither a nor b

144. A currency depreciation will begin to improve the trade balance immediately
(a) If the demand for imports and exports are inelastic.
(b) If the demand for imports and exports are elastic.
(c) If imports decrease and exports decrease.
(d) none of the above

145. When a country's currency depreciates against the currencies of major trading partners,
(a) The country's exports tend to rise and imports fall.
(b) The country's exports tend to fall and imports rise.
(c) The country's exports tend to rise and imports rise.
(d) The country's exports tend to fall and imports fall.

146. A depreciation will begin to improve the trade balance immediately if
(a) Imports and exports are responsive to the exchange rate changes.
(b) Imports and exports are inelastic to the exchange rate changes.
(c) Consumers exhibit brand loyalty and price inelasticity.
(d) None of the above

147. In the long run, both exports and imports tend to be
(a) Unresponsive to changes in exchange rates.
(b) Responsive to changes in exchange rates.
(c) both of the above
(d) None of the above

148. The Central Bank of India is
(a) The Bank of India.
(b) The Reserve Bank of India.
(c) The EXIM bank.
(d) The State Bank of India.

149. The vast majority of the foreign-exchange reserves held by central banks are denominated in
(a) Local currencies (b) U.S. dollars
(c) Yen (d) Euro

For questions 152-155, the notation is
Y = GNP = national income,
C = consumption, I = private investment, G = government spending, X = exports, M = imports, S= savings

150. The current account balance is given by
(a) C + I + G + X + M (b) X – M
(c) I + X + M (d) M – X

151. The difference between a country's savings and investment is given by
(a) S – I (b) I x S
(c) X – M (d) GNP – Y

152. If the difference between tax revenue and government expenditures is negative, it implies that
(a) Tax revenue is insufficient to cover government spending.
(b) A government budget deficit exists.
(c) The government will be issuing new debt securities.
(d) all of the above

153. National income, or Gross National Product is given by
(a) $GNP \equiv Y \equiv C + I + G + X + M$
(b) $GNP \equiv Y \equiv C + I + G + X - M$
(c) $GNP \equiv I \equiv C + Y + G + X - M$
(d) $GNP \equiv Y \equiv C + I + X + M - G$

154. Minimum Support Prices are fixed on the recommendations of
(a) Ministry of Consumer Affairs, Food and Public Distribution
(b) Planning Commission
(c) Ministry of Agriculture
(d) Commission for Agricultural Costs and Prices

155. Minimum Support Price are declared for ....crops
(a) 24 (b) 25
(c) 27 (d) 28

156. MSP for jute in 2016-17 was
(a) Rs 3200/ quintal
(b) Rs 3400/ quintal
(c) Rs 3500/quintal
(d) Rs 3600/quintal

157. CACP recommended 2016-17 MSP for tur at ....per quintal. The government, however, fixed the same at ....per quintal
(a) 3850, 4300 (b) 3850, 3850
(c) 4300, 4300 (d) 4625, 5050

158. Which of the following is a zaid crop?
(a) Watermelon
(b) Soybean
(c) Maize
(d) Sesame seed

159. Golden Rays Project was launched in ....to increase the productivity of ....
(a) 2009, Maize (b) 2008, Rice
(c) 2009, Barley (d) 2011, Sorghum

160. National Agriculture Policy envisages to achieve agricultural growth rate of
(a) 2% (b) 3%
(c) 4% (d) 4.5%

161. Minimum Support prices were first announced in
(a) 1980-81 (b) 1975-76
(c) 1966-67 (d) 1970-71

162. MSP was first announced by the government of India in 1966-67, for which of the following crop/s
(a) Wheat
(b) Wheat, rice
(c) Rice
(d) Wheat, rice, maize, and soybean

163. Heckscher-Ohlin theory explains
(a) The law of comparative advantage
(b) Pattern of trade in terms of relative factor endowments of countries
(c) The law of comparative cost
(d) None of the above

164. Uruguay Round of multilateral trade negotiations
(a) Aimed at reversing the rising trend of non-tariff trade barriers
(b) Replaced the GATT with WTO
(c) Brought services and agriculture into WTO
(d) All of the above

165. Hedging and speculation in financial markets
(a) Are opposite activities
(b) Parallel activities
(c) Same activities
(d) None of these

166. Public-Private Partnership projects in India are mostly related to
(a) Airports
(b) Ports
(c) Railway
(d) Roads

167. RashtriyaKrishiVikasYojana was launched by GOI in August 2007 with the objective(s) of
(a) To incentivise the states to increase public investment in agriculture and allied sectors
(b) To ensure that local needs/crops/priorities are better reflected
(c) To achieve the goal of reducing the yield gaps in important crops
(d) All of the above

168. The total central pool stock of foodgrains (Rice and Wheat) at the beginning of November 2016 stood at
(a) 548.10 lakh MT
(b) 313.62 lakh MT
(c) 453.10 lakh MT
(d) 662.80 lakh MT

169. Public Private Partnership came to be introduced in India to
(a) Mitigate the financial burden of the governments
(b) Respond to the technological change that allow unbundling of infrastructure
(c) Alter the character of goods and services from pure public to private ones
(d) All of the above

170. Privatization refers to a range of policies including
A) Changing part of a public enterprise's ownership to the private sector
B) Liberalization of entry into activities previously restricted to the public sector
C) Two infant industries merging into a monopoly
D) Franchising or contracting public services or leasing public assets to the private sector
(a) C only.
(b) D only.
(c) A, B and D only.
(d) None of the above.

171. Which of the following is not a public policy to promote the private sector?
(a) Promoting market monopolies and cartel formation.
(b) Promoting investment in market research
(c) Providing adequate infrastructure for public and private agencies
(d) Developing markets.

172. The market efficiently allocates scarce resources among alternative ends such that
(a) Consumers receive goods for which they are willing to pay.
(b) Production resources hire out to maximize income.
(c) The market determines available labor and capital.
(d) All of the above

173. The real exchange rate is
(a) The effective exchange rate
(b) The nominal exchange rate adjusted for relative inflation rates at home and abroad.
(c) The USD value divided by WPI
(d) The CPI divided by USD

174. Overvaluation of Rupee relative to the US Dollar will
(a) Always leads to a balance of trade.
(b) Discourage import substitution and exports.
(c) Encourage exports and discourage imports.
(d) Create a trade surplus in the local market.

175. The buffer stock management helps in
(a) Buying and Selling of commodities to maintain prices within a certain range.
(b) Stabilizing exports.
(c) Subsidizes commodities that are important for national security reasons.
(d) None of the above

176. The Organization of Petroleum Exporting Countries (OPEC) is a ____ whose members have agreed to limit output and fix prices.
(a) Duopoly.
(b) Generalized system of tariff preferences.
(c) Monopoly
(d) Cartel.

177. If export prices increase7 percent and import prices 15 percent, the commodity terms of trade is
(a) 2.14. (b) 0.93.
(c) 0.25. (d) 0.46

178. The commodity terms of trade equals the
(a) (exports prices minus import prices)/ exchange rates.
(b) exchange rates of country i divided by exchange rates of country *j*.
(c) external balances/ balance of payments.
(d) price index of exports divided by the price index of imports.

179. The selling by Hyundai of its car cheaper in the United States than at home (Korea) is known as
(a) Subsidized imports.
(b) Dumping.
(c) Internal economies of scale.
(d) Export substitution.

180. The infant industry arguments refers to a tariff designed to
(a) Help foreign industries establish themselves in the local market.
(b) Protect young manufacturing products from foreign competition.
(c) Help consumers enjoy a variety of products in the local market.
(d) Provide incentives for established local manufacturing firms to venture in foreign markets.

181. International trade and specialization are determined by
(a) Absolute advantage.
(b) Comparative advantage.
(c) Absolute costs.
(d) Production possibility frontier.

182. Some economists and third world policy makers criticize MNCs, arguing that they have a negative effect on the developing country because they
(a) Increase the LDC's technological dependence on foreign sources, resulting in less technological innovation by local workers.
(b) Hamper local entrepreneurship and investment in infant industries.
(c) Increase unemployment rates from unsuitable technology.
(d) All of the above

183. MNCs can help the developing country to
I Finance a savings gap or balance of payments deficit.
II Obtain foreign technology and innovative methods of increasing productivity.
III Generate appropriate technology by adapting existing processes.
IV Employ domestic labor, especially in skilled jobs.
(a) I and II only
(b) III and IV only
(c) I, II and III only
(d) All of the above

184. The greenhouse effect is the phenomenon by which
(a) The earth's atmosphere traps infrared radiation.
(b) The earth's atmosphere traps carbon do oxide.
(c) The earth's atmosphere traps chloflurocarbons.
(d) None of the above

185. An economy is said to be known as capitalistic
(a) Where private owners of capital make decisions based on profit.
(b) Government interventionis more in the means of production.
(c) Which is more or less the same as prevails in the United States
(d) Which is more or less the same as prevails in India

186. Monopoly emerges when a firm has
(a) Access to more economic information than competitors.
(b) Superior access to training and education.
(c) A lower discount of future earnings.
(d) All of the above.

187. Which of the following was a socialist country?
(a) Holland
(b) China
(c) Soviet Union
(d) Britain.

188. Which of the following is a form of urban bias?
I Government may set price floors on food and price ceilings for industrial goods.
II Tax incentives and subsidies to infant-industry.
III Tariff and quota protection for industry.
IV Spending more for education, training, housing, health and transport in urban areas than in rural areas.
(a) I, II, III only.
(b) I, II and IV only.
(c) II, III and IV only.
(d) I, II, III and IV

189. The goal of integrated pest management (IPT) is to
(a) Reduce yield losses by pests while minimizing the negative effects of pest control.
(b) Have year-round plantings of a single crop.
(c) Undertake monoculture pest planning.
(d) Encompass biological control through fertilizers.

190. Economists in India, Pakistan, the Philippines, and Mexico argue that foodgrain growth would not have kept up with population growth in the last four decades without

I the improved packages of high-yielding seed varieties.

II fertilizers, pesticides and irrigation.

III improved transportation.

IV better extension service.

(a) I and II only.
(b) II and III only.
(c) I, II and III only.
(d) I, II, III and IV.

191. Following are the international agricultural research centers

I International Center for the Improvement of Maize and Wheat (CIMMYT).

II International Rice Research Institute (IRRI).

III Synthetic Rubber Research Institute (SRRI).

IV Center for International Agricultural Production Control (CIAPC).

(a) I only.
(b) I and II only.
(c) III and IV only.
(d) IV only.

192. The development of high yielding varieties (HYVs) of wheat and rice is known as

(a) The agribusiness revolution.
(b) Farming system theory.
(c) The Green Revolution.
(d) agri-R&D.

193. About __________of the world's population lives in LDCs.

(a) 80%.
(b) 50%.
(c) 25%.
(d) 35%.

194. The vicious circle theory states that

(a) The factors of production are underutilized.
(b) Low income levels restrict consumption
(c) Low income levels promote consumption of substitutes
(d) Low per capita income creates low savings that results in low investments

195. Which of the following statement is true about low-income countries?

(a) Less than 10% of the labor force is in agriculture.
(b) Large population share a small income from agriculture
(c) Agriculture labor is more than 50%
(d) Share of labor force is about 30%.

196. Dual economies are countries

(a) with double capital and labor.
(b) With a modern manufacturing sector as well as traditional agriculture sector.
(c) That specializes in labor-intensive products more than capital-intensive products.
(d) With foreign-owned and domestically-owned capital.

197. The SverigesRiksbank Prize in Economic Sciences in Memory of Alfred Nobel 2013 was awarded to

(a) Eugene F. Fama, Lars Peter Hansen and Robert J. Shiller
(b) Peter A. Diamond, Dale T. Mortensen and Christopher A . Pissarides
(c) Leonid Hurwicz, Eric S. Maskin and Roger B. Myerson
(d) Robert J. Aumann and Thomas C. Schelling

198. The Sveriges Riks bank Prize in Economic Sciences in Memory of Alfred Nobel 2013 was awarded to Eugene F. Fama, Lars Peter Hansen and Robert J. Shiller, for :
(a) empirical analysis of asset prices
(b) analysis of markets with search frictions
(c) having laid the foundations of mechanism design theory
(d) having enhanced our understanding of conflict and cooperation through game-theory analysis

199. The Food and Agriculture Organization of the UN advises that an efficient and adequate marketing system is a .... for agricultural diversification, providing better prices to producers and the availability of competitively priced produce to consumers.
(a) precondition
(b) precursor
(c) predecessor
(d) preference

200. Rural markets form part of a trade network; they are normally arranged on a ....basis on specific weekdays and are commonly organised at a central place in a village or district centre.
(a) ad hoc
(b) casual
(c) irregular
(d) periodic

201. .... wholesale and semi-wholesale markets are located within or near major cities (usually with populations exceeding 0.5 million).
(a) Seasonal (b) Subsidized
(c) Temporary (d) Terminal

202. Retail markets directly serve consumers in main urban areas; although primarily retail, they may have some semi-wholesale functions, particularly if they allow producers to trade – in which case, they are often called .... markets.
(a) direct (b) farmers
(c) open-air (d) produce

203. Internal factors for change include increasing volumes of produce to be handled; alterations to commercial practices and trading patterns; and the .... of professional specialised wholesalers.
(a) convergence
(b) conveyance
(c) emergence
(d) reliance

204. Which of the following can be the underlying for a commodity derivative contract?
(a) Interest Rate (b) Euro-Indian Rupee
(c) Guarseed (d) NIFTY

205. Daily mark to market is done ________________
(a) Till the date of expiry of the contract
(b) On the last day of week
(c) On the last trading day of the month
(d) All of the above

206. The actual process of exchanging money and goods is termed as.................
(a) Transfer (b) Clearing
(c) Settlement (d) Netting

207. ___________ work at making profits by taking advantage of discrepancy between prices of the same product across different markets.
(a) Arbitragers
(b) Speculators
(c) Exchange
(d) Hedgers

208. Commodity exchanges enable producers and consumer to hedge their _______ given the uncertainty of the future.
(a) price risk (b) seasonal risk
(c) weather risk (d) production risk

209. _________can only trade through their account or on account of their clients.
(a) Trading Member
(b) Trading cum Clearing Member
(c) Both of the above
(d) None of the above

210. The cash settlement is only for the incremental gain/ loss as determined on the basis of ______.
(a) Final settlement price
(b) Average price for the day
(c) Opening price.
(d) Last traded price

211. __________ is the last day on which the futures contract will be traded, at the end of which it will cease to exist.
(a) Pay off Date
(b) Exercise Date
(c) Expiry Date
(d) Maturity Date

212. "A ________ option gives the holder the right but not the obligation to buy an asset by a certain date for a certain price."
(a) Put
(b) At the money
(c) In the money
(d) Call

213. An ________ option is an option that would lead to a zero cash flow to the holder if it were exercised immediately.
(a) Call
(b) At the money
(c) In the money
(d) Put

214. The total number of outstanding contracts (long/short) at any point in time is called ____________.
(a) Hedge Limit
(b) Open Interest
(c) Volume
(d) Delivery Lot

215. On introduction of new contracts, the base price is the ________ of the underlying commodity in the prevailing spot markets.
(a) Last open price
(b) Last closing price
(c) Decided by participants
(d) Decided by exchange

216. ____ refers to issue of physical delivery against the credit in the demat account
(a) Re-materialisation
(b) Securitisation
(c) De-materialisation
(d) Liquidation

217. The difference between futures contracts and other contract arrangements is that futures
(a) are standardized, negotiable, and traded on an organized exchange like the NCDEX
(b) are offered at any local APMC market
(c) are available only for metals and energy
(d) none of these

218. A Basis
(a) determines the outcome of the hedge position when a futures position is liquidated
(b) is the difference between a futures contract price and the local cash price for the commodity
(c) None of the above
(d) a & b

219. The amount of the basis
(a) is most critical as the contract approaches maturity
(b) is affected by local conditions
(c) None of the above
(d) a & b

220. The characteristic of a futures market and cash market price relationship that makes hedging feasible is
(a) basis is always constant throughout the life of the contract
(b) losses in the cash market are offset by losses in the futures market
(c) their tendency to converge as the contract approaches maturity
(d) all of the above

221. A difference in using options for price protection compared with hedging or cash contracting is
(a) an option position establishes a minimum selling price or maximum purchase price, but leaves the buyer in a position to benefit from favorable price changes
(b) hedging and cash contracting set an approximate selling price, regardless of later price changes
(c) there is no cost in using options unlike hedging or cash contracting
(d) a and b

222. Option premiums are
(a) paid up front
(b) made up of intrinsic value and time value
(c) the cost of buying specific options
(d) all of the above

223. The Agricultural and Processed Food Products Export Development Authority Act came into effect on
(a) 13th March 1986
(b) 13th August 1986
(c) 13th February, 1986
(d) 13th November, 1986

224. The Agricultural and Processed Food Products Export Development Authority (APEDA) , replaced........in 1986
(a) Processed Food Export Promotion Council (PFEPC)
(b) Agricultural and Processed Food Products Export Development Council (APEDC)
(c) Agricultural and Processed Food Promotion Council (APFPC)
(d) Agricultural and Processed Food Export Promotion Council (APFEPC)

225. Major function of APEDA, include
(a) Improving of packaging of the Scheduled products
(b) Improving of marketing of the Scheduled products outside India
(c) Promotion of export oriented production and development of the Scheduled products
(d) All of the above

226. APEDA has been entrusted with the responsibility to monitor import of ....
(a) Wheat (b) Crude vegetable oil
(c) Coarse grains (d) Sugar

227. The EU banned the import of Indian Alphonso Mangoes with effect from May 1, 2014 due to
(a) High chloropyriphos levels
(b) Contamination with fruit flies
(c) Low demand
(d) Poor quality of product

228. Which country had the largest share in Fresh Mango imports from India in 2015-16
(a) United Arab Emirates
(b) Pakistan
(c) Australia
(d) United States

229. In 2014-15, which of the following agricultural produce earned the maximum foreign exchange for India?
(a) Buffalo meat (b) Non Basmati Rice
(c) Basmati Rice (d) Guargum

230. In 2015-16, which of the following agricultural produce earned the maximum foreign exchange for India?
(a) Meat and Preparations
(b) Guargum
(c) Non Basmati Rice
(d) Cereals

231. Which of the following financial schemes is launched by APEDA?
(a) Schemes for Market Development
(b) Schemes for Infrastructure Development
(c) Schemes for Quality Development
(d) All of the above

232. Which of the following belt is an agri export zone for oranges in Madhya Pradesh?
(a) Chhindwara, Hoshangabad, Betul
(b) Shivpuri, Guna, Vidisha, Raisen, Narsinghpura, Chhindwara
(c) Guna, Mandsaur, Ujjain, Rajgarh, Ratlam, Shajapur and Neemuch
(d) Indore, Dhar, Shajapur and Dewas / Bhopal

233. Top five importers of walnuts from India are (in 2015-16):
(a) United Kingdom, Egypt Arab Republic, Netherland, Germany, United States.
(b) United Arab Emirates, United Kingdom, Saudi Arabia, Qatar, Kuwait
(c) Germany, UK, Netherland, Egypt Arab Republic, United States
(d) United States, United Arab Emirates, Australia, Saudi Arabia, Bangladesh

234. The term agribusiness was coined by
(a) John Herbert Davis & Ray Allan Goldberg
(b) John Herbert Davis
(c) Ray Allan Goldberg
(d) John Wilkinson

235. A country focus on agribusiness, because
(a) Post liberalization, emphasis is upon bringing dynamism to agriculture sector
(b) Limited presence of agro industry results in low value addition in agricultural commodities making a country less competitive internationally
(c) Strong forward and backward linkages in agriculture, along with right and timely policy intervention is required to address food security issues.
(d) All of the above

236. Which of these is not an external factor affecting business environment
(a) Social
(b) Physiological
(c) Legal
(d) Technological

237. Legal Factors affecting business environment include
(a) discrimination law
(b) consumer law
(c) employment law
(d) all of the above

238. The environment matters for conducting a business because
(a) The environment provides resources that an organization needs in order to create goods and services.
(b) The environment is a source of opportunities and threats for an organization
(c) The environment shapes the various strategic decisions that executives make as they attempt to lead their organizations to success
(d) All of the above

239. India is a
(a) Middle income country
(b) Low income country
(c) High income country
(d) Average income country

240. As per the current EXIM policy of the government of India, the registered quantity for export of sugar from India is upto
(a) 25 thousand metric tons
(b) 35 thousand metric tons
(c) 50 thousand metric tons
(d) There is no limit on the registered quantity for export of sugar

241. Following quantities of pulses is being allowed for export to the Republic of Maldives under bilateral trade agreement between Government of India and Government of Maldives during the period 2015-16
(a) 87.85 metric tons
(b) 96.63 metric tons
(c) 106.29 metric tons
(d) 110.25 metric tons

242. As per EXIM policy 2009-2014, the prohibition on export of pulses did not apply to:
(a) Kabuli Chana
(b) 10,000 MTs of organic pulses and lentils per annum.
(c) Both of the above
(d) None of the above

243. As per EXIM policy 2009-2014, export of organic pulses and lentils was subjected to following conditions
(a) Quantity limit shall be 10,000 MTs per annum; It should be duly certified by APEDA as being organic pulses and lentils
(b) Export contracts should be registered with APEDA, New Delhi prior to shipment
(c) Exports shall be allowed only from Customs EDI Ports
(d) All of the above

244. On 30 April, 2014, the government of India reduced the Minimum Export Price on export of edible oils in branded consumer packs of upto 5 Kgs to USD 1100 per MT. Earlier it was
(a) USD 1400 per MT
(b) USD 1200 per MT
(c) USD 1300 per MT
(d) USD 1500 per MT

245. Funds to states are provided for Development of complementary infrastructure such as, roads connecting the production centres with the ports, setting up of Inland Container Depots and Container Freight Stations, under the scheme called
(a) Scheme for Assistance to States for Developing Export Infrastructure and Allied Activities(ASIDE)
(b) Market Access Initiative
(c) Market Development Assistance
(d) Towns of Export Excellence

246. Under Market Development Scheme (MDS), financial assistance includes, amongst others, participation in
(a) Trade Fairs and Buyer Seller meets abroad or in India
(b) Export promotion seminars
(c) Both of the above
(d) None of the above

247. IBEF (originally called India Brand Equity Fund and Quality later renamed as India Brand Equity Foundation) was setup by the Ministry of Commerce on
(a) 26 January 1989
(b) 21 September 2001
(c) 11 July 1996
(d) 15 August 2003

248. Objective of VisheshKrishi Gram UpajYojana is to promote export of
(a) Agricultural Produce and their value added products
(b) Minor Forest Produce and their value added variants
(c) Forest Based Products
(d) All of the above

249. Which of the following falls among the top ten commodities that India imports
(a) Cashewnut (b) Apples
(c) Vegetable oils (d) Pulses

250. The ministry of food processing industry of the government of India was set up in
(a) 1987 (b) 1988
(c) 1999 (d) 1997

251. The cabinet minister for Ministry of Food Processing in the Government of India is
(a) Mrs. Sushma Swaraj
(b) Mrs. Najma Heptulla
(c) Bibi Harsimrat Kaur Badal
(d) Mrs SmritiIrani

252. Which of these does not fall under Ministry of Food Processing Industries?
(a) NIFTEM (b) NMPPB
(c) FSSAI (d) IICPT

253. What was the total number of factories in registered Food Processing Industries unit in 2010-11
(a) 35,838 (b) 36,838
(c) 36,800 (d) 34,000

254. Total exports from FPI sector in 2012-13 in million US$ was
(a) 36,057 (b) 31,762
(c) 20,427 (d) 25,762

255. Share of exports from FPI in total national exports in 2012-13 was
(a) 13% (b) 12%
(c) 12.50% (d) 12.25%

256. In 2012-13, the share of FDI in food processing industries to the total national FDI was
(a) 1.79%
(b) 1.89%
(c) 1.75%
(d) 1.80%

257. Which of the following emphasizes upon the thrust areas of the national seeds policy, 2002
(a) Promotion of domestic seed industry
(b) Export of seeds
(c) Import of seeds and planting material
(d) All of the above

258. Seeds of transgenic plant varieties for research purposes can be imported only through ....as per the EPA, 1986.
(a) The National Bureau of Plant Genetic Resources (NBPGR)
(b) Directorate of plant protection, quarantine and storage
(c) APEDA
(d) None of the above

259. The National Seeds Policy was announced in
(a) 1999
(b) 2008
(c) 2002
(d) 2005

260. National Mission on seeds was launched for
(a) 11th five year plan
(b) 10th five year plan
(c) 12th five year plan
(d) 2012-13 Annual plan

261. The total approved outlay for the National Mission on seeds is
(a) Rs. 3,773.40 crore for five years
(b) Rs. 3,880.00 crore for five years
(c) Rs. 2,980.50 crore for five years
(d) Rs. 3,573.40 crore for five years

262. At present, private sector contributes ......percent of the total seed supply in the country
(a) 70 (b) 80
(c) 40 (d) 50

263. What 'NiryatBandhu' scheme is all about?
(a) Mentoring first generation entrepreneurs
(b) Mentoring interested individuals in the arena of international business
(c) Both of the above
(d) None of the above

264. The following categories of export products / sectors are ineligible for Duty Credit Scrip, under Focus Market Scheme
(a) Export of Milk and Milk Products covered under ITC HS Codes 0401 to 0406, 1 9011 001, 1 9011 010, 2105 & 3501.
(b) Cereals, of all types;
(c) Sugar, of all types and in all forms;
(d) All of the above

265. The chairman of 14th Finance Commisision of India is
(a) Dr. Vijay Kelkar
(b) Dr. C. Rangrajan
(c) Dr. Y.V. Reddy
(d) Dr. RaghuramRajan

266. The former name of State Bank of India was
(a) Central Bank of India
(b) United Bank of India
(c) Imperial Bank of India
(d) People's Bank of India

267. On July 12, 1982, the ARDC was merged into
(a) RBI (b) NABARD
(c) EXIM Bank (d) None of the above

268. How many banks were nationalized in 1969?
(a) 16
(b) 14
(c) 15
(d) 20

269. Regional Rural Banks were established in
(a) 1897
(b) 1975
(c) 1965
(d) 1975

270. Which was the first super market in India
(a) Subhiksha
(b) Food World
(c) Giant
(d) Nilgiris

271. The first ever super market 'Food World' was set up in ....in the city of ....
(a) 1997, Chennai
(b) 1998, Hyderabad
(c) 1996, Bangalore
(d) 1996, Coimbatore

272. In March 2014, Which of the following had struck a joint venture deal with India's Tata Group to become the first foreign supermarket to enter the country's US$500-billion retail sector
(a) TESCO, Britain
(b) Walmart, US
(c) Amazon, US
(d) Home Depot, US

273. Which of the following is not a part of Integrated scheme for agricultural marketing
(a) Marketing Research and Information Network ?
(b) Strengthening of Agmark Grading Facilities
(c) Agri-Business Development through Venture Capital Assistance and Project Development Facility
(d) VisheshKrishi Gram UpajYojana (VKGUY)

274. According to the third advanced estimates (2015-16) for production of foodgrains, released in May 2016, the total foodgrain production is estimated at.......million tons against the target.....million tons
(a) 264.38, 259.00
(b) 264.38, 257.13
(c) 252.23, 264.10
(a) 259.00, 257.13

275. Which of the following emerged as a leader in Agricultural exports from India for the year 2015-16 (in value terms), contributing around 15 percent to total exports?
(a) Meat and preparations
(b) Basmati Rice
(c) Marine products
(d) Cotton

276. According to the third estimates of horticultural crops for the year 2015-16, India's total fruit production is estimated at
(a) 90.41 million tons
(b) 91.44 million tons
(c) 93.41 million tons
(d) 94.41 million tons

277. According to the third estimates of horticultural crops for the year 2015-16, India's total vegetable production is estimated at
(a) 170.25 million tons
(b) 166.60 million tons
(c) 171.25 million tons
(d) 174.25 million tons

278. According to the third estimates of horticultural crops for the year 2015-16, India's total fruits and vegetables production is estimated at
(a) 254.66 million tons
(b) 258.05 million tons
(c) 254.55 million tons
(d) 253.55 million tons

279. National production target for foodgrains for the year 2015-16 is
(a) 264 million tons
(b) 259 million tons
(c) 257 million tons
(d) 262 million tons

280. National production target for oilseeds for the year 2015-16 is
(a) 31 million tons
(b) 32 million tons
(c) 33 million tons
(d) 34 million tons

281. National production target for cotton for the year 2015-16 is
(a) 31 million bales
(b) 34 million bales
(c) 35 million bales
(d) 32 million bales

282. National production target for coarse cereals for the year 2015-16 is
(a) 40.00 million tons
(b) 41.50 million tons
(c) 43.20 million tons
(d) 39.00 million tons

283. Who is the current agricultural minister in the Government of India?
(a) Shri Radha Mohan Singh
(b) Shri ArunJaitley
(c) Smt. SushmaSwaraj
(d) Shri Venkaiah Naidu

284. Which of the following is not a component of National Mission on Sustainable Agriculture?
(a) Rainfed Area Development
(b) On Farm Water Management
(c) Soil Health Management
(d) Agricultural Market Management

285. The XII plan outlay for Integrated Scheme for Agricultural Marketing is
(a) Rs. 4548.00 crore
(b) Rs. 4540.00 crore
(c) Rs. 4556.00 crore
(d) Rs. 4550.00 crore

286. India's foreign exchange rate system is
(a) Free float (b) Managed float
(c) Fixed (d) Fixed target of band

287. India's first port-based Special Economic Zone named International Container Transshipment Terminal (ICTI) is being set-up at
(a) Kandla (b) Kochi
(c) Goa (d) Tuticorin

288. India GDP growth rate at factor cost for Quarter I of 2016-17 at constant (2011-12) prices has registered a growth of …….% over corresponding period of 2015-16
(a) 4.6 % (b) 7.1%
(c) 6.1% (d) 6.0%

289. 'agriculture, forestry and fishing' sector in first quarter of 2016-17 has shown a growth rate in GVA of
(a) 4.4%
(b) 1.8%
(c) 3.4%
(d) 4.7%

290. Food Corporation of India was established in
(a) 1964
(b) 1966
(c) 1965
(d) 1963

291. Which was the first state to launch a policy to establish Special Agricultural Zones
(a) Gujarat
(b) Tamil Nadu
(c) Uttarakhand
(d) Andhra Pradesh

292. When economists say that developing countries have a 'dualistic' economy, the 2 parts referred to are
(a) Rich people and poor people.
(b) Modern sector and traditional sector.
(c) Town and country.
(d) Men and women.

293. Which of the following is NOT one of the problems facing most developing nations?
(a) Labour shortages
(b) Foreign debt
(c) Rapid population growth
(d) Food shortages

294. Low agricultural productivity in developing countries is mainly caused by
(a) A lack of effective demand for food products.
(b) Shortages of inputs, including land.
(c) Migration from rural areas to urban areas.
(d) An over-investment in farm equipment.

295. One of the benefits of putting resources into agricultural projects is that
(a) Successful agricultural projects produce surplus food to support urban development.
(b) Agricultural projects usually have very low import requirements.
(c) Export prices for agricultural products are more stable than those for industrial products.
(d) Agricultural investment will prevent the flight of capital abroad.

296. Which of the following factors has been suggested as an explanation for the lack of economic growth in many poor nations?
(a) A steady rate of capital formation.
(b) An adequate level of social overhead capital.
(c) The supply of human resources is too high.
(d) The constraints imposed by dependency on the already-developed nations.

297. Which of the following is not a member of the WTO?
(a) United States (b) China
(c) Russia (d) Iran

298. Members of the World Trade Organization are required to
(a) Eliminate all tariffs.
(b) Allocate quotas on a first-come-first-served basis.
(c) Use the WTO's definitions of products, such as what can be called "beer."
(d) Enforce patent rights for patent holders from other countries.

299. A situation where any advantage given by one member of the WTO to another member must be extended to all WTO members.
(a) The trade diversion principle
(b) The comparative advantage principle
(c) The intra-regional principle
(d) The most favoured nation principle

300. A form of protection which is often used to protect the agricultural sector in EU countries
(a) Exchange controls
(b) Tariffs on imports
(c) Public sector contracts
(d) Voluntary export restraints

301. What is the underlying characteristic of the WTO?
(a) It facilitates economic co-operation between different countries
(b) It resolves disputes between economic trade blocks
(c) It facilitates the development of less developed countries
(d) It acts as an umbrella institution that regulates the agreements concluded at the Uruguay round, the organisation's ultimate goal being the promotion of free international trade

302. Does the WTO come with its own institutional framework?
(a) No, the WTO depends on the relevant frameworks of national governments
(b) No, the WTO provides certain institutional arrangements but only on an ad hoc basis
(c) Yes, the WTO provides a certain institutional framework which changes depending on the nature of free trade agreements
(d) Yes, the WTO provides a common institutional framework for the implementation of free trade agreements

303. What are the three key 'components' of the WTO?
(a) A Senate, a Judiciary and a Directorate
(b) A Trade Commission, a Dispute Settlement Body and a Council of Ministers
(c) An Executive apparatus, a Legislative apparatus and an Enforcement apparatus
(d) A Board of Governors, the Assembly of Member States and a Steering Committee

304. Does the WTO apply its framework in exactly the same fashion throughout the world?
(a) Yes, the WTO applies its framework in exactly the same way throughout the world
(b) Yes, the WTO applies its framework in exactly the same way around the world, unless it deals with trade blocks
(c) No, the WTO makes allowance for regional variation
(d) No, the WTO makes allowance for variation in the case of economically stronger countries

305. What are the two main functions of the WTO?
(a) The promotion of free trade and economic liberalism
(b) The administration of the WTO agreements and the resolution of international trade disputes
(c) The promotion of world peace, economic stability and financial co-operation
(d) The administration of the WTO's institutional framework and the promotion of economic co-operation amongst its members

306. Theoretically, what is the most significant organ of the WTO?
(a) The Committee on Trade and Development
(b) The Council for Trade in Goods
(c) The General Council
(d) The Ministerial Conference

307. Practically, what is the most significant organ of the WTO?
(a) The Committee on Trade and Development
(b) The Council for Trade in Goods
(c) The General Council
(d) The Ministerial Conference

308. Where is provision made with regard to the WTO's institutional arrangements?
(a) In the WTO Agreement
(b) In the GATT
(c) In the UN Charter
(d) In the WTO Agreement on Subsidies and Countervailing Measures

309. With regard to the WTO, subsidies are covered in two separate international economic law agreements. Which are they?
(a) The GATT and the Agreement on Subsidies and Countervailing Measures
(b) The Reform Treaty and the Treaty of Rome
(c) The North American Free Trade Agreement and the Organization of American States Charter
(d) The International Monetary Fund's Articles of Agreement and the European Central Bank's Statute

310. What are countervailing measures?
(a) Measures which a State may initiate against another State when that other State has legally subsidised any of its domestic industries
(b) Measures which a national government may initiate against another State when that other State has not subsidised any of its domestic industries
(c) Measures which a State may initiate against another State when that other State has illegally subsidised any of its domestic industries
(d) Measures which a national government may initiate against another State when that other State has not illegally subsidised any its domestic industries

311. Which of the following is not an economic rationale for trade intervention?
(a) employment
(b) balance of payments considerations
(c) protection to domestic industry
(d) preservation of national identity

312. Direct payments made by the government to domestic companies to encourage exports or to protect them from imports are known as
(a) aids
(b) subsidies
(c) export tariffs
(d) voluntary export restraints

313. A quantitative import or export restriction which prohibits or limits the quantity of a product being exported or imported is called a(n)
(a) Restraint
(b) embargo
(c) import tariff
(d) quota

314. Bilateral agreements that restrict exports are called
(a) Transit tariffs
(b) Voluntary export restraints
(c) Export quotas
(d) Orderly marketing arrangement

315. The WTO was born out of
(a) Uruguay Round
(b) Tokyo Round
(c) Geneva Round
(d) Torquay Round

316. Which of the following is referred to as a predecessor to WTO?
(a) OPEC (b) GATT
(c) IMF (d) World Bank

317. The quota that restricted trade in textiles was known as
(a) MFA (b) GATS
(c) TRIPS (d) TRIMS

318. The current round of negotiation in the WTO is known as the
(a) Cancun Meet
(b) Torquay Round
(c) Doha Development Round (DDA)
(d) Seattle Summit

319. "Bhagwa" is an Indian variety of
(a) mango (b) pomegranate
(c) kiwi (d) orange

320. Economic globalization is characterized by
(a) cross-cultural flows of ideas
(b) international trade and investment flows
(c) international migration
(d) rise of regionalism

321. Order the following events in global trade and finance
i. Establishment of electronic stock exchange
ii. Formation of WTO
iii. The first McDonalds
(a) i, ii, iii
(b) ii, iii, i
(c) iii, i, ii
(d) iii, ii, i

322. What is 'The Quad' at the World Trade Organization (WTO)?
(a) The coffee shop at the WTO.
(b) The decision-making structure composed of the US, the EU, Brazil and India.
(c) The four most powerful countries within the WTO.
(d) The four department directors within the WTO.

323. When the housing market boom first began to unravel globally in 2007 banks discovered their over-exposure to what?
(a) Derivatives.
(b) Shoddy buildings.
(c) Politically motivated loans.
(d) 'Toxic assets' of mortgage-backed securities

324. General Agreement on Trade in Services will not be applicable to
(a) Services supplied from one country to another – cross border supply
(b) Transaction of goods across the border – Export Import
(c) Individuals traveling from own country to supply services in another – presence of natural persons.
(d) Consumers/firms making use of a service in another country – consumption abroad.

325. As a part of WTO guidelines, Agreement on Agriculture (AOA) doesn't consider
(a) Direct payments to farmers are permitted.
(b) Indirect assistance and support to farmers including R & D support by govt. are not permitted.
(c) Domestic policies which directly effect on production and trade have to be cut back.
(d) Least developed countries do not need to make any cuts.

326. Quantitative restrictions refer to limit set by countries to curb
(a) Imports
(b) Exports
(c) Imports & exports
(d) None of the above

327. Consider the following statements regarding WTO agreements

1. Genetically Modified Organisms (GMO s) is covered under the Agreement on the Application of Sanitary and Phyto-Sanitary Measures (SPS)
2. Under the SPS agreement, the WTO sets constraints on member-states' policies relating to food safety as well as animal and plant health about imported pests and diseases

Which of the statements given above is/are correct?
(a) Only 1 (b) Only 2
(c) Both 1 and 2 (d) Neither 1 nor 2

328. Consider the following statements

1. Geographical indication is a name or sign used on certain products which corresponds to a specific geographical location or origin.
2. Geographical indication acts as a certification that the product possesses certain qualities or reputation due to its geographical origin.

Which of the statements given above is/are correct?
(a) Only 1 (b) Only 2
(c) Both 1 and 2 (d) Neither 1 nor 2

329. Consider the following statements

1. Countervailing duties are trade import duties imposed under the United Nations Organisation.
2. Countervailing duties are imposed under World Trade Organisation (WTO) Rules to neutralize the negative effects of subsidies.

Which of the statements given above is/are correct?
(a) Only 1 (b) Only 2
(c) Both 1 and 2 (d) Neither 1 nor 2

330. Consider the following statements

1. MFN is a status or level of treatment accorded by one state to another in international trade.
2. MFN is one of the cornerstones of World Trade Organisation trade law.

Which of the statements given above is/are correct?
(a) Only 1 (b) Only 2
(c) Both 1 and 2 (d) Neither 1 nor 2

331. Consider the following barriers

1. Anti-dumping practice
2. Packaging, labeling, and marking standards
3. Taxes on imports
4. Rules of origin
5. Sanitary and photo-sanitary conditions

Which of the barriers given above is/are come under the Non-Tariff Barriers (NTBs) of WTO?
(a) 1, 2, 3 and 5 (b) 1, 3, 4 and 5
(c) 2, 3, 4 and 5 (d) 1, 2, 4 and 5

332. Which trade theory holds that nations can increase their economic well-being by specializing in the production of goods they produce more efficiently than anyone else?
(a) The Theory of absolute advantage
(b) The theory of comparative advantage
(c) The factor endowment theory
(d) The International Product Life Cycle Theory

333. The major differences between internal and international trade are based on
(a) Different national policies
(b) Adjustment of Balance of payments
(c) Both of them
(d) None of them

334. Among the following which one is the best advantage of international trade
(a) Benefit of price stability
(b) Benefit of demand and supply equilibrium for commodities equilibrium
(c) Benefit of achieving equilibrium in foreign exchange reserve
(d) Benefits of specialization

335. Mercantilism held that
(a) Silver alone was the mainstay of national wealth
(b) Gold alone was the mainstay of national wealth
(c) Silver and Gold were the mainstay of national wealth
(d) Silver and gold were not important for the national wealth of a country

336. The principle of absolute cost advantage is given by
(a) Adam Smith (b) Samuelson
(c) Robertson (d) Ricardo

337. According to David Home, under gold standard
(a) A perpetual surplus is impossible
(b) A perpetual deficit is impossible
(c) A perpetual surplus is impossible, but a perpetual deficit is possible
(d) A perpetual deficit is impossible but a perpetual surplus is possible

338. Who written the book 'Principles of Economics'?
(a) Afred Marshall (b) David Ricardo
(c) Paul Samuelson (d) Amartya Sen

339. International Trade is a branch of
(a) International Economics
(b) International Economics
(c) International Business
(d) International Marketing

340.. Inter-regional Trade may be differentiated from International trade on the basis of
(a) Factors mobility
(b) Different currencies
(c) Different national groups
(d) All of the above

341. The National Wealth was reflected in a country's holding of precious metal was the view of
(a) Mercantilists
(b) Asian Traders
(c) Industrialists from the emerging economies
(d) Americans

342. Tariffs, quotas and other commercial policies were proposed by
(a) Mercantilists
(b) Classical Economists
(c) Modern Economists
(d) None of the above

343. One of the earliest attack on mercantilism was raised by
(a) David Hume (b) Adam Smith
(c) Paul Samuelson (d) David Ricardo

344. Doctrine of Price-Specie Flow mechanism in his Political Discourses was given by
(a) Adam Smith
(b) David Hume
(c) Paul Samuelson
(d) David Ricardo

345. Mercantilists view of fixed wealth of the world was:
(a) Adam Smith (b) David Hume
(c) Paul Samuelson (d) David Ricardo

346. Mercantilism was based on the ideology of
(a) Globalisation
(b) Nationalization
(c) Regionalization
(d) Privatization

347. Who wrote, "The wealth of nations."
(a) Adam Smith in 1776
(b) David Ricardo in 1784
(c) Paul Samuelson in 1776
(d) Robertson in 1784

348. Given a natural or acquired advantage, with free trade, nation could concentrate their production on the goods that they could make more cheaply with all the consequent benefits of division of labour. This is
(a) Theory of absolute cost advantage
(b) Theory of comparative cost advantage
(c) Theory of cost advantage
(d) Theory of international trade

349. The theory of absolute cost advantage "seems convincing but it is not very deep," was the view of
(a) Ellsworth
(b) Ricardo
(c) Sodersten
(d) None of the above

350. Which of the following is true about International Economics?
(a) It broadens mental outlook
(b) It widens outlook of citizens
(c) It solves complex economic problems
(d) All of the above

351. Comparative advantage is one of the few things in economics which is true, but not obvious. Who said it?
(a) Adam Smith
(b) David Ricardo
(c) Paul Samuelson
(d) None of the above

352. In Ricardo's view, which of the following was the cause of comparative cost differences
(a) Technological advancement
(b) Profit differences
(c) Difference in labor cost of production
(d) All the above

353. Ricardian theory of comparative cost advantage suggests that trade between two countries is desirable
(a) If one has absolute advantage in the production of both the commodities and the other absolute disadvantage in the production of both the commodities.
(b) If one has absolute advantage in the production of at least one commodity and the other absolute disadvantage in the production of both the commodities.
(c) If one has absolute advantage in the production of both the commodities and the other absolute disadvantage in the production of at least one commodity.
(d) If one has comparative advantage in the production of both the commodities, without considering the status of the other country.

354. The theory of Opportunity cost is considered a new life to the theory of comparative advantage. Who introduced this theory of opportunity cost?
(a) Samuelson (b) Keynes
(c) Abbas Lerner (d) G. Haberler

355. The other names of opportunity cost theory are
(a) Substitution curve
(b) Transformation curve or production possibility curve
(c) Production frontier or production indifference curve
(d) All of the above

356. Autarky is a state of
(a) Trade between friendly nations
(b) No trade between nations
(c) Trade between neighbours
(d) Trade with all nations

357. Who wrote the famous book 'Inter-Regional and International Trade (1933)'?
(a) Bertil Ohlin
(b) G. Haberler
(c) David Ricardo
(d) Eli Hecksher

358. Factor endowment theory of international trade is given by
(a) Hecksher-Ohlin
(b) Haberler
(c) Prebisch Singer
(d) P. Samuelson

359. Leontief paradox on the factor composition of U.S. imports and exports
(a) Supports Hecksher-Ohlin model
(b) Refutes Hecksher-Onlin model
(c) Supports Comparative advantage theory
(d) Refutes Comparative advantage theory

360. According to Leontif's study
(a) U.S. exports consisted of capital intensive goods
(b) U.S. exports consisted of labour-intensive goods
(c) U.S. imports consisted of labour-intensive goods
(d) None of the above

361. The Hecksher-Ohlin theory postulates that as a result of trade, the difference in factor prices between nations
(a) Diminishes
(b) Increases
(c) Remains unchanged
(d) First increases then decreases

362. The gains from trade refer to the
(a) Net benefits or increase in goods that a country gets by trading with other countries
(b) Gains from exports only
(c) Gains from imports only
(d) None of the above

363. Which factor/s determine the size of the gains from international trade?
(a) Differences in cost ratios, size of country
(b) Nature of terms of trade, technological conditions
(c) Productive efficiency, nature of commodities involved in trade
(d) All of the above

364. Terms of trade can be defined as
(a) Quantity of domestic goods that must be given in exchange for one unit of imported goods
(b) Quantity of goods to be exported to balance the deficit in the foreign exchange reserves with respect to a country
(c) Quantity of goods to be exported in the exchange of one unit of exported goods
(d) Ratio between goods exported and goods imported in a given year

365. There have been substantial shift in pattern and composition of trade items of developing countries. Now days, in world markets, developing countries tend to rely primarily on exports of
(a) Manufactured goods
(b) Agricultural goods
(c) Services
(d) Primary goods

366. Terms of trade of the developing countries are unfavorable due to the fact that whenever income in developed countries increases, the demand for primary goods reduces because
(a) Income elasticity of demand for primary goods is high
(b) Income elasticity of demand for primary goods is low
(c) Income elasticity of demand is neutral towards primary goods
(d) None of the above

367. Free trade refers to the trade
(a) That is free from all artificial barriers like tariffs, quantitative restrictions, exchange control and others
(b) Lets nations enjoy lower prices and higher levels of output, income and consumption that could not be achieved in isolation
(c) Both of the above
(d) None of the above

368. The infant industry argument was supported by
(a) Alexander Hamilton (USA)
(b) F. List (Germany)
(c) J.S. Mill (England)
(d) All the above

369. M. Keynes pointed out for an economy which is neither in equilibrium nor in sight of equilibrium,
(a) Protection and not free trade is the most rational trade policy
(b) Free trade and not protectionism is the most rational policy
(c) Neither protectionism nor free trade will work
(d) None of the above

370. The arguments for free trade are
(a) Maximization of output and economies of scale
(b) Lower prices of imported goods and better reward for factors
(c) Increase in real income
(d) All of the above

371. The arguments for protection can be classified into three broad categories
(a) Fallacious, non economic, economic
(b) Fallacious, social, socio economic
(c) Fallacious, social, economic
(d) Fallacious, technological, economic

372. What is the non economic argument for protection?
(a) Infant industry argument
(b) Anti-dumping argument
(c) Defense argument
(d) Terms of trade argument

373. Protection encourages
(a) Laissez faire
(b) Competition
(c) Self-reliance
(d) Division of labour

374. The Dutch disease model was developed by:
(a) W. Max Corden and Peter Neary
(b) Gunnar Myrdal
(c) Raul Prebisch
(d) Hans Singer

375. The Dutch disease is a
(a) Relationship between the exploitation of natural resources and a decline in the manufacturing sector
(b) Relationship between the exploitation of natural resources and an increase in the manufacturing sector
(c) Relationship between the underuse of natural resources and a decline in the manufacturing sector
(d) Relationship between the underuse of natural resources and an increase in the manufacturing sector

376. An economic loss which a country suffers on account of its newly exploited natural resources is called
(a) Leontief paradox
(b) Dead Weight Loss
(c) The Dutch disease
(d) None of the above

377. On the basis of purpose, tariffs are classified as
(a) Specific tariff, Ad valorem tariff, Compound tariff
(b) Export duties, Import duties, Transit duties
(c) Single-column tariff, Multiple column tariff, Traditional or conventional tariff
(d) Revenue tariff, protective tariff, countervailing and antidumping duties

378. On the basis of quantification, tariffs are classified as
(a) Specific tariff, Ad valorem tariff, Compound tariff
(b) Export duties, Import duties, Transit duties
(c) Single-column tariff, Multiple column tariff, Traditional or conventional tariff
(d) Revenue tariff, protective tariff, countervailing and antidumping duties

379. On the basis of origin and destination of goods crossing national boundaries, tariffs are classified as
(a) Specific tariff, Ad valorem tariff, Compound tariff
(b) Export duties, Import duties, Transit duties
(c) Single-column tariff, Multiple column tariff, Traditional or conventional tariff
(d) Revenue tariff, protective tariff, countervailing and antidumping duties

380. On the basis of its application between two different countries, tariffs are classified as
(a) Specific tariff, Ad valorem tariff, Compound tariff
(b) Export duties, Import duties, Transit duties
(c) Single-column tariff, Multiple column tariff, Traditional or conventional tariff
(d) Revenue tariff, protective tariff, countervailing and antidumping duties

381. A specific tariff is
(a) Any tax on a particular imported good
(b) An import tax that must be paid in kind
(c) A requirement to pay the government a specified fraction of the monetary value of an imported good.
(d) A tax on imports defined as an amount of currency per unit of the good.

382. A tariff on imports benefits domestic producers of the imported good because
(a) They get the tariff revenue.
(b) It raises the price for which they can sell their product on the domestic market.
(c) It prevents imports from rising above a specified quantity.
(d) It reduces their producer surplus, making them more efficient.

383. When a large country levies a tariff on imports
(a) The world price falls.
(b) Foreigners are hurt.
(c) The domestic price rises by less than the tariff.
(d) All of the above.

384. Suppose that the tariff on shirts is 20% while the tariff on the cloth used to make the shirts is also 20% and there is no tariff on any of the inputs needed to produce cloth. The tariff on cloth is now reduced to 10%. Which of the following, if any, is not true?
(a) The nominal rate of protection on cloth is reduced.
(b) The effective rate of protection on cloth is reduced.
(c) The nominal rate of protection on shirts is unchanged.
(d) The effective rate of protection on shirts is unchanged

385. The WTO's Agreement on Textiles and Clothing promised
(a) To prevent job losses in these industries in developed countries.
(b) To phase out all quotas on textiles and apparel by Dec. 31, 2004.
(c) To eliminate tariffs on these products in the next round of trade negotiations.
(d) To help developing countries escape from these dead-end industries.

386. A uniform rate of duty for all commodities imposed by law without any discrimination between countries is known as
(a) Single column tariff
(b) Multiple column tariff
(c) Traditional or conventional tariff
(d) Import duty

387. When two or more duties are imposed by law on each class of commodity, such tariff is called:
(a) Single column tariff
(b) Multiple column tariff
(c) Traditional or conventional tariff
(d) Import duty

388. When a basic duty is imposed by law for each class of the commodity with the provision that such duty may be reduced reciprocally under international agreements, such tariff is called
(a) Single column tariff
(b) Multiple column tariff
(c) Traditional or conventional tariff
(d) Import duty

389. When a tariff is imposed on imported goods, it:
(a) Affects the domestic consumption of an imported good
(b) Affects the domestic production of goods that compete with the imported good
(c) Affects the foreign production of the imported good
(d) Changes the structure of the domestic economy
(e) All of the above

390. The effects of tariff could be
(a) Protective effect
(b) Consumption effect
(c) Revenue effect and Redistribution effect
(d) Balance of payments effect
(e) All of the above

391. The optimum tariff rate is the rate that
(a) Maximizes country's welfare
(b) Which the community indifference curve of a nation is tangent to the foreign country's offer curve
(c) Both a and b
(d) Neither a nor b

392. The most important non-tariff barriers are:
(a) Transport regulations
(b) Health regulations
(c) Quotas
(d) Packaging regulations

393. Which of the following is a non tariff barrier?
(a) Voluntary export restraint
(b) Health and product standard
(c) Environmental protection laws
(d) Ad valorem duties

394. Kindleberger's formula to measure optimum tariff
(a) t=[1/e-1] (b) t= [1-e/e-1]
(c) t=[1-e/ (1/e)] (d) t= e

395. The main difference between a tariff and a quota is
(a) A tariff raises the price of imports more than a quota.
(b) A quota does not harm domestic consumers.
(c) A tariff does not harm foreign producers.
(d) A tariff generates government revenue, while a quota, unless it is sold, does not.

396. A government procurement regulation or practice constitutes a nontariff barrier when
(a) Government agencies are required to purchase from the lowest bidder.
(b) Government shows a preference for domestic sellers over foreign sellers.
(c) Government requires that goods that it purchases meet a uniform safety standard.
(d) Government purchases are financed by tax receipts.

397. Import quotas are most commonly administered
(a) By permitting all imports until the quota is filled for the year, then none after that.
(b) By auctioning import licenses to the highest bidder.
(c) By granting import rights to domestic firms.
(d) By granting import rights to foreign firms or governments.

398. Which of the following will cause the tariff equivalent of a quota to increase in a small country?
(a) A decrease in domestic demand (the demand curving shifting left).
(b) A decrease in domestic supply (the supply curving shifting left).
(c) A rise in the world price.
(d) A rise in the quantity of imports permitted by the quota

399. What happens if a quota has been keeping the domestic price above the world price, but then the world price rises above what has been the domestic price?
(a) The country begins to export the good.
(b) The tariff-equivalent of the quota becomes negative.
(c) The domestic price rises above the world price, to keep the tariff-equivalent of the quota constant.
(d) Quota rents become zero.

400. Under unilateral quota:
(a) A country fixes an absolute limit on the importation of a commodity during a given period
(b) It is fixed by law or decree without prior negotiation with foreign governments
(c) The autonomously fixed quota may be either global or allocated
(d) All of the above

401. Under bilateral quota:
(a) Quotas are determined by mutual negotiations and agreement
(b) As the shipment of imported goods is spread evenly over the entire quota period, price fluctuations are minimum
(c) Both a and b
(d) Neither a nor b

402. Quota regulations are generally administered by means of
(a) Import licensing
(b) Export subsidies
(c) Both a and b
(d) Neither a nor b

403. Which is incorrect? Import licensing system
(a) Checks smuggling
(b) Promotes free trade
(c) Increases possibility of black marketing
(d) Increases possibility of corruption

404. Which of the following is a free-trade agreement between the United States, Canada, and Mexico?
(a) GATT
(b) WTO
(c) NAFTA
(d) EU

405. The ratio of export price to import price is called:
(a) Terms of trade
(b) Price ratio
(c) Tariff level
(d) The size of the quote

406. The _______ begins at the country's point of complete specialization and then runs outward at the slope that corresponds to the world price ratio.
(a) Production possibility frontier
(b) Consumption possibility curves
(c) Demand curve
(d) Supply of imports curve

407. A tariff is a .... A quota is a.........
(a) tax levied on imports; limit on the quantity of imports
(b) tax levied on imports; tariff that exists only on exported goods
(c) quota placed only on imported goods; limit on the quantity of imports
(d) limit on the quantity of imports; tax levied on imports

408. A tariff that is so high that it forces imports to zero is called
(a) Choking
(b) Damaging
(c) Integrating
(d) Prohibitive

409. The anti dumping measures are
(a) Tariff duty
(b) Import Quotas
(c) Import Embargo
(d) Voluntary Export Restraint
(e) All of the above

410. The drawback of exchange control is:
(a) It leads to contraction in International trade
(b) It leads to emergence of economic nationalism
(c) It leads to creation of multilateral trade
(d) It leads to disequilibrium in the country's balance of payments

411. Which of the following represents the stage where economic integration is most complete?
(a) Economic union (b) Customs union
(c) Monetary union (d) Common market

412. Which of the following represents the stage where economic integration is least complete?
(a) Free trade area
(b) Monetary union
(c) Common market
(d) Customs union

413. Customs union theory reasons that the formation of a customs union will decrease members' real welfare when the
(a) Trade diversion effect exceeds the trade creation effect
(b) Trade production effect exceeds the trade consumption effect
(c) Trade consumption effect exceeds the trade production effect
(d) Trade creation effect exceeds the trade diversion effect

414. Which economic integration scheme is solely intended to abolish trade restrictions among member countries, while setting up common tariffs against nonmembers?
(a) Economic union
(b) Common market
(c) Free trade area
(d) Customs union

415. By 1992 the European Union had become a full-fledged
(a) Economic union
(b) Monetary union
(c) Common market
(d) Fiscal union

416. Which organization of nations permits free trade among its members in industrial goods, while each member maintains freedom in its trade policies toward non-member countries?
(a) European Union
(b) Benelux
(c) Council for Mutual Economic Assistance
(d) North American Free Trade Association

417. Which form of economic integration occurs when participating countries abolish tariffs on trade among themselves, establish a common tariff on imports from nonmembers, and permit free movement of capital and labor within the organization?
(a) Free trade area
(b) Economic union
(c) Common market
(d) Monetary union

418. Which organization was founded in 1957 whose objective was to create an economic union among its members?
(a) General Agreements on Tariffs and Trade
(b) Organization of Economic Cooperation and Development
(c) European Union
(d) Latin American Free Trade Association

419. Which nation is not a member of the North American Free Trade Association?
(a) Canada
(b) Greenland
(c) Mexico
(d) United States

420. NAFTA is a:
(a) Monetary union
(b) Free trade area
(c) Common market
(d) Customs union

421. When the United States, Canada, and Mexico form a free trade area, and Mexico begins importing a product from Canada rather than from the lowest cost world producer
(a) Trade diversion occurs
(b) Trade creation occurs
(c) World welfare rises
(d) World welfare falls to zero

422. When the formation of a free trade area results in the reduction of trade with nonmember nations in favor of member countries, ________ occurs.
(a) Trade devaluation
(b) Trade revaluation
(c) Trade creation
(d) Trade diversion

423. The implementation of a common market involves all of the following except
(a) Elimination of trade restrictions among member countries
(b) A common tax system and monetary union
(c) Prohibition of restrictions on factor movements
(d) A common tariff levied in imports from nonmembers

424. Suppose that steel from Japan faces a 20 percent tariff in France and a 25 percent tariff in Italy, while France and Italy maintain free trade between each other. France and Italy are therefore part of a (an)
(a) Free trade area
(b) Customs union
(c) Common market
(d) Economic union

425. Suppose that Mexico and Canada form a free-trade area and Canada begins importing steel from Mexico rather than from Germany. There occurs
(a) Trade diversion
(b) Trade creation
(c) Trade destruction
(d) Trade exhaustion

426. Suppose that Mexico and Canada form a free-trade area. Mexicans then decrease auto manufacturing and increase imports of autos from Canada, while the Canadians decrease computer production and import more computers from Mexico. This is an example o
(a) Trade diversion
(b) Trade creation
(c) Trade destruction
(d) Trade exhaustion

427. If the United States and Canada abolish all tariffs on each other's goods and implement a common tariff on goods imported from other countries, there occurs a (an):
(a) Free-trade area
(b) Customs union
(c) Common market
(d) Economic union

428. Suppose that the United Kingdom and Italy abolish all tariffs on each other's goods and all restrictions on movements of factors of production between them. They also implement a common protectionist policy toward other countries. This is an example of a (an):
(a) Free-trade area
(b) Customs union
(c) Common market
(d) Economic union

429. Who is the author of the book, " The Customs Union Issues"?
(a) Jacob Viner
(b) Thomas Graham
(c) Kindleberger
(d) Harry Johnson

430. Which of the following is the correct sequence in increasing degree of inbtegration between member countries?
(a) FTA-EU-Common Market-Customs Union
(b) Common market-Customs Union-FTA-EU
(c) FTA-Customs Union-Common market-EU
(d) Common market-EU-FTA-Customs Union

431. International business firms have the fundamental goals of:
(a) Expanding market share
(b) Expanding sales revenue
(c) Expanding profits
(d) All of the above

432. What factors are taken into consideration to analyze the alternative foreign markets
(a) Current and potential size of the alternative markets?
(b) Level of competition the firm will face in each of these alternative markets
(c) Legal and political environments
(d) Socio cultural environment
(e) All of the above

433. Which of the following is not a locational factor for a firm while entering a market?
(a) Cheap land acquisition cost
(b) Political stability
(c) Logistic Requirements
(d) Ownership

434. Different modes to enter in International business are:
(a) Exporting (b) Licensing
(c) Franchising (d) FDI
(e) All of the above

435. First mover advantage suggests that:
(a) Pioneering businesses are able to obtain higher profits and other benefits as the consequence of early market entry.
(b) First mover firms expand more rapidly in international markets than late movers in international markets.
(c) Competing multinational firms enter an important market when a market is growing very fast.
(d) Multinational firms with first mover advantages have greater strategic incentives for investing in technical innovations than late movers.

436. Which of the following is not a mode of entry into the foreign market?
(a) Export
(b) Franchise
(c) Turnkey projects
(d) Internationalization

437. Franchising means
(a) The transfer of patented information and trademarks, information and know-how as well as information needed to sell a product or service
(b) The use of franchising for licensing new technologies in global markets
(c) The transfer of a business concept, with corresponding operational guidelines, to non-domestic parties for a fee
(d) Greenfield investment in a completely new facility, or acquisition of or merger with an already established local firm.

438. Horizontal and Vertical are types of:
(a) Mergers and Acquisitions
(b) Greenfield Strategy
(c) Direct Exporting
(d) FDI

439. Selling of products by a company to its affiliated company in host country is called:
(a) Intracorporate Transfers
(b) Direct Exporting
(c) Indirect Exporting
(d) None of the above

440. Disadvantage of Franchising is:
(a) Leakage of trade secrets
(b) Low cost
(c) Risk of product failure
(d) None of the above

441. The major benefits of trade include:
(a) Reduction in transportation cost
(b) Increases in incomes and employment
(c) Advances in technical methods
(d) b and c

442. Faster growth in emerging economies:
(a) Has resulted in higher inflation
(b) Has reduced globalization efforts
(c) Has put pressure on the supply of raw materials
(d) Has reduced global imbalances

443. The Final Act of the Uruguay Round established:
(a) Reduction in global trade barriers and a multilateral framework of discipline for trade in services
(b) Reduction in global trade barriers and protection of trade-related intellectual property rights
(c) A multilateral framework of discipline for trade in services and protection of trade-relatedintellectual property rights
(d) Reduction in global trade barriers, a multilateral framework of discipline for trade inservices, and protection of trade-related intellectual property rights

444. What was the primary focus of the Doha Round?
(a) Reduction of dumping by emerging economies in other countries
(b) Reduction of trade distorting agricultural subsidies by developing countries
(c) Reduction of trade distorting agricultural subsidies by developed countries
(d) All of the above

445. Which of the following is not part of the four Pacific Rim countries?
(a) Hong Kong (b) South Korea
(c) China (d) Taiwan

446. The European Free Trade Area, formed in 1960, has the following countries as members
(a) Austria, Denmark, Norway, Portugal
(b) Austria, Spain, Czech Republic, Portugal
(c) Austria, Denmark, Norway, Spain
(d) Austria, Norway, Portugal, Spain

447. Which of the following is not required in a partnership?
(a) Share in profits
(b) Share of bank accounts
(c) Written agreement
(d) All of the above

448. The difference between a systematic approach and a reactive approach is
(a) The systematic approach involves the selection of a product based on immediate market need, whereas the reactive approach is based on overall market demand.
(b) A systematic approach involves the selection of a service based only on immediate market need, whereas a reactive approach is based on the selection based on overall market demand.
(c) A systematic approach involves the selection of a product and service based on overall market demand, whereas a reactive approach is based on the selection on the basis of immediate market demand.
(d) None of the above

449. Lack of focus on a given product or market and absence of a long-term relationship with an
(a) importer are disadvantages of using:
(b) Systematic approach
(c) Reactive approach
(d) a and b
(e) None of the above

450. Countries with high inflation rates should be carefully considered because they may limit:
(a) The volume of imports by restricting the availability of foreign exchange
(b) The volume of exports by restricting the availability of foreign exchange
(c) The volume of money supply by restricting the availability of foreign exchange
(d) The volume of exports, imports, and money supply by restricting the availability of foreign exchange

451. Who is the Commerce Minister of India?
(a) Nirmala Sitharaman
(b) Radha Mohan Singh
(c) Arun Jaitly
(d) Sushman Swaraj

452. Which of the following is not a commercial document?
(a) Commercial Invoice
(b) Shipping Bill
(c) Bill of Lading
(d) Bill of Exchange

453. Which of the following are Principal Commercial Documents?
(a) Commercial Invoice
(b) Certificate of Origin
(c) Shipment Advice
(d) Certificate of Insurance
(e) All of the above

454. Which of the following are Auxilliary Commercial Documents?
(a) Pro-forma Invoice
(b) Insurance Declaration
(c) Mate's Receipt
(d) Insurance Declaration
(e) All of the above

455. Documents that help is effecting physical transfer of goods and their 'title' from the exporter to the im-porter are called:
(a) Commercial Documents
(b) Regulatory Documents
(c) Ancillary Documents
(d) None of the above

456. The declaration issued by an officer of a a vessel whereby he indicates that the goods have been received on board on his vesseel for shipment to destined country is called
(a) Shipping bill
(b) Bill of Lading
(c) Mate's Receipt
(d) Shipping order

457. A ....is a contract of affreightment between the exporter/ consignee of the goods and the courier/ shipping company
(a) Bill of Lading
(b) Shipping order
(c) Mate's receipt
(d) Commercial invoice

458. ....means that the price quoted by the vendor includes the price of the goods and all expenses up to and inclusing the cost of loading the goods on to the vessel
(a) FOB
(b) CIF
(c) Freight and Insurance
(d) Cost and Insurance

459. In the case of ...., the price quoted includes the cost of the goods, the cpst of insuring goods to destination, and the freight or cost of transport
(a) FOB
(b) CIF
(c) Freight and Insurance
(d) Cost and Insurance

460. The exporter uses...................in financing the transactions for pre-shipment credit
(a) Letter of credit
(b) Mate's receipt
(c) Bill of Lading

461. Which of the following is not true about airway bill?
(a) It cannot be used as collection instrument for receiving payments
(b) It is non negotiable
(c) It carries the title of the goods
(d) All of the above are true

462. Which of the following are regulatory documents in International Trade
(a) Shipping Bill
(b) Dock Challan
(c) Freight payment certificate
(d) Insurance Premium payment certificate
(e) All of the above

463. The MSP for paddy (Grade A) for the year 2016-17 is
(a) Rs 1570 per quintal
(b) Rs 1510 per quintal
(c) Rs 1470 per quintal
(d) Rs 1475 per quintal

464. For 2016-17, a bonus of RS 425 per quintal is payable over and above the MSP Rs 4800 per quintal on
(a) Moong
(b) Urad
(c) Tur
(d) None of the above

465. In terms of value, four largest exports from India in 2015-16 were (in descending order)
(a) Marine products, Buffalo Meat, Basmati Rice, Spices
(b) Marine Products, Basmati Rice, Buffalo Meat, Spices
(c) Buffalo Meat, Marine Products, Basmati Rice, Spices
(d) Buffalo Meat, Basmati Rice, Marine Products, Spices

466. In terms of quanity, four largest exports from India in 2015-16 were (in descending order)
(a) Non Basmati Rice, Basmati Rice, Sugar, Oil Meals
(b) Non Basmati Rice, Basmati Rice, Oil Meals, Sugar
(c) Basmati Rice, Oil Meals, Sugar, Non Basmati Rice
(d) Basmati Rice, Sugar, Non Basmati Rice, Oil Meals

467. Who is the Director General of the WTO?
(a) Roberto Azevedo
(b) Pascal Lamy
(c) Ban ki Moon
(d) None of the above

ANSWER KEY (Agribusiness Policy)

1. (a)	**2.** (a)	**3.** (c)	**4.** (d)	**5.** (b)	**6.** (b)	**7.** (c)	**8.** (a)	**9.** (c)	**10.** (b)
11. (a)	**12.** (a)	**13.** (b)	**14.** (d)	**15.** (d)	**16.** (d)	**17.** (b)	**18.** (a)	**19.** (b)	**20.** (b)
21. (c)	**22.** (d)	**23.** (d)	**24.** (b)	**25.** (b)	**26.** (d)	**27.** (d)	**28.** (a)	**29.** (c)	**30.** (a)
31. (d)	**32.** (d)	**33.** (c)	**34.** (a)	**35.** (c)	**36.** (b)	**37.** (c)	**38.** (d)	**39.** (b)	**40.** (c)
41. (d)	**42.** (b)	**43.** (c)	**44.** (b)	**45.** (a)	**46.** (b)	**47.** (c)	**48.** (a)	**49.** (c)	**50.** (d)
51. (a)	**52.** (d)	**53.** (d)	**54.** (c)	**55.** (c)	**56.** (a)	**57.** (d)	**58.** (c)	**59.** (a)	**60.** (b)
61. (b)	**62.** (a)	**63.** (b)	**64.** (c)	**65.** (c)	**66.** (b)	**67.** (b)	**68.** (b)	**69.** (c)	**70.** (c)
71. (b)	**72.** (b)	**73.** (d)	**74.** (d)	**75.** (c)	**76.** (c)	**77.** (c)	**78.** (a)	**79.** (b)	**80.** (d)
81. (a)	**82.** (b)	**83.** (c)	**84.** (d)	**85.** (d)	**86.** (a)	**87.** (b)	**88.** (b)	**89.** (d)	**90.** (c)
91. (b)	**92.** (c)	**93.** (b)	**94.** (b)	**95.** (c)	**96.** (a)	**97.** (c)	**98.** (c)	**99.** (d)	**100.** (b)
101. (a)	**102.** (a)	**103.** (b)	**104.** (b)	**105.** (d)	**106.** (b)	**107.** (a)	**108.** (d)	**109.** (b)	**110.** (a)
111. (a)	**112.** (c)	**113.** (b)	**114.** (d)	**115.** (c)	**116.** (a)	**117.** (d)	**118.** (b)	**119.** (c)	**120.** (c)
121. (c)	**122.** (c)	**123.** (d)	**124.** (c)	**125.** (a)	**126.** (a)	**127.** (a)	**128.** (b)	**129.** (c)	**130.** (b)
131. (d)	**132.** (d)	**133.** (a)	**134.** (d)	**135.** (a)	**136.** (d)	**137.** (c)	**138.** (c)	**139.** (c)	**140.** (c)
141. (b)	**142.** (c)	**143.** (a)	**144.** (b)	**145.** (a)	**146.** (a)	**147.** (b)	**148.** (b)	**149.** (b)	**150.** (b)
151. (a)	**152.** (d)	**153.** (b)	**154.** (d)	**155.** (b)	**156.** (a)	**157.** (d)	**158.** (a)	**159.** (a)	**160.** (c)
161. (c)	**162.** (a)	**163.** (b)	**164.** (d)	**165.** (a)	**166.** (d)	**167.** (d)	**168.** (b)	**169.** (a)	**170.** (c)
171. (a)	**172.** (c)	**173.** (b)	**174.** (b)	**175.** (a)	**176.** (d)	**177.** (b)	**178.** (d)	**179.** (b)	**180.** (b)

181. (b) **182.** (d) **183.** (d) **184.** (a) **185.** (a) **186.** (d) **187.** (c) **188.** (c) **189.** (a) **190.** (d)
191. (b) **192.** (c) **193.** (c) **194.** (d) **195.** (b) **196.** (b) **197.** (a) **198.** (a) **199.** (a) **200.** (d)
201. (d) **202.** (b) **203.** (c) **204.** (c) **205.** (a) **206.** (c) **207.** (a) **208.** (a) **209.** (a) **210.** (a)
211. (c) **212.** (d) **213.** (b) **214.** (b) **215.** (b) **216.** (a) **217.** (a) **218.** (d) **219.** (d) **220.** (c)
221. (d) **222.** (d) **223.** (c) **224.** (a) **225.** (d) **226.** (d) **227.** (b) **228.** (a) **229.** (a) **230.** (d)
231. (d) **232.** (a) **233.** (c) **234.** (a) **235.** (d) **236.** (b) **237.** (d) **238.** (d) **239.** (b) **240.** (c)
241. (b) **242.** (c) **243.** (d) **244.** (a) **245.** (a) **246.** (c) **247.** (c) **248.** (d) **249.** (c) **250.** (b)
251. (c) **252.** (c) **253.** (a) **254.** (a) **255.** (b) **256.** (a) **257.** (d) **258.** (a) **259.** (c) **260.** (c)
261. (a) **262.** (c) **263.** (c) **264.** (d) **265.** (c) **266.** (c) **267.** (b) **268.** (b) **269.** (d) **270.** (b)
271. (c) **272.** (a) **273.** (d) **274.** (c) **275.** (c) **276.** (b) **277.** (b) **278.** (b) **279.** (a) **280.** (c)
281. (c) **282.** (c) **283.** (a) **284.** (d) **285.** (a) **286.** (b) **287.** (b) **288.** (b) **289.** (b) **290.** (c)
291. (c) **292.** (b) **293.** (a) **294.** (a) **295.** (b) **296.** (b) **297.** (d) **298.** (d) **299.** (d) **300.** (b)
301. (d) **302.** (c) **303.** (c) **304.** (c) **305.** (d) **306.** (d) **307.** (c) **308.** (a) **309.** (a) **310.** (c)
311. (d) **312.** (b) **313.** (d) **314.** (b) **315.** (a) **316.** (b) **317.** (a) **318.** (c) **319.** (b) **320.** (b)
321. (c) **322.** (b) **323.** (d) **324.** (b) **325.** (b) **326.** (c) **327.** (c) **328.** (c) **329.** (b) **330.** (c)
331. (d) **332.** (b) **333.** (c) **334.** (d) **335.** (c) **336.** (a) **337.** (a) **338.** (a) **339.** (a) **340.** (d)
341. (a) **342.** (a) **343.** (a) **344.** (b) **345.** (a) **346.** (b) **347.** (a) **348.** (b) **349.** (c) **350.** (d)
351. (c) **352.** (c) **353.** (b) **354.** (d) **355.** (d) **356.** (b) **357.** (a) **358.** (a) **359.** (b) **360.** (b)
361. (c) **362.** (a) **363.** (d) **364.** (a) **365.** (d) **366.** (b) **367.** (b) **368.** (d) **369.** (a) **370.** (d)
371. (a) **372.** (c) **373.** (c) **374.** (a) **375.** (a) **376.** (c) **377.** (d) **378.** (a) **379.** (b) **380.** (c)
381. (d) **382.** (b) **383.** (d) **384.** (d) **385.** (b) **386.** (a) **387.** (b) **388.** (c) **389.** (e) **390.** (e)
391. (c) **392.** (b) **393.** (a) **394.** (b) **395.** (d) **396.** (b) **397.** (d) **398.** (b) **399.** (d) **400.** (c)
401. (a) **402.** (b) **403.** (c) **404.** (c) **405.** (a) **406.** (b) **407.** (a) **408.** (d) **409.** (e) **410.** (a)
411. (c) **412.** (a) **413.** (a) **414.** (d) **415.** (c) **416.** (d) **417.** (c) **418.** (c) **419.** (b) **420.** (b)
421. (a) **422.** (d) **423.** (d) **424.** (a) **425.** (a) **426.** (b) **427.** (b) **428.** (c) **429.** (a) **430.** (c)
431. (d) **432.** (e) **433.** (d) **434.** (e) **435.** (a) **436.** (d) **437.** (c) **438.** (a) **439.** (a) **440.** (a)
441. (d) **442.** (c) **443.** (d) **444.** (c) **445.** (c) **446.** (a) **447.** (c) **448.** (c) **449.** (c) **450.** (a)
451. (a) **452.** (b) **453.** (e) **454.** (e) **455.** (a) **456.** (c) **457.** (a) **458.** (a) **459.** (b) **460.** (a)
461. (c) **462.** (e) **463.** (b) **464.** (a) **465.** (a) **466.** (a) **467.** (a)

CHAPTER 11

11. Project Management and Entrepreneurship Development

1. A set of organized activities is called as
   (a) Project (b) Programme
   (c) Policy (d) Rule
2. Combination of many projects
   (a) Project (b) Programme
   (c) Policy (d) Rule
3. The end phase of project life cycle is called as
   (a) Conception Stage
   (b) Design Stage
   (c) Implementation Stage
   (d) Commissioning Stage
4. Odd man out
   (a) Setting up a cement plant and construction of a highway are two different project
   (b) A project is not rigid in its life time
   (c) The activities of a project are unique and routine
   (d) In conception stage, project ideas are conceived
5. Maximum flexibility has been observed in
   (a) Crash projec
   (b) Normal project
   (c) Both
   (d) None
6. The cost relationship between Normal project and Crash project is
   (a) Both the cases costs are same
   (b) Normal project cost is greater than crash project cost
   (c) Crash project cost is greater than Normal project
   (d) There is no such relationship
7. Small projects are those on which investment on plant and machinery is
   (a) Greater than 10 cr
   (b) Less than 1 cr
   (c) Greater than 100 cr
   (d) In between 1 cr to 100 cr
8. A project in which organization moves more towards customer is
   (a) Backward integration project
   (b) Forward integration project
   (c) Both
   (d) None
9. A production firm is going to process its produce is an example of
   (a) Backward integration project
   (b) Forward integration project
   (c) Both
   (d) None
10. Molasses is derived from
    (a) Paddy (b) Sugarcane
    (c) Cotton (d) Ground Nut
11. Kum kum is derived from
    (a) Cardamom (b) Turmeric
    (c) Ginger (d) Onion
12. Sagu is derived from
    (a) Turmeric (b) Tapioca
    (c) Tamarind (d) Rose grass
13. Hydraulic brake fluid is derived from
    (a) Castor seed (b) Paddy
    (c) Rose grass (d) Cardamom
14. Oleoresin is derived from
    (a) Sugarcane (b) Onion
    (c) Chilly (d) Sundra wood

15. Lycopene pigment is found in
(a) Castor (b) Tomato
(c) Tobacco (d) Onion

16. Linamarin is found in
(a) Maize
(b) Tapioca meal
(c) Turmeric
(d) Ground nut oil cake

17. Tartaric acid is derived from
(a) Paddy (b) Tamarind
(c) Rubber (d) Sundra wood

18. Papain is found in
(a) Tomato (b) Persimon
(c) Papaya (d) Chilli

19. Acidic soil mostly found in
(a) Maharashtra (b) West Bengal
(c) Gujarat (d) Tamil Nadu

20. Red soil is mostly found in
(a) Maharashtra (b) West Bengal
(c) Gujarat (d) Tamil Nadu

21. Black soil is mostly found in
(a) Maharashtra (b) West Bengal
(c) Gujarat (d) Tamil Nadu

22. Saline soil is found in
(a) Maharashtra (b) West Bengal
(c) Gujarat (d) Tamil Nadu

23. Parent group of BPL company
(a) Mittal (b) Nambiar
(c) Nanda (d) Nadar

24. Parent group of Hero company
(a) Mittal (b) Munjal
(c) Jhawar (d) Modi

25. Parent group of Escorts company
(a) Mittal
(b) Nanda
(c) Gupta
(d) Dhoot

26. Publishing Home of Frontline Magazine
(a) Living Media
(b) Hindu Group
(c) Ministry of Information and Broadcasting
(d) Bennet Coleman & Co

27. Publishing Home of Yojana Magazine
(a) Living Media
(b) Hindu Group
(c) Ministry of Information and Broadcasting
(d) Bennet Coleman & Co

28. Publishing Home of 'Filmfare Magazine'
(a) Living Media
(b) Hindu Group
(c) Ministry of Information and Broadcasting
(d) Bennet Coleman & Co

29. Publishing Home of 'The Economic Times'
(a) Living Media
(b) Hindu Group
(c) Ministry of Information and Broadcasting
(d) Bennet Coleman & Co

30. Head quarter of Asian Development Bank
(a) Jakarta, Indonesia
(b) Manila, Philippines
(c) Kuwait
(d) Geneva

31. Head quarter of International Monetary Fund
(a) Jakarta, Indonesia
(b) Manila, Philippines
(c) Washington, DC, USA
(d) New York, USA

32. Head quarter of International Finance Corporation
(a) Jakarta, Indonesia
(b) Manila, Philippines
(c) Washington, DC, USA
(d) New York, USA

33. Lac Bangles industry is seen in maximum at
(a) Kolkata (b) Jodhpur
(c) Bhubaneswar (d) Rajkot

34. Jute mills are found in maximum at
(a) Kolkata
(b) Kanpur
(c) Chennai
(d) Mumbai

35. IDBI came into existence
(a) 1955 (b) 1964
(c) 1948 (d) 1989

36. SIDBI came into existence
(a) 1955 (b) 1964
(c) 1948 (d) 1989

37. National Small Industry Corporation came into existence
(a) 1955 (b) 1964
(c) 1948 (d) 1989

38. Industrial Finance Corporation of India came into existence
(a) 1955 (b) 1964
(c) 1948 (d) 1989

39. Small Industry Development Organization came into existence
(a) 1951 (b) 1961
(c) 1901 (d) 1991

40. Khadi and Village Industry Commission was established in the year
(a) 1966 (b) 1976
(c) 1956 (d) 1970

41. Entrepreneurship development institute of India was set up in
(a) 1955 (b) 1975
(c) 1983 (d) 1991

42. The extent of refinance for Pilot rainfed farming project by NABARD is
(a) 50% (b) 100%
(c) 75% (d) 25%

43. The extent of refinance for Bio gas Scheme by NABARD is
(a) 50% (b) 100%
(c) 75% (d) 25%

44. The extent of refinance for Farm Mechanization Scheme by NABARD is
(a) 50% (b) 100%
(c) 75% (d) 25%

45. The extent of refinance for Rural electrification by NABARD is
(a) 50% (b) 100%
(c) 75% (d) 25%

46. The head quarter of SIDBI
(a) Mumbai (b) Lucknow
(c) Hyderabad (d) New Delhi

47. The headquarter of BIRD (Bankers Institute for Rural Development)
(a) Mumbai (b) Lucknow
(c) Hyderabad (d) New Delhi

48. Dhara Oilseed is the product of
(a) Agrotech foods ltd
(b) Panak foods
(c) NDDB
(d) Marico Indus

49. Sundrop Oilseed is the product of
(a) Agrotech foods ltd
(b) Panak foods
(c) NDDB
(d) Marico Indus

50. Gemini Oilseed is the product of
(a) Agrotech foods ltd
(b) Panak foods
(c) NDDB
(d) Marico Indus

51. Saffola Oilseed is the product of
(a) Agrotech foods ltd
(b) Panak foods
(c) NDDB
(d) Marico Indus

52. Fortune Oilseed is the product of
(a) Ruchisoya
(b) Adani wilmar ltd
(c) HLL
(d) Kaneria Oils

53. Flora oilseed is the product of
(a) Ruchisoya
(b) Adani wilmar ltd
(c) HLL
(d) Kaneria Oils

54. Soyum oilseed is the product of
(a) Ruchisoya (b) Adani wilmar ltd
(c) HLL (d) Kaneria Oils

55. Rani oilseed is the product of
(a) Ruchisoya (b) Adani wilmar ltd
(c) HLL (d) Kaneria Oils

56. Nature fresh oilseed is the product of
(a) Ruchisoya (b) Cargil
(c) Kaneria Oils (d) Chambal

57. Geepee Ceval oilseed is the product of
(a) Ruchisoya (b) Cargil
(c) Kaneria Oils (d) Chambal

58. Annapurna Atta is the product of
(a) NDDB
(b) HUL
(c) General Mills India
(d) ITC

59. Pillsbury Chhaki fresh Atta is the product of
(a) NDDB
(b) HUL
(c) General Mills India
(d) ITC

60. Captain Cook salt is the product of
(a) HUL
(b) Tata
(c) DCW Home products Ltd
(d) Dabur India

61. Nutra salt is the product of
(a) HUL
(b) Tata
(c) DCW Home products Ltd
(d) Dabur India

62. Hide & Seek biscuits is the product of company
(a) Brittania (b) Cadbury
(c) Parle (d) ITC

63. Sun feast biscuit is the product of
(a) Brittania (b) Cadbury
(c) Parle (d) ITC

64. Oreo biscuit brand is in the name of company
(a) Brittania (b) Cadbury
(c) Parle. (d) ITC

65. Largest Market share of company in biscuit segment
(a) Brittania (b) Cadbury
(c) Parle (d) ITC

66. Nutri Choice biscuit is the brand of
(a) Brittania (b) Cadbury
(c) Parle (d) ITC

67. Little Heart biscuit is the product of
(a) Brittania (b) Cadbury
(c) Parle (d) ITC

68. Mother dairy is the brand name product of
(a) NDRI (b) NDDB
(c) VITA (d) GCMMF

69. VITA is the cooperative society of which state
(a) Odisha
(b) Uttar Pradesh
(c) Haryana
(d) Himachal Pradesh

70. Syngenta AG is global agribusiness company in the field of seeds and agrochemicals. Its head quarter is at
(a) New York
(b) Seol, South Korea
(c) Basel, Switzerland
(d) Mumbai, India

71. Head quarter of Monsanto
(a) New York,USA
(b) Seoul, South Korea
(c) Basel, Switzerland
(d) Creve Coeur, Missouri

72. Industrial Estate Programme was set up in the year
(a) 1955 (b) 1973
(c) 1982 (d) 1965

73. Integrated Infrastructural development Scheme was set up in the year
(a) 1991 (b) 1993
(c) 1995 (d) 1994

74. Industrial Growth Centres Scheme was set up in
(a) 1948 (b) 1968
(c) 1988 (d) 1998

75. Khadi and Village Industries Commission Act was put forth in
(a) 1965 (b) 1956
(c) 1975 (d) 1982

76. Micro, Small and Medium Enterprises Development Act came in the year
(a) 2001 (b) 2005
(c) 2006 (d) 2004

77. According to MSME Development Act,2006 , investment limit for micro enterprises in manufacturing category is
(a) upto Rs 25 lakh
(b) Rs 25 lakh to Rs 5 crore
(c) Rs 5 crore to Rs 10 crore
(d) More than 10 crore

78. According to MSME Development Act, 2006 , investment limit for small enterprises in manufacturing category is
(a) upto Rs 25 lakh
(b) Rs 25 lakh to Rs 5 crore
(c) Rs 5 crore to Rs 10 crore
(d) More than 10 crore

79. According to MSME Development Act, 2006, investment limit for Medium enterprises in manufacturing category is
(a) upto Rs 25 lakh
(b) Rs 25 lakh to Rs 5 crore
(c) Rs 5 crore to Rs 10 crore
(d) More than 10 crore

80. According to MSME Development Act,2006 , investment limit for micro enterprises in service category is
(a) upto Rs 10 lakh
(b) Rs 10 lakh to Rs 2 crore
(c) Rs 2 crore to Rs 5 crore
(d) More than 5 crore

81. According to MSME Development Act,2006 , investment limit for small enterprises in service category is
(a) upto Rs 10 lakh
(b) Rs 10 lakh to Rs 2 crore
(c) Rs 2 crore to Rs 5 crore
(d) More than 5 crore

82. According to MSME Development Act,2006 , investment limit for Medium enterprises in service category is
(a) upto Rs 10 lakh
(b) Rs 10 lakh to Rs 2 crore
(c) Rs 2 crore to Rs 5 crore
(d) More than 5 crore

83. Mahila Vikas Nidhi and Mahila Udyam Nidhi are the scheme of
(a) NABARD (b) SIDBI
(c) IDBI (d) ICICI

84. The Head quarter of Indian Trade Promotion Organisation (ITPO) is in
(a) Nasik, Moharastra
(b) Kanpur, UP
(c) Pragati Maidan, New Delhi
(d) Hyderabad, Andhra Pradesh

85. All India Khadi and Village Board was set up in
(a) 1951 (b) 1953
(c) 1955 (d) 1957

86. Manufacture of Cane Sugar and Khandsari is under the purview of
(a) Ministry of Agriculture
(b) Ministry of Khadi and Village Industries
(c) Ministry of Commerce
(d) Ministry of Rural Development

87. Manufacture of Gums and Resins come under
(a) Ministry of Agriculture
(b) Ministry of Khadi and Village Industries
(c) Ministry of Commerce
(d) Ministry of Rural Development

88. Processing of Maize and Ragi come under
(a) Ministry of Agriculture
(b) Ministry of Khadi and Village Industries
(c) Ministry of Commerce
(d) Ministry of Rural Development

89. Sick Industrial Company Act came in the year
(a) 1985
(b) 1955
(c) 2000
(d) 2005

90. Full form of BIFR
(a) Business of Industrial and Financial Reconstruction
(b) Bureau of Industrial and Financial Reconstruction
(c) Board of Industrial and Financial Reconstruction
(d) None

91. Amalgamation of sick unit with health unit came after the recommendation of
(a) Tiwari Committee
(b) Chokshi Committee
(c) Khusro Committee
(d) Nasrsimah Committee

92. Rastriya krishi Vikas Yojana came in the year
(a) 2000-01
(b) 2005-06
(c) 2007-08
(d) 2008-09

93. From Kharif 2010, pulses component of Integrated Scheme of Oilseeds Pulses Oil palm and Maize (ISOPOM) has been merged under
(a) Rastriya Krishi vikas yojana
(b) National Food Security Mission
(c) Accelerated Pulses Production Programme
(d) Initiation of Management Information System (MIS)

94. According to National Mission on Micro Irrigation, subsidy for small and marginal farmer
(a) 50% of the total cost
(b) 60% of the total cost
(c) 70% of the total cost
(d) 75% of the total cost

95. According to National Mission on Micro Irrigation, subsidy for general category of farmers
(a) 50% of the total cost
(b) 60% of the total cost
(c) 70% of the total cost
(d) 75% of the total cost

96. Under Development of Commercial Horticulture through Production and Post-Harvest Management scheme back ended capital investment subsidy
(a) not exceeding 20% of the total project cost with a maximum limit of Rs.25.00 lakh per project
(b) not exceeding 20% of the total project cost with a maximum limit of Rs.50.00 lakh per project
(c) not exceeding 25% of the total project cost with a maximum limit of Rs.25.00 lakh per project
(d) not exceeding 25% of the total project cost with a maximum limit of Rs.50.00 lakh per project

97. Under Investment Subsidy for Construction/Expansion/Modernization of Cold Storages and Storages for Horticulture Produce scheme back-ended capital investment subsidy
(a) 50% of the project cost
(b) 40% of the project cost
(c) 60% of the project cost
(d) 75% of the project cost

98. Under Investment Subsidy for Construction/Expansion/Modernization of Cold Storages and Storages for Horticulture Produce scheme back-ended capital investment subsidy for N-E region
(a) 50% of the project cost
(b) 40% of the project cost
(c) 60% of the project cost
(d) 55% of the project cost

99. Under Investment Subsidy for Construction/Expansion/Modernization of Cold Storages and Storages for Horticulture Produce, the capacity of cold storage must be
(a) 5000 MT per project
(b) 10,000 MT per project
(c) 20,000 MT per project
(d) 25,000 MT per project

100. Coconut Development Board was established in the year
(a) 1975
(b) 1979
(c) 1981
(d) 1985

101. Central Institute of Horticulture is at
(a) Mumbai (b) Kohima
(c) Nagaland (d) New delhi

102. Under NATIONAL MISSION ON OILSEEDS & OIL PALM , assistance for production of certified and foundation seed is
(a) Rs 500 per quintal
(b) Rs 1000 per quintal
(c) Rs 1500 per quintal
(d) Rs 2000 per quintal

103. Under NATIONAL MISSION ON OILSEEDS & OIL PALM , assistance for distribution of certified and foundation seed is
(a) Rs 500 per quintal
(b) Rs 1000 per quintal
(c) Rs 1200 per quintal
(d) Rs 1500 per quintal

104. Back ended credit support to Agri clinic and agri Business centre is
(a) Subsidy upto 25%
(b) Subsidy upto 36%
(c) Subsidy upto 44%
(d) Subsidy upto 50%

105. Back ended credit support to Agri clinic and agri Business centre for SC/ST/ women/NE region/other hill areas
(a) Subsidy upto 25%
(b) Subsidy upto 36%
(c) Subsidy upto 44%
(d) Subsidy upto 50%

106. Toll free telephone number for Kisan Call Center is
(a) 18001801551 (b) 18001801555
(c) 18001801550 (d) 18001801591

107. National Plant Protection Training Institute is at
(a) New Delhi (b) Pune
(c) Hyderabad (d) Chandigarh

108. Under Development of Market Infrastructure, Grading and Standardization, the subsidy is
(a) 50% of Capital Cost
(b) 25% of Capital cost
(c) 60% of Capital Cost
(d) 75% of Capital Cost

109. Subsidy provided by NABARD for establishment/ modernization of rural slaughter house
(a) 25% of total cost
(b) 50% of total cost
(c) 75% of total cost
(d) 44% of total cost

110. "Innovation and Entrepreneurship" is a book written by
(a) Philip Kotler
(b) C K Prahlad
(c) Pradeep Khandewala
(d) Peter F Drucker

111. Sick Industries is referred to which of the following bodies for rehabilitation
(a) SIDBI (b) SIDO
(c) BIFR (d) NSIC

112. Which of the following organisation is meant for promoting small scale industries at district level
(a) SIDBI (b) DIC
(c) SFC (d) NABARD

113. The measure taken by Government to promote small scale industries
(a) Provision of land
(b) Provision of Marketing Facility
(c) Arrangement of credit and raw material
(d) All

114. Strategies formulated to convert a sick unit to healthy is referred to
(a) Turnaround
(b) Expansion
(c) Diversification
(d) Stabilisation

115. Which is an international credit rating agency?
(a) Mc Kinsey
(b) Standard and Poor
(c) IMF
(d) BCG

116. Which among the following will suit the best as location for a chemical plant
(a) Desert Area (b) Coastal Area
(c) Platen (d) Hill Area

117. The future value of present sum is called as
(a) Discounting
(b) Compounding
(c) Opportunity cost of capital
(d) Annuity

118. The present value of future sum is called
(a) Discounting
(b) Compounding
(c) Opportunity cost of capital
(d) Annuity

119. A stream of payments or returns over time is called as
(a) Discounting
(b) Compounding
(c) Opportunity cost of capital
(d) Annuity

120. If future value of present sum is denoted as A, Principal amount invested is P, interest rate in per cent is I, number of years is t, then formula for compound interest is
(a) $A = P\ (1 - i)^t$
(b) $A = P\ (1 + i)^t$
(c) $A = P\ (1/i)^t$
(d) $A = P\ (1 \times i)^t$

121. If present worth of future money is PW, money value in future is P, rate of interest is I, project life period in years is t, then Present value is
(a) $PW = P\ 1 - \frac{(1 + i)^{-t}}{i}$
(b) $PW = P\ 1 + \frac{(1 + i)^{-t}}{i}$
(c) $PW = P\ 1 + \frac{(1 - i)^{-t}}{i}$
(d) $PW = P\ 1 + \frac{(1 \times i)^{-t}}{i}$

122. The investment criteria that have taken time value into consideration
(a) Payback Period
(b) Accounting rate of Return
(c) Internal Rate of Return
(d) Discounted Pay Back Period

123. Marginal efficiency of capital is also called as
(a) NPV (b) ARR
(c) IRR (d) PBP

124. IRR is the discount rate at which the present value of the net cash flows i.e. NPV is equal to
(a) 1
(b) 0
(c) -1
(d) 2

125. The relationship between Profitability Index (PI), Net Present Value (NPV) and Initial Capital Expenditure(Co) is
(a) PI = NPV/Co
(b) PI = NPV x Co
(c) PI = NPV + Co
(d) PI = NPV – Co

126. The technique that calculate the number of years required to recover the original cash outlay invested in a project is called as
(a) Benefit-Cost ratio
(b) ARR
(c) Pay Back Period
(d) IRR

127. The capital budgeting technique that takes accounting information into consideration is called as
(a) IRR
(b) PBC
(c) ARR
(d) NPV

128. It is widely used as a measure of social benefit in project analysis is
(a) PBP
(b) IRR
(c) B-C Ratio
(d) NPW

129. In which of the following measures, project is made break even, through manipulating the discount factor is
(a) PBP
(b) NPW
(c) IRR
(d) ARR

130. Adjusting the discounted measures in accordance with the variations in project costs and output prices is called as
(a) Market Price
(b) Sensitivity Analysis
(c) Equilibrium Price
(d) NPW

131. The price which is otherwise called as accounting price and represents the true and intrinsic value of a product is called as
(a) Minimum support Price
(b) Market Price
(c) Equilibrium Price
(d) Shadow Price

132. If the change in discounted cash inflow and outflow is positive,
(a) It will be better not to invest
(b) It will be tough situation to decide
(c) It will be wise to go for investment
(d) You may accept the project

133. The ratio of average income with average investment is called as
(a) IRR
(b) PBP
(c) B-C ratio
(d) ARR

134. Calculation of risk on investment by analyzing the impact of alternative variables is called as
(a) Sensitivity Analysis
(b) Simulation Analysis
(c) Scenario Analysis
(d) Market Analysis

135. Calculation of risk on investment by considering interaction among variables and probability of change in variables is called as
(a) Sensitivity Analysis
(b) Simulation Analysis
(c) Scenario Analysis
(d) Market Analysis

136. The entrepreneur who imitates techniques and technology innovate by others is
(a) Innovative entrepreneur
(b) Imitative entrepreneur
(c) Fabian entrepreneur
(d) Drone entrepreneur

137. An entrepreneur who is very cautious in taking the decisions
(a) Innovative entrepreneur
(b) Imitative entrepreneur
(c) Fabian entrepreneur
(d) Drone entrepreneur

138. An entrepreneur who is conservative or orthodox in outlook and feel comfortable with their old-fashioned technology is
(a) Innovative entrepreneur
(b) Imitative entrepreneur
(c) Fabian entrepreneur
(d) Drone entrepreneur

139. Enterprising couples who work in and share ownership is called
(a) Copreneur
(b) Intrapreneur
(c) Drone entrepreneur
(d) Fabian entrepreneur

140. An organizational entrepreneur is also called
(a) Copreneur
(b) Intrapreneur
(c) Drone entrepreneur
(d) Fabian entrepreneur

141. Homemaker entrepreneur is otherwise called
(a) Women entrepreneur
(b) Hompreneur
(c) Mompreneur
(d) E-entrepreneur

142. National Institute for Entrepreneurship and Small Business development is at
(a) Delhi
(b) Ahmedabad
(c) Mumbai
(d) Hyderabad

143. Rural Entrepreneurship Development Institute is at
(a) Kanpur (b) Bhubaneswar
(c) Ranchi (d) Pune

144. Which city is more favourable for selling electric blanket
(a) Mumbai (b) Chennai
(c) Shillong (d) Pune

145. Global compact relates to
(a) Human Rights
(b) Social responsibility
(c) Environment Protection
(d) Women rights

146. Global compact is an initiative of
(a) UN (b) WTO
(c) IMF (d) ILO

147. The criteria for small business is
(a) Capital invested
(b) Number of managers
(c) Number of employees
(d) Types of business

148. The most appropriate goal of corporate social responsibility is centered around
(a) Philanthropic activities
(b) Voluntary activities
(c) Ethical based activities
(d) Legalistic activities

149. Technology Development and Modernization Fund is provided by
(a) SIDBI (b) NABARD
(c) KVIC (d) SIDO

150. Which of the following organizations provide institutional support to SSI?
(a) Indian Institute of Foreign Trade
(b) Nationalized Banks
(c) Tool Design Institute of SIDO
(d) Ministry of Commerce

151. Egalitarian behaviour refers to
(a) Natural behavior
(b) Moral behavior
(c) Social behavior
(d) Ethical behavior

152. Which of the following is the cause of sickness of an enterprise
(a) Lack of adequate capital
(b) Lack of competition
(c) Lack of demand of the products
(d) All of the above

153. In India the first report on corporate governance was given by
(a) Narayan Murthy committee
(b) Cadbury's committee
(c) Naresh Chandra committee
(d) Kumarmangalam Birla committee

154. Agenda 21 relates to
(a) Child labour
(b) Human rights
(c) Environment
(d) None of these

155. Small and Medium enterprises are
(a) Labour intensive
(b) Capital intensive
(c) Market leader
(d) Industry price determiner

ANSWER KEY (Project Management and Entrepreneurship Development)

1. (a) **2.** (b) **3.** (d) **4.** (c) **5.** (b) **6.** (c) **7.** (b) **8.** (b) **9.** (b) **10.** (b)
11. (b) **12.** (b) **13.** (a) **14.** (c) **15.** (b) **16.** (b) **17.** (b) **18.** (c) **19.** (b) **20.** (d)
21. (a) **22.** (c) **23.** (b) **24.** (b) **25.** (b) **26.** (b) **27.** (c) **28.** (d) **29.** (d) **30.** (b)
31. (c) **32.** (c) **33.** (b) **34.** (a) **35.** (b) **36.** (d) **37.** (a) **38.** (c) **39.** (a) **40.** (c)
41. (c) **42.** (b) **43.** (c) **44.** (a) **45.** (a) **46.** (b) **47.** (b) **48.** (c) **49.** (a) **50.** (b)
51. (d) **52.** (b) **53.** (c) **54.** (a) **55.** (d) **56.** (b) **57.** (d) **58.** (b) **59.** (c) **60.** (c)
61. (d) **62.** (c) **63.** (d) **64.** (b) **65.** (c) **66.** (a) **67.** (a) **68.** (b) **69.** (c) **70.** (c)
71. (d) **72.** (a) **73.** (d) **74.** (c) **75.** (b) **76.** (c) **77.** (a) **78.** (b) **79.** (c) **80.** (a)

81. (b)	**82.** (c)	**83.** (b)	**84.** (c)	**85.** (b)	**86.** (b)	**87.** (b)	**88.** (b)	**89.** (a)	**90.** (c)
91. (b)	**92.** (c)	**93.** (b)	**94.** (b)	**95.** (a)	**96.** (a)	**97.** (b)	**98.** (d)	**99.** (a)	**100.** (b)
101. (c)	**102.** (b)	**103.** (c)	**104.** (b)	**105.** (c)	**106.** (a)	**107.** (c)	**108.** (a)	**109.** (b)	**110.** (d)
111. (c)	**112.** (b)	**113.** (d)	**114.** (a)	**115.** (b)	**116.** (b)	**117.** (b)	**118.** (a)	**119.** (d)	**120.** (b)
121. (a)	**122.** (c)	**123.** (c)	**124.** (b)	**125.** (a)	**126.** (c)	**127.** (c)	**128.** (c)	**129.** (c)	**130.** (d)
131. (d)	**132.** (c)	**133.** (d)	**134.** (c)	**135.** (b)	**136.** (b)	**137.** (c)	**138.** (d)	**139.** (a)	**140.** (b)
141. (c)	**142.** (a)	**143.** (c)	**144.** (c)	**145.** (b)	**146.** (a)	**147.** (a)	**148.** (d)	**149.** (a)	**150.** (c)
151. (b)	**152.** (d)	**153.** (d)	**154.** (c)	**155.** (a)					

CHAPTER 12

12. Research Methodology

1. Data that are collected at fresh or for the first time is
(a) Primary data (b) Secondary data
(c) Both (d) None

2. Data which have been collected by someone else is
(a) Primary data (b) Secondary data
(c) Both (d) None

3. The source that is not categorized in secondary data collection method is by
(a) Collection from web portal
(b) Collection from books
(c) Collection through schedules
(d) Collection from trade journal

4. The demerit of data collection through observation method is
(a) It is expensive
(b) Unforeseen factors may interfere with observations
(c) Emotional attachment of the participates may loss the objectivity
(d) All of the above

5. Interview where only predetermined questions have being asked in a highly standardized technique is
(a) Structured Interview
(b) Unstructured Interview
(c) Both
(d) None

6. The interview that is concerned with broad underlying feeling of individual's life experience is called as
(a) Focused interview
(b) Clinical interview
(c) Non-directive interview
(d) All

7. The form of question in questionnaire where the respondents are invited for free response is
(a) Close question
(b) Open question
(c) Both
(d) None

8. Proforma containing a set of questions are being filled by the enumerator is
(a) Questionnaire (b) Schedules
(c) Invoice (d) Warranty cards

9. The device of data collection that records extent of body excitement as a result of the visual stimulus is
(a) Eye Camera
(b) Motion Picture Camera
(c) Pupilometric Camera
(d) Psychogalvanometer

10. The device of data collection that is mostly use as designing advertising material is
(a) Eye Camera
(b) Motion Picture Camera
(c) Pupilometric Camera
(d) Psychogalvanometer

11. The test that helps the respondents to create inference from the picture is
(a) Rosenzweig test
(b) Thematic Apperception Test
(c) Rorschach test
(d) Holtzman Inkblot test

12. The part of the population on the basis of which a judgment can be made is
(a) Sample (b) Sample frame
(c) Census (d) Universe

13. Total of the items in any field of enquiry is
(a) Universe (b) Population
(c) Both (d) None

14. A list of sampling unit of the population is
(a) Sampling design
(b) Sample frame
(c) Sampling analysis
(d) Response

15. A definite plan for obtaining a sample from the sampling is
(a) Sampling analysis
(b) Sampling design
(c) Sample survey
(d) Sample Check out

16. A characteristic of a sample is
(a) Statistics (b) Parameter
(c) Both (d) None

17. A characteristic of population is
(a) Statistics (b) Parameter
(c) Both (d) None

18. Types of sampling error is
(a) Frame error
(b) Chance error
(c) Response error
(d) All

19. More the size of the sample, more will be
(a) Sampling error
(b) Non sampling error
(c) Response error
(d) All

20. The standard deviation of sampling distribution of a statistic is known as
(a) Sampling error
(b) Non-sampling error
(c) Standard error
(d) None of the above

21. In sampling technique in which every item of the universe has an equal chance of inclusion is
(a) Probability sampling
(b) Non-probability sampling
(c) Both
(d) None

22. Probability sampling does not include
(a) Simple Random Sampling
(b) Stratified Sampling
(c) Quota Sampling
(d) Systemic Random Sampling

23. Non-Probability sampling does not include
(a) Purposive Sampling
(b) Judgment Sampling
(c) Cluster Sampling
(d) Quota Sampling

24. Who among the following scientists have prepared random number table?
(a) L.H.C. Tippett
(b) Francis Yates
(c) R.A. Fisher
(d) All of the above

25. Types of sampling when an enumerator choose every item at a particular interval from the sampling frame is called as
(a) Stratified Sampling
(b) Systemic Sampling
(c) Cluster Sampling
(d) Sequential Sampling

26. If ten items have to be chosen from a list of two hundred sample units, then according to systematic sampling technique , first unit should be in between
(a) 1-10 (b) 1-20
(c) 1-30 (d) 1-40

27. Find the correct statement
(a) Elements within the strata are homogenous in nature
(b) Elements between the strata are heterogeneous in nature
(c) The strata should be mutually exclusive and collectively exhaustive
(d) All are correct

28. According to researcher, the number of strata should be
(a) Within 5
(b) Within 6
(c) Within 10
(d) No limit

29. Find the wrong one
(a) The elements within the cluster are homogenous in nature
(b) Two clusters are homogenous to each other
(c) Objectives of cluster sampling to increase the sampling efficiency
(d) Probability of selection has to be done in two stage cluster sampling

30. Area sampling is a type of
(a) Quota Sampling
(b) Cluster Sampling
(c) Stratified Sampling
(d) Simple Random Sampling

31. A non-probability sampling technique in which samples are collected in waves by obtaining referrals from referrals are
(a) Quota Sampling
(b) Judgmental Sampling
(c) Snowball Sampling
(d) Convenient Sampling

32. The scaling technique that provide convenient ways of keeping track of people, object and events by assigning number symbols to them is called as
(a) Nominal Scale (b) Ordinal Scale
(c) Interval Scale (d) Ratio Scale

33. The scaling technique that ranks items from highest to lowest is
(a) Nominal Scale (b) Ordinal Scale
(c) Interval Scale (d) Ratio Scale

34. The scaling technique that establishes meaning of difference between two variable is
(a) Nominal Scale (b) Ordinal Scale
(c) Interval Scale (d) Ratio Scale

35. The scale that represents the actual amounts of variables as it is the absolute or true value of measurement is
(a) Nominal Scale (b) Ordinal Scale
(c) Interval Scale (d) Ratio Scale

36. Multiplication and division of the variables can only be managed through
(a) Nominal Scale (b) Ordinal Scale
(c) Interval Scale (d) Ratio Scale

37. The Nominal Scale is represented through
(a) Number
(b) Alphabet
(c) Symbol
(d) All of the above

38. A comparative scaling technique in which respondents are presented with several objects simultaneously and asked to order or rank them according to some criteria
(a) Paired comparision scaling
(b) Rank order scaling
(c) Constant sum scaling
(d) Q-sort scaling

39. A comparative scaling technique in which respondents are required to allocate a constant sum of units such as points, dollars among a set of stimulus objects with respect to some criteria
(a) Paired comparision scaling
(b) Rank order scaling
(c) Constant sum scaling
(d) Q-sort scaling

40. A comparative scaling technique that uses a rank order procedure to sort objects based on similarity with respect to some criteria
(a) Paired comparision scaling
(b) Rank order scaling
(c) Constant sum scaling
(d) Q-sort scaling

41. Non-comparative scaling technique, where respondents rate the object by placing a mark at the appropriate position is
(a) Continuous rating scale
(b) Likert Scale
(c) Semantic Differential Scale
(d) Stapel Scale

42. Graphic rating scale is also called as
(a) Continuous rating scale
(b) Likert Scale
(c) Semantic Differential Scale
(d) Stapel Scale

43. Itemised rating scale that categorized into five responses ranging from strongly disagree to strongly agree is known as
(a) Likert Scale
(b) Semantic Differential Scale
(c) Stapel Scale
(d) Thurstone Scale

44. A 7-point rating scale with end points associated with bipolar adjectives like 'cold' and 'warm' is
(a) Likert Scale
(b) Semantic Differential Scale
(c) Stapel Scale
(d) Thurstone Scale

45. Differential scale technique in which selection of items is made by a panel of judges is
(a) Likert Scale
(b) Semantic Differential Scale
(c) Stapel Scale
(d) Thurstone Scale

46. A scale for measuring attitudes that consists of a single adjective in the middle of an even numbered range of values, from -5 to +5 without a neutral point is
(a) Likert Scale
(b) Semantic Differential Scale
(c) Stapel Scale
(d) Thurstone Scale

47. Stapel scale is usually presented
(a) Horizontally (b) Vertically
(c) Slanting manner (d) Circularly

48. Types of research design, where prior formulation of hypothesis has not been developed is known as
(a) Descriptive research
(b) Exploratory research
(c) Causal research
(d) None of the above

49. Flexible and unstructured research process is the characteristics of
(a) Descriptive research
(b) Exploratory research
(c) Causal research
(d) None of the above

50. Types of research design that is marked by a clear statement of the problem, specific hypothesis and detail information needs is
(a) Descriptive research
(b) Exploratory research
(c) Causal research
(d) None of the above

51. Research design that is established by the assumption of cause and effect relationship among the variables is
(a) Descriptive research
(b) Exploratory research
(c) Causal research
(d) None of the above

52. Research design that involve the collection of information from any given sample of population elements only once is
(a) Cross sectional design
(b) Longitudinal design
(c) Both
(d) None

53. Types of research design involving a fixed sample of population elements that is measured repeatedly is
(a) Cross sectional design
(b) Longitudinal design
(c) Both
(d) None

54. A research process starts from
(a) Formulation of hypothesis
(b) Define research problem
(c) Design research
(d) Collection of data

55. Independent variable that is not related to the purpose of study is
(a) Controlled variable
(b) Explicit Variable
(c) Extraneous variable
(d) Confounded Variable

56. According to Professor R A Fisher, the principle that is not relevant in research design is
(a) Principle of replication
(b) Principle of reflection
(c) Principle of randomization
(d) Principle of local control

57. 'An experiment should be repeated more than once' is the basic principle comes under
(a) Principle of replication
(b) Principle of randomization
(c) Principle of local control
(d) None of the above

58. The principle that states that we conduct an experiment against the effect of extraneous factors is
(a) Principle of replication
(b) Principle of randomization
(c) Principle of local control
(d) None of the above

59. The principle that measure and remove the extraneous factor, the known source of variability from the research design is
(a) Principle of replication
(b) Principle of randomization
(c) Principle of local control
(d) None of the above

60. Principle of local control is not found in
(a) Completely randomized design
(b) Randomized block design
(c) Latin square design
(d) All of the above

61. The experimental design where we find homogeneous experimental materials is
(a) Completely randomized design
(b) Randomized block design
(c) Latin square design
(d) All of the above

62. In Randomized Block Design, if the number of subjects in each block is 5 then the number of treatment it will be allocated is
(a) 3
(b) 4
(c) 5
(d) 6

63. Find the wrong statement
(a) In RBD, the subject material within the blocks are homogenous in nature
(b) In RBD, the number of treatments should be equal to number of subjects in each block
(c) In RBD, treatments are applied randomly in each block
(d) In RBD, the number of treatment is equal to number of blocks

64. Experimental design that is very frequently used in agriculture is known as
(a) Completely randomized design
(b) Randomized block design
(c) Latin square design
(d) All of the above

65. The experimental design that is always balanced with both rows and column
(a) Completely randomized design
(b) Randomized block design
(c) Latin square design
(d) All of the above

66. When there is more number of rows than column then that matrix is called as
(a) Horizontal Matrix
(b) Vertical Matrix
(c) Both
(d) None

67. If in a matrix number of rows is equal to number of column, it is called
(a) Horizontal matrix
(b) Vertical matrix
(c) Square matrix
(d) Row matrix

68. A square matrix, in which all the elements, except those in the leading diagonal are zero, is called as
(a) Horizontal matrix
(b) Vertical matrix
(c) Scalar matrix
(d) None

69. A square matrix in which all the elements in the leading diagonal are one and remaining elements are zero, is called as
(a) Row matrix
(b) Vertical matrix
(c) Unit matrix
(d) Null matrix

70. When the determinant of a matrix is zero, it is called as
(a) Null matrix
(b) Singular matrix
(c) Nonsingular matrix
(d) Diagonal matrix

71. When one matrix of 3 rows and 2 column is multiplied with another matrix of 2 rows and 3 column, the new matrix that is formed having
(a) 2 rows and 2 column
(b) 3 rows and 2 column
(c) 2 rows and 3 column
(d) 3 rows and 3 column

72. A square matrix where aij =aji, then it is called as
(a) Singular matrix
(b) Symmetric matrix
(c) Skew symmetric matrix
(d) Null matrix

73. A matrix is obtained by interchanging rows and column is called as
(a) Singular matrix
(b) Symmetric matrix
(c) Skew symmetric matrix
(d) Transpose matrix

74. If the product of matrix A and its transpose A1 is an identity matrix, it is called as
(a) Hermitian matrix
(b) Diagonal matrix
(c) Orthogonal matrix
(d) Idempotent matrix

75. Mathematical diagram that display values for two variables for a set of data is
(a) Scalar diagram
(b) Gantt Chart
(c) Scatter Diagram
(d) P chart

76. The statistical relationship between two or more variables that shows degree of association is
(a) Causation
(b) Correlation
(c) Regression
(d) All of the above

77. Types of correlation where we study the relationship between two variable by eliminating the effect of other variable is
(a) Multiple correlation
(b) Spurious correlation
(c) Partial correlation
(d) Bivariate correlation

78. Coefficient of correlation mainly is denoted as
(a) a (b) b
(c) r (d) r^2

79. Coefficient of correlation ranges from
(a) 0 to 1 (b) -1 to 0
(c) -1 to 1 (d) $-\infty$ to ∞

80. Regression coefficient ranges from
(a) 0 to 1 (b) -1 to 0
(c) -1 to 1 (d) $-\infty$ to ∞

81. If the regression coefficient is the coefficient of determination, then it ranges from
(a) 0 to 1 (b) -1 to 0
(c) -1 to 1 (d) $-\infty$ to ∞

82. The value $R_{1.23}$ lies between
(a) 0 to 1 (b) -1 to 0
(c) -1 to 1 (d) $-\infty$ to ∞

83. Correlation coefficient indicates
(a) Magnitude of relationship
(b) Direction of relationship
(c) Both
(d) None

84. In scatter diagram when the relationship between variables is curvilinear then coefficient of correlation is
(a) -1
(b) 0
(c) 1
(d) Positive

85. The statement that is false
(a) Coefficient of correlation is unit less and pure number
(b) Coefficient of correlation ranges from -1 to1
(c) Coefficient of correlation is independent for change in origin but depend on change in scale
(d) $r_{xy} = r_{yx}$

86. If coefficient of correlation is 0.9 then the coefficient of determination is
(a) 0.9 (b) 0.91
(c) 0.81 (d) 0.18

87. Coefficient of determination varies in the range of
(a) 0 to 1 (b) -1 to 0
(c) -1 to 1 (d) $-\infty$ to ∞

88. Regression concept is discovered by
(a) Karl Pearson (b) Francis Galton
(c) Robert Fisher (d) Mendel

89. If two regression line intersects at right angle then coefficient of correlation is
(a) -1
(b) 0
(c) 1
(d) Negative

90. If two regression lines coincide then coefficient of correlation is
(a) -1
(b) 1
(c) 0
(d) Either a or b

91. The relationship between correlation coefficient and regression coefficient is
(a) Correlation coefficient is the arithmetic mean between the regression coefficient
(b) Correlation coefficient is the geometric mean between the regression coefficient
(c) Correlation coefficient is the harmonic mean between the regression coefficient
(d) No relationship

92. Find the wrong statement
(a) Regression coefficient is the independent of change of origin but change in scale
(b) Arithmetic mean of regression coefficient is smaller than the correlation coefficient
(c) If one of the regression coefficient is greater than unity, the other must be less than unity
(d) If two regressions make a very small acute angle, regression coefficient is very high

93. Independent variable is also called as
(a) Criterion variable
(b) Predictor variable
(c) Both
(d) None

94. Dependent variable is also called
(a) Criterion variable
(b) Predictor variable
(c) Both
(d) None

95. The term variance first used by
(a) Karl Pearson
(b) Francis Galton
(c) Robert Fisher
(d) Professor Snedecor

96. The term variance first coined by
(a) Karl Pearson
(b) Francis Galton
(c) Robert Fisher
(d) Professor Snedecor

97. Nature of the independent variable in ANOVA is
(a) Metric (b) Categorical
(c) Both (d) None

98. Nature of the dependent variable in ANOVA is
(a) Metric (b) Categorical
(c) Both (d) None

99. Nature of the dependent variable in regression is
(a) Metric (b) Categorical
(c) Both (d) None

100. Nature of the independent variable in regression is
(a) Metric
(b) Categorical
(c) Both
(d) None

101. Find the wrong one
(a) R^2 tells how the regression line predicts the estimates to actual value
(b) R^2 justifies the proportion of the variation in Y being explained by the variation in X
(c) R^2 tells the proportion of the variation of independent variables explained by the dependent variable
(d) $R^2 = \frac{\sum (\text{Estimated value} - \text{Mean of dependent variable})^2}{\sum (\text{Actual value} - \text{Mean of dependent variable})^2}$

102. Residual Sum of Square is otherwise called as
(a) Sum of square due to regression
(b) Explained sum of square
(c) Sum of Square due to error
(d) Total sum of square

103. Explained sum of Square is otherwise called as
(a) Sum of square due to regression
(b) Residual Sum of Square
(c) Sum of Square due to error
(d) Total sum of square

104. If number of observation is 'N' and number of explanatory variable is 'K', then degree of freedom 'df' is
(a) N + K
(b) N – K + 2
(c) N - K – 1
(d) N + K + 1

105. If sum of square within group is 42, sum of square between group is 6, then total sum of square is
(a) 36
(b) 7
(c) 48
(d) 252

106. Adding more independent variables to a multiple regression procedure does not mean the regression will be better or offer better predictions; in fact it can make things worse. This is called as
(a) Partial collenearity
(b) Multi collenearilty
(c) Overfitting
(d) All of the above

107. The addition of more independent variables creates more relationships among them. So not only are the independent variables related to the dependent variable, they are also potentially related to each other. When this happens, it is called as
(a) Partial collenearity
(b) Multi collenearity
(c) Multiple Regression
(d) Overfitting

108. The average distance of data point from regression line in dependent variables unit is called as
(a) Variance
(b) Standard Error of regression
(c) Adjusted R^2
(d) Degree of Freedom

109. When there is one independent variables, the relationship between p- value of independent value coefficient and p- value of ANOVA is
(a) Both are equal
(b) p- value of independent value coefficient is greater
(c) p- value of independent value coefficient is smaller
(d) no relationship

110. If there is no relationship between independent variables among themselves, then Variance Inflation Factor (VIF) is equal to
(a) 1 (b) 1
(c) 5 (d) > 10

111. If numbers of predictors is 3, then Mallows Cp should be
(a) 3 (b) 4
(c) 5 (d) 0

112. Factor analysis is otherwise called as
(a) Data Reduction Technique
(b) Dimension Reduction Technique
(c) Both
(d) None

113. Mallows Cp is an indicator to understand multicollenearity among independent variables. When Number of Predictors are P then Mallows Cp should be
(a) P – 1 (b) P + 1
(c) 2P + 1 (d) 2P – 1

114. If number of variables are 10, then it is advisable that numbers of respondents should be in the range of
(a) 10-20 (b) 20-30
(c) 30-40 (d) 40-50

115. KMO is an indicator to understand data adequacy in
(a) Discriminant Analysis
(b) Factor Analysis
(c) Cluster Analysis
(d) Conjoint Analysis

116. Range of KMO is in between
(a) -1 to +1 (b) 0 to 1
(c) $-\infty$ to $+\infty$ (d) 0 to $+\infty$

117. Bartlett's test of Sphericity is a test to understand any correlation among the variables in
(a) Discriminant Analysis
(b) Factor Analysis
(c) Cluster Analysis
(d) Conjoint Analysis

118. Eigen Value that represents total variance explain by each factor, should be accepted in value
(a) 0 (b) 0 to 1
(c) >1 (d) Negative

119. Wilk's Lambda is an indicator for
(a) Discriminant Analysis
(b) Factor Analysis
(c) Cluster Analysis
(d) Conjoint Analysis

120. Range of Wilk's Lambda is
(a) -1 to +1 (b) 0 to 1
(c) $-\infty$ to $+\infty$ (d) 0 to $+\infty$

121. Wilk's Lambda tells about
(a) Explained Variance/ Total Variance
(b) Unexplained Variance/ Total Variance
(c) Only Total Variance
(d) Only Between group Variance

122. BOX's test of equality of Covariance matrices mainly taken in to consideration in
(a) Discriminant Analysis
(b) Factor Analysis
(c) Cluster Analysis
(d) Conjoint Analysis

123. Canonical Correlation is an indicator to be observed in
(a) Discriminant Analysis
(b) Factor Analysis
(c) Cluster Analysis
(d) Conjoint Analysis

124. Eigen Value tells about
(a) Explained Variance/ Total Variance
(b) Explained Variance/ Unexplained Variance
(c) Unexplained Variance/ Explained Variance
(d) Unexplained Variance/ Total Variance

125. Hierarchical tree is a nexted set of partition of data that is going to be represented by a tree diagram in
(a) Discriminant Analysis
(b) Factor Analysis
(c) Cluster Analysis
(d) Conjoint Analysis

126. Difference between Cluster analysis and Decision tree is that
(a) In cluster analysis, there is no objective function
(b) In cluster analysis, there is no dependent variables
(c) Both
(d) None

127. Shortest distance between two points which are elements of two different cluster is called as
(a) Single Linkage
(b) Double Linkage
(c) Complete Linkage
(d) Multi Linkage

128. Farthest distance between two points which are elements of two different cluster is called as
(a) Single Linkage
(b) Double Linkage
(c) Complete Linkage
(d) Multi Linkage

129. K-mean Clustering analysis is a
(a) Hierarchical Clustering technique
(b) Non Hierarchical Clustering technique
(c) Both
(d) None

130. Nature of the independent variable in discriminant analysis is
(a) Metric (b) Categorical
(c) Both (d) None

131. Nature of the dependent variable in discriminant analysis is
(a) Metric
(b) Categorical
(c) Both
(d) None

132. A statistical tool- Eigen value helps in variance analysis, mainly used in
(a) Factor Analysis
(b) Cluster Analysis
(c) Conjoint Analysis
(d) Multidimensional scaling

133. Scree plot is a statistical tool used in
(a) Factor Analysis
(b) Cluster Analysis
(c) Conjoint Analysis
(d) Multidimensional scaling

134. Icicle plot is a statistical tool used in
(a) Factor Analysis
(b) Cluster Analysis
(c) Conjoint Analysis
(d) Multidimensional scaling

135. Index number is always expressed in
(a) Square
(b) Rate
(c) Percentage
(d) Decimal

136. The reference period for comparison between two year has been written in
(a) Numerator
(b) Denominator
(c) Both
(d) None

137. Inflation is measured in India on the basis of which index
(a) Wholesale Price Index
(b) Retail Price Index
(c) Consumer Price Index
(d) All of the above

138. Hypothesis is a
(a) Theory
(b) Technique
(c) Assumption
(d) Model

139. The shape of the normal distribution curve is
(a) Square shape
(b) Curvilinear shape
(c) Bell shape
(d) Circular shape

140. An error in rejecting the null hypothesis when it is true is
(a) Type I error
(b) Type II error
(c) Both
(d) None

141. An error in accepting the null hypothesis when it is false is
(a) Type I error
(b) Type II error
(c) Both
(d) None

142. Type I error is also called
(a) Level of determinant
(b) Degree of freedom
(c) Level of significance
(d) All of the above

143. X^2 distribution curve is
(a) Skewed to the left
(b) Skewed to the right
(c) Normally distributed
(d) None of the above

144. The probable value in x2 test is
(a) Negative
(b) Zero
(c) Positive
(d) All of the above

145. T test for hypothesis testing is proposed by
(a) Francis Galton
(b) W S Gosset
(c) Fisher
(d) Karl Pearson

146. Find the wrong statement
(a) t-distribution is family of curves
(b) In t-distribution, the population standard deviation is not known
(c) The smaller the sample, the greater will be the critical t-score value
(d) The critical t-score value is smaller than z-score value for a given level of significance

147. Which is the non-parametric test among following
(a) Z-test (b) Run test
(c) T-test (d) F test

148. In research report, the font size of the letter should be
(a) 10 (b) 11
(c) 12 (d) 14

149. In research report, the standard theme font of the letter should be
(a) Ariel Black
(b) Calibri
(c) Times New Roman
(d) Aharoni

150. In research report, the standard spacing between the lines should be
(a) 1.0 spacing
(b) 1.5 spacing
(c) 2.0 spacing
(d) 2.5 spacing

151. The correct margin for a standard research report is
(a) At least 1 and 11/2 inch left hand margin
(b) At least 1 inch right hand margin
(c) Both the above
(d) None

152. The correct sequence for writing bibliography for books and pamphlets are

(1) Name of the author, last name

(2) Title, underlined to indicate italics

(3) Place, publisher and date of publication

(4) Number of volumes
(a) 1243 (b) 2143
(c) 1234 (d) 4321

153. The correct sequence for writing bibliography for magazines and newspaper are

(1) Name of the author, last name first

(2) Title of the article, in quotation mark

(3) Name of periodical, underlined to indicate italics

(4) The volume or volume and number

(5) The date of the issue

(6) The pagination
(a) 123654
(b) 123564
(c) 652341
(d) 123456

154. In research report, the table is represented through
(a) Arabic number
(b) Title
(c) Both
(d) None

155. In table of a research report, the information cannot be incorporated should be explained by
(a) Headings
(b) Stubs
(c) Footnotes
(d) All

156. Designation placed over the vertical column of the table is called
(a) Headings (b) Stubs
(c) Both (d) None

157. Designation placed over the left hand column of the table is called
(a) Headings (b) Stubs
(c) Both (d) None

158. The footnotes in the table of a research report is represented through
(a) Numbers (b) Symbols
(c) Letters (d) Both b and c

159. The footnotes should come
(a) After the main table
(b) Before the source note
(c) Both
(d) None

160. In a research report, the abbreviation, 'cf.,' means
(a) chapter (b) compare
(c) column (d) correction

161. In a research report, the abbreviation, 'eng.,' means
(a) chapter (b) compare
(c) column (d) correction

162. In a research report, the abbreviation, 'cf.,' means
(a) long (b) line
(c) letter (d) last

163. In a research report, the abbreviation, 'n.p.,' means
(a) no place
(b) no date
(c) no publisher
(d) note well

ANSWER KEY (Research Methodology)

1. (a) **2.** (b) **3.** (c) **4.** (d) **5.** (a) **6.** (b) **7.** (b) **8.** (b) **9.** (b) **10.** (a)
11. (b) **12.** (a) **13.** (c) **14.** (b) **15.** (b) **16.** (a) **17.** (b) **18.** (d) **19.** (b) **20.** (c)
21. (a) **22.** (c) **23.** (c) **24.** (d) **25.** (b) **26.** (b) **27.** (d) **28.** (b) **29.** (a) **30.** (b)
31. (c) **32.** (a) **33.** (b) **34.** (c) **35.** (d) **36.** (d) **37.** (d) **38.** (b) **39.** (c) **40.** (d)
41. (a) **42.** (a) **43.** (a) **44.** (b) **45.** (d) **46.** (c) **47.** (b) **48.** (b) **49.** (b) **50.** (a)
51. (a) **52.** (a) **53.** (b) **54.** (b) **55.** (c) **56.** (b) **57.** (a) **58.** (b) **59.** (c) **60.** (a)
61. (a) **62.** (c) **63.** (d) **64.** (c) **65.** (c) **66.** (b) **67.** (c) **68.** (c) **69.** (c) **70.** (b)
71. (d) **72.** (b) **73.** (d) **74.** (c) **75.** (c) **76.** (b) **77.** (c) **78.** (c) **79.** (c) **80.** (d)
81. (a) **82.** (a) **83.** (c) **84.** (b) **85.** (c) **86.** (c) **87.** (a) **88.** (b) **89.** (b) **90.** (d)
91. (b) **92.** (b) **93.** (b) **94.** (a) **95.** (c) **96.** (d) **97.** (b) **98.** (a) **99.** (a) **100.** (a)
101. (c) **102.** (c) **103.** (a) **104.** (c) **105.** (c) **106.** (c) **107.** (b) **108.** (b) **109.** (a) **110.** (b)
111. (b) **112.** (c) **113.** (b) **114.** (d) **115.** (b) **116.** (b) **117.** (b) **118.** (c) **119.** (a) **120.** (b)
121. (b) **122.** (a) **123.** (a) **124.** (b) **125.** (c) **126.** (c) **127.** (a) **128.** (c) **129.** (b) **130.** (d)
131. (c) **132.** (c) **133.** (a) **134.** (b) **135.** (d) **136.** (c) **137.** (a) **138.** (b) **139.** (a) **140.** (a)
141. (b) **142.** (c) **143.** (b) **144.** (c) **145.** (c) **146.** (c) **147.** (a) **148.** (b) **149.** (c) **150.** (b)
151. (c) **152.** (b) **153.** (d) **154.** (b) **155.** (c) **156.** (c) **157.** (c) **158.** (c) **159.** (c) **160.** (b)
162. (b) **163.** (b)

13. General Study in Agribusiness Management

1. When was ICAR established?
(a) 16 July, 1929
(b) 16th July 1949
(c) 19th July 1926
(d) 19th July 1949

2. Which committee recommended RAWE programme?
(a) Randhawa Committee
(b) Gajendra Gadkar Committee
(c) Mohan Singh Mehta Committee
(d) Khusro Committee

3. When was ASRB established?
(a) 1st Nov, 1983
(b) 1st Nov, 1973
(c) 1st Oct, 1973
(d) 1st Oct, 1983

4. Which committee recommended ASRB as an independent recruitment agency?
(a) Randhawa Committee
(b) Gajendra Gadkar Committee
(c) Mohan Singh Mehta Committee
(d) Khusro Committee

5. Who is the Director General of ICAR
(a) Dr. S. Ayyapan
(b) Dr. Gurbachan Singh
(c) Dr. T. Mohapatra
(d) Dr. S.K. Bandyopadhyay

6. Who is the chairman of ASRB
(a) Dr. S. Ayyapan
(b) Dr A K Misra
(c) Dr. V.N. Sharda
(d) Dr. S.K. Bandyopadhyay

7. Headquarter of ICAR is located at
(a) Mumbai (b) New Delhi
(c) Kolkata (d) Chennai

8. The Indian Council of Agricultural Research (ICAR) is an autonomous organization under
(a) Department of Agricultural Research and Education (DARE)
(b) Department of Agriculture and Cooperation
(c) Department of Animal Husbandry, Dairying & Fisheries
(d) All of the above

9. First Agricultural University was established in India at_______ in ______
(a) Hisar, 1956 (b) Pantnagar, 1960
(c) Pusa, 1960 (d) Hyderabad, 1956

10. Central Agricultural University is located at
(a) New Delhi (b) Varanasi
(c) Jhansi (d) Imphal

11. First Krishi Vigyan Kendra was established at ____________ in _______
(a) Madras, 1974
(b) Puducherry, 1964
(c) Puducherry, 1974
(d) Madras, 1964

12. Total No. of Krishi Vigyan Kendras (KVKs) as of date (November 2016) is
(a) 467 (b) 367
(c) 537 (d) 651

13. KVK was established on the recommendation of
(a) Randhawa Committee
(b) Gajendra Gadkar Committee
(c) Mohan Singh Mehta Committee
(d) Khusro Committee

14. National Agricultural Technology Project (NATP) was launched in the year
(a) 1998 (b) 1999
(c) 2000 (d) 2001

15. National Agricultural Innovation Project (NATP) was launched in the year
(a) 2003 (b) 2004
(c) 2005 (d) 2006

16. Who is the governer of RBI
(a) Raghuram Rajan
(b) d. Subbarao
(c) Y.c. Reddy
(d) Montek Singh Ahluwalia

17. Who is the Chairman of 'Commission on Agricultural Costs & Prices'?
(a) Dr. Ashok Gulati
(b) Dr. Y.K.Alagh
(c) Dr. Manmohan Singh
(d) Dr. Vijay Paul Sharma

18. Total no. of All India Coordinated Research Projects of ICAR is
(a) 40 (b) 50
(c) 60 (d) 70

19. Total No. of Deemed Universities under ICAR is
(a) 2 (b) 3
(c) 4 (d) 5

20. Total no. of Agricultural Technology Information Centres is
(a) 33 (b) 44
(c) 55 (d) 66

21. Central Institute of Fisheries Education is located in
(a) New Delhi (b) Mumbai
(c) Kolkata (d) Chennai

22. Central Institute on Post Harvest Engineering and Technology is located at
(a) Jhansi (b) Ludhiana
(c) Karnal (d) Dehradun

23. National Academy of Agricultural Research & Management is located in
(a) New Delhi
(b) Bangaluru
(c) Coimbatore
(d) Hyderabad

24. National Centre for Agricultural Economics & Policy Research is located in
(a) New Delhi (b) Mumbai
(c) Kolkata (d) Chennai

25. Directorate of Knowledge Management in Agriculture is located in
(a) New Delhi (b) Mumbai
(c) Kolkata (d) Chennai

26. World Food is celebrated on
(a) 15th October
(b) 16th October
(c) 15th November
(d) 16th NOvember

27. World Forestry Day is celebrated on
(a) 21st February (b) 21st March
(c) 22 February (d) 22 March

28. World Day for water is celebrated on
(a) 21st February
(b) 21st March
(c) 22 February
(d) 22 March

29. Earth Day is celebrated on
(a) 21st April
(b) 21st March
(c) 22 April
(d) 22 March

30. World Zoonoses day is celebrated on
(a) July 6 (b) August 6
(c) July 9 (d) September 9

31. World Environment Day is celebrated on
(a) May 5 (b) June 5
(c) July 5 (d) August 5

32. World Veterinary Day is celebrated on
(a) Last Saturday of April
(b) Last Saturday of May
(c) Last Friday of April
(d) Last Friday of May

33. National Initiative on Climate Resilient Agriculture (NICRA) was launched on
(a) 2nd Feb, 2011
(b) 3rd March, 2012
(c) 2nd Feb 2011
(d) 3rd March 2011

34. CeRA stands for
(a) Consortium for electronic Reply in Agriculture
(b) Consortium for e-Resources in Agriculture
(c) Centre for Resource allocation
(d) Centre for Resilient Agriculture

35. Full Text database of Indian Agricultural Doctoral Dissertations is
(a) Krishi Kiran (b) E-Prabha
(c) Kishi Prabha (d) E-Krishikiran

36. Nodal Centre for Krishi Prabha is
(a) HAU, Hisar
(b) NDRI, Karnal
(c) IARI, New Delhi
(d) IVRI, Izzatnagar

37. India stands at __________ position in the world with respect to milk production
(a) First
(b) Second
(c) Third
(d) Fourth

38. India stands at __________ position in the world with respect to fruit & Vegetable production
(a) First (b) Second
(c) Third (d) Fourth

39. India stands at __________ position in the world with respect to pulse production
(a) First (b) Second
(c) Third (d) Fourth

40. India stands at __________ position in the world with respect to egg production
(a) First (b) Second
(c) Third (d) Fourth

41. India stands at __________ position in the world with respect to fish production
(a) First (b) Second
(c) Third (d) Fourth

42. Share of Agriculture towards India's GDP is (in 2015-16)
(a) 10.75% (b) 15.35%
(c) 17.37% (d) 15.97%

43. Growth rate of Agriculture in the year 2015-16 is estimated to be
(a) 2.6% (b) 3.6%
(c) 8.3% (d) 5.6%

44. India's milk production in the year 2014-15 is estimated to be reached the level of
(a) 130.3 MMT (b) 135.3 MMT
(c) 137.3 MMT (d) 146.3 MMT

45. Share of Scheduled Commercial Bank towards Agricultural Credit is
(a) 10% (b) 15%
(c) 75% (d) 80%

46. Share of Regional Rural Bank towards Agricultural Credit is
(a) 10% (b) 15%
(c) 75% (d) 80%

47. Share of Co-operative Bank towards Agricultural Credit is
(a) 10% (b) 15%
(c) 75% (d) 80%

48. Under Agri-clinic & Agribusiness centre scheme, individuals are entitled to financial assistance in the form of loan upto
(a) `5 lakhs (b) `10 lakhs
(c) `20 lakhs (d) `50 lakhs

49. Under Agri-clinic & Agribusiness centre scheme, a group of individuals is entitled to financial assistance in the form of loan upto
(a) `5 lakhs (b) `20 lakhs
(c) `50 lakhs (d) `100 lakhs

50. Maximum limit of loan under microfinance is
(a) `5 lakhs (b) `10 lakhs
(c) `25 lakhs (d) `50 lakhs

51. Use of Farm mechanization equipment in India stands at
(a) 40-45% (b) 55-60%
(c) 35-40% (d) 65-70%

52. Highest mechanization on farm is in
(a) Harvesting and Threshing
(b) Irrigation
(c) Plant Protection
(d) Seedling and Planting

53. Highest penetration of micro-irrigation system is
(a) Rajasthan (b) Punjab
(c) Haryana (d) Gujarat

54. NMSA refers to
(a) National Mission on Sustainable Agriculture
(b) National Mission of Specialized Agriculture
(c) Natural Mission for sustainable agriculture
(d) None of these

55. State with largest area covered under micro-irrigation is...
(a) Rajasthan (b) Punjab
(c) Haryana (d) Gujarat

56. Among the crops, maximum adoption of drip irrigation system was in
(a) Vegetable Crops
(b) Medicinal Crops
(c) Ornamental Crops
(d) Fruits Crops

57. As of 2015, area covered under micro-irrigation is
(a) 6 million hectares
(b) 6.5 million hectares
(c) 7.73 million hectares
(d) 8.5 million hectares

58. Area covered under drip irrigation system
(a) 20% (b) 50%
(c) 43.62% (d) 45%

59. Area covered under sprinkler irrigation system
(a) 56.38% (b) 59%
(c) 65% (d) 69%

60. Rice is cultivated in how many countries all over the world?
(a) 117 (b) 118
(c) 119 (d) 120

61. Area under paddy cultivation worldwide (in million hectare)
(a) 162.7 (b) 162.8
(c) 162.6 (d) 162.9

62. Yield loss (in%) in transplanted rice due to weed infestation
(a) 10 (b) 14
(c) 12 (d) 16

63. Full form of HMNEH is_____________.
(a) Horticulture Mission for North East and Himalayan state
(b) Horticulture Ministry for North East and Himalayan state
(c) Himalayan Ministry for North East area of horticulture
(d) Horticulture Mission for North East and Himachal Pradesh

64. Gujarat stands first in production of ________.
(a) Cauliflower (b) Onion
(c) Chilli (d) Tomato

65. Horticulture Contribute __________ million-hectare area in overall agriculture production of Gujarat State.
(a) 1.40 (b) 1.45
(c) 1.47 (d) 1.49

66. Centre of Excellence - Date palm is located at_________
(a) Kukma (b) Talala
(c) Pokhrej (d) Prantij

67. Which one among these is major export destination for the Indian dairy products?
(a) Switzerland (b) China
(c) Egypt (d) Australia

68. India's international trade in meat and meat products is
(a) 1% (b) 13%
(c) 40% (d) 82%

69. Punjab State Cooperative Milk Producers Federation (MILKFED) was established as a ____
(a) Three tier system
(b) Two tier system
(c) Four tier system
(d) None of the above

70. Government of India approved NAM in the year
(a) 2012 (b) 2014
(c) 2015 (d) 2016

71. How many markets are targeted to be covered under NAM by 2017-18
(a) 585 (b) 535
(c) 485 (d) 435

72. The lead promoter of National Agricultural Market is
(a) ISAP (b) SFAC
(c) FICCI (d) CII

73. Forward Market commission was merged with SEBI in the year
(a) 2000 (b) 2005
(c) 2010 (d) 2015

74. Which is the highest oilseed producing country?
(a) India (b) Brazil
(c) China (d) USA

75. Which state is highest producer of rapeseed-mustard in India?
(a) Haryana (b) Punjab
(c) Rajasthan (d) Uttar Pradesh

76. Highest production of Ground nut in ____
(a) Haryana (b) Punjab
(c) Gujarat (d) Uttar Pradesh

77. Which country is largest importer of oil meals from India?
(a) Japan
(b) Australia
(c) USA
(d) China

78. India imports palm oil from
(a) Indonesia and Malaysia
(b) Australia
(c) USA
(d) Russia

79. Indian vegetable oil economy is world's fourth largest, which accounts for.......of world oilseed output.
(a) 8% (b) 9%
(c) 7% (d) 5%

80. How much percentage contribution of sugarcane to the nation GDP during 2014 15?
(a) 0.97 (b) 0.7
(c) 0.92 (d) 0.68

81. Growing area of sugarcane is .... in India.
(a) 2.56% (b) 2.98%
(c) 3.4% (d) 4%

82. Sugarcane which type crop?
(a) Long Duration
(b) Short Duration
(c) Both
(d) None of the above

83. Sugarcane is ....most important source of per capital dietary energy supply.
(a) 1st (b) 2nd
(c) 3rd (d) 4th

84. Who is first most important source of energy?
(a) Sugarcane (b) Cereals
(c) Oilseed (d) None of the above

85. Molasses use________
(a) Ethanol Production
(b) Livestock Feed
(c) Yeast Production
(d) All of the above

86. Which state was announced to be first organic state in January 2016 in India?
(a) Maharashtra
(b) Tamilnadu
(c) Odisha
(d) Sikkim

87. Which of the following is the largest fertilizer producer in India?
(a) Coromandel International LTD
(b) Indian farmer fertilizer corporation LTD
(c) Gujarat state fertilizer fertilizer & chemicals LTD
(d) Rashtriya chemicals & fertilizer LTD

88. KCC stands for
(a) Kisan Call Center
(b) Kisan Calling Center
(c) Kisan Common Center
(d) None of the above

89. VKC Stands for
(a) Village Knowledge Center
(b) Village Kalyan Center
(c) Village Kisan Center
(d) Village Kamal Center

90. SMART ________ can help improve water management and reduce water consumption.
(a) Barley (b) Wheat
(c) Rice (d) Cotton

91. SMART Barley was launched at Kisan Kaushal Sangosthi at Sirsa in Year _______.
(a) March 2016 (b) May 2016
(c) May 2017 (d) March 2017

92. Which is the largest rice producing country in world?
(a) China (b) India
(c) Japan (d) United States

93. Which company has the largest market share in tractor market segments?
(a) John Dheere
(b) Mahindra and Mahindra
(c) Force Motors
(d) TAFE

94. Which state was providing maximum loan disbursement to farmers?
(a) Maharashtra (b) Uttar Pradesh
(c) Punjab (d) Madhya Pradesh

95. Which variety is the high yielding variety among following varieties of Basmati rice?
(a) Pusa 1211 (b) Pusa 1121
(c) Pusa 1322 (d) Pusa 1112

96. How much pulse's share to total food grain production in India?
(a) 1-5% (b) 10-15%
(c) 6-7% (d) 20-25%

97. Which is the major export destination country of India for processed food products?
(a) UK (b) Europe
(c) USA (d) Russia

98. Which one among these is major export destination for the Indian dairy products?
(a) Switzerland (b) China
(c) Egypt (d) Australia

99. Which is the major state for clumping of maximum poultry industry in India?
(a) Assam (b) Odisha
(c) Andhra Pradesh (d) Karnataka

100. Which is the state number one in potato productivity?
(a) Uttar Pradesh (b) West Bengal
(c) Gujarat (d) Bihar

ANSWER KEY (General Study in Agribusiness Management)

1.	(a)	**2.**	(a)	**3.**	(b)	**4.**	(b)	**5.**	(c)	**6.**	(b)	**7.**	(b)	**8.**	(a)	**9.**	(b)	**10.**	(d)
11.	(c)	**12.**	(d)	**13.**	(c)	**14.**	(a)	**15.**	(c)	**16.**	(a)	**17.**	(d)	**18.**	(c)	**19.**	(c)	**20.**	(b)
21.	(b)	**22.**	(b)	**23.**	(d)	**24.**	(a)	**25.**	(a)	**26.**	(b)	**27.**	(b)	**28.**	(d)	**29.**	(c)	**30.**	(a)
31.	(b)	**32.**	(a)	**33.**	(a)	**34.**	(b)	**35.**	(c)	**36.**	(a)	**37.**	(a)	**38.**	(b)	**39.**	(a)	**40.**	(c)
41.	(c)	**42.**	(b)	**43.**	(c)	**44.**	(d)	**45.**	(c)	**46.**	(a)	**47.**	(b)	**48.**	(c)	**49.**	(d)	**50.**	(c)
51.	(a)	**52.**	(a)	**53.**	(c)	**54.**	(a)	**55.**	(a)	**56.**	(d)	**57.**	(c)	**58.**	(c)	**59.**	(a)	**60.**	(b)
61.	(a)	**62.**	(b)	**63.**	(a)	**64.**	(b)	**65.**	(c)	**66.**	(a)	**67.**	(c)	**68.**	(d)	**69.**	(a)	**70.**	(c)
71.	(a)	**72.**	(b)	**73.**	(d)	**74.**	(d)	**75.**	(c)	**76.**	(c)	**77.**	(a)	**78.**	(a)	**79.**	(c)	**80.**	(b)
81.	(a)	**82.**	(a)	**83.**	(c)	**84.**	(b)	**85.**	(d)	**86.**	(d)	**87.**	(a)	**88.**	(a)	**89.**	(a)	**90.**	(a)
91.	(a)	**92.**	(a)	**93.**	(b)	**94.**	(b)	**95.**	(b)	**96.**	(c)	**97.**	(c)	**98.**	(c)	**99.**	(c)	**100.**	(c)

Practice Set 1

1. During cost analysis process, When average cost is at its minimum, then
   (a) Marginal cost is equal to Average cost
   (b) Marginal cost is less than Average cost
   (c) Marginal cost is more than Average cost
   (d) None of the above

2. Output under Price discrimination will show a characteristics like
   (a) Generally be larger than under simple monopoly
   (b) Be inderminate because, we cannot know what prices can be charged
   (c) Usually be the same as under perfect competition
   (d) Be produced at higher average cost than under simple monopoly

3. Other way to describe Opportunity cost is
   (a) Cost related to an optimum level of production
   (b) Variable Cost
   (c) Short-run Cost
   (d) Cost of one product in terms of production of other forgone

4. When marginal propensity to consume (MP(c) is one, then multiplier value (K) is
   (a) Zero
   (b) One
   (c) Infinity
   (d) None of the above

5. Decision of Monopolist relies on
   (a) Price
   (b) Quantity
   (c) Both
   (d) Either Price or Quantity

6. Sometimes your behaviour results in a congruent response from other person. Such behaviour is referred to as
   (a) Perceptual defense
   (b) Perceptual Bias
   (c) Misperception
   (d) Response evocation

7. The change related phenomenon resisted by the organisations in order to maintain an equilibrium, is known as
   (a) Conformity to norms
   (b) Insecurity and regression
   (c) Homoeostasis
   (d) Self-distrust

8. The view that the conflict should be accepted and encouraged, corresponds to which of the following views
   (a) Traditional view
   (b) Behavioural View
   (c) Interactionist View
   (d) Both A and B

9. A creativity technique based on "the fitting together of diverse elements" principle for the generation of alternative solutions is referred to as
   (a) Synectics
   (b) Brainstorming
   (c) Nominal Grouping
   (d) Creative thinking

10. The concept of task oriented and people oriented leadership was developed by
    (a) John Berry
    (b) Maxwell
    (c) Blake and Mouton
    (d) Hertzberg

11. A broad statement of the purpose, duties and responsibilities of a job or position, may be referred to as
(a) Job analysis
(b) Task analysis
(c) Job description
(d) Job specification

12. BARS stands for
(a) Behavioural Abnormality Rating Scales
(b) Behavioural Abnormality Response Scales
(c) Behavioural Anchored Response Scales
(d) Behavioural Anchored Rating Scales

13. The art of judging, character of a person from appearance specially the face, based on the assumption that face is an index of mind, is known as
(a) Graphology
(b) Physiognomy
(c) Phrenology
(d) None of the above

14. The wage which has to be paid by an employer to his workers, irrespective of his ability, is known as
(a) Living wage
(b) Minimum wage
(c) Fair wage
(d) None of the above

15. Which of the following machineries have been provided by the Industrial Disputes Act for settlement of disputes between the employee and the employer
(a) Conciliation
(b) Arbitration
(c) Adjudication
(d) All of the above

16. Leaders who rely forever on the same internal advisers run the risk of being betrayed due to
(a) Lack of trust
(b) Improper distance from authority
(c) Closer leader-advisor relationship
(d) Lack of impartiality of the opinion

17. A ltd. Has beta (β) of 1.25. The risk free rate of return is 0.08 and Rm = 0.15. Share price is Rs 30. The expected return is
(a) 10%
(b) 8%
(c) 15%
(d) 16.75%

18. In financial evaluation of an acquisition, the value of merger focusses on
(a) Corporate value
(b) Shareholder value
(c) Stakeholder value
(d) None of the above

19. Interest is paid on a deep discount bond
(a) Monthly
(b) Annually
(c) Half yearly
(d) At the end of life of bond

20. Liberal credit policy may enhance the sales but will lead to
(a) Faster collection
(b) Higher turnover
(c) Increase in debts
(d) None of the above

21. The term marketing may be described as the performance of business activities that
(a) Gives profit the organisation
(b) Directs flow of goods from producer to consumer
(c) Directs selling
(d) Aims at satisfying the consumers

22. In 'Marketing Myopia' the tendency of firms is to take a
(a) Broad view of the market
(b) Narrow view of the market
(c) Narrow view of MIS
(d) Broad view of MIS

23. Marketers can do on-line marketing by cheating an electronic presence on
(a) The internet, placing ads online
(b) Participating in forums, news groups, web communities
(c) Using web casting and email
(d) All of the above

24. Green marketing principles simulates to
(a) Agriculture marketing
(b) Preservation of environment
(c) Social marketing environment
(d) Organic Product marketing

25. Which one of the following is not coming under principles of personal selling?
(a) Prospecting
(b) Presentation and Demonstration
(c) Promoting
(d) Follow-up and Maintenance

26. In L.P.P, a basic solution is called Degenerate if
(a) None of the variables vanish
(b) All basic variables have infinite values
(c) L.P.P has a finite no. of solution
(d) One or more of the basic variables vanish

27. In queuing theory M/M/1/∞/FIFO stands for
(a) Multiple channel, Multiple criteria, single line, infinite population first in first out
(b) Multiple arrivals, Multiple channels, single distribution infinite population, first in first out
(c) Multiple single channel, Poison arrival, exponential service, infinite queue length, first in first out
(d) Poisson arrivals, exponential service, single channel, infinite queue length, first in first out

28. In North-West Corner rule for solving a transportation problem
(a) The first assignment is made in the cell occupying the lower left- hand corner of the transportation table
(b) The last assignment is made in the cell occupying the upper left hand corner of the transportation table
(c) The last assignment is made in the cell occupying the lower left hand corner of the transportation table
(d) The first assignment is made in the cell occupying the upper left- hand corner of the transportation table

29. The main advantage of the Delphi technique over other judgmental forecasting technique is
(a) The opinion of others are not allowed to influence indivisual experts
(b) It requires that a consensus is reached
(c) It is based on the majority decision
(d) For this less information is required

30. The pessimistic time estimate is
(a) Maximum possible time it could take to accomplish the job
(b) Minimum possible time it could take to accomplish the job
(c) The total time estimate of the critical path
(d) None of the above

31. If $r_{12} = 0.86$, $r_{13} = 0.65$, $r_{23} = 0.72$ then $r_{12.3}$ is
(a) 0.743
(b) 0.529
(c) 0.65
(d) 0.86

32. A company is said to be multinational, if
(a) Production and marketing are done in many countries
(b) Domestically produced items are sold globally
(c) Raw materials are procured from many countries
(d) Workers are hired globally

33. Ansoff's growth vector sometimes simulates to
(a) Generic Strategy
(b) Product- Market strategy
(c) Both of the above
(d) None of the above

34. Five forces model was developed by
(a) Clayton Paul Alderfer
(b) Abraham Harold Maslow
(c) Michael Porter
(d) Elton Mayo

35. Related aspect of TRIPS that mainly considered
(a) Patents
(b) Trades in Services
(c) Globalisation of Business
(d) Labour Practices

36. Small scale Industry is defined in terms of
(a) Volume of production
(b) Number of employees
(c) Amount of investment in plants and machineries
(d) Sales turnover

37. If each one of the partners of a firm pulls out of a business and decides to separate, it is called
(a) Dissolution of partnership
(b) Dissolution of the firm
(c) Lockout of the firm
(d) Bankruptcy of the firm

38. Overdraft is a short term finance measure to
(a) Pay taxes
(b) Repay loan
(c) Purchase plant and equipment
(d) Meet working capital requirement

39. Rural Infrastructure Development Fund is provided by
(a) NABARD
(b) SIDBI
(c) ICICI
(d) Nationalized Bank

40. Which of the following is a measure taken for revival of sick SSI?
(a) Margin money scheme
(b) Reservation of items
(c) Export promotion
(d) Tax holiday

41. "Ecology Paradigm" of management refers to
(a) Environmental management
(b) Strategic Choice Framework
(c) Stockholders' Management
(d) None of the above

42. Strategizing within value framework is an approach of
(a) Governance
(b) Management
(c) Administration
(d) None of the above

43. Economic value added is the measure of
(a) Shareholders wealth
(b) Strategy evaluation
(c) Stake holders
(d) None of the above

44. Egalitarian behaviour refers to
(a) Natural behaviour
(b) Moral behaviour
(c) Ethical behaviour
(d) Social beheviour

45. Organisational value refers to
(a) Indivisual beliefs
(b) Group belief
(c) Corporate belief and ethos
(d) None of the above

46. In the law of variable proportions, phase 'two' indicates
(a) Increasing returns
(b) Constant returns
(c) Diminishing returns
(d) Negative returns

47. The demand curve facing the pure monopolist is
(a) Positively sloped
(b) Negatively sloped
(c) Horizontal
(d) None of the above

48. Total consumer expenditure on a good falls as its price falls, this indicates that its price elasticity is
(a) One
(b) Less than One
(c) More than One
(d) None of the above

49. National Income is
(a) NDP at Factor cost
(b) NNP at Factor cost
(c) NNP at market price
(d) GNP at factor cost

50. "Supply creates its own demand" is known as
(a) Keynesian Law
(b) Veblen Law
(c) Say's Law
(d) None of the above

51. An organisation that has developed capacity to adapt and change is described as
(a) Virtual Organisation
(b) Learning Organisation
(c) Developing Organisation
(d) Changing Organisation

52. The personality theory that emphasizes the unconscious determinants of behaviour is known as
(a) Cognitive theory
(b) Psychodynamic theory
(c) Perpetual theory
(d) Refreezing

53. 'Grapevine' is the term used in relation to
(a) Formal communication
(b) Informal communication
(c) Both Formal communication and Informal communication
(d) None of the above

54. Which theory of motivation places a premium on the effect-reward relationship
(a) Vroom's (b) McClelland's
(c) Equity (d) Herzberg

55. ____________ is the third person acting as a catalyst, who assets in the achievement of the smooth introduction and implementation of change
(a) Change agent
(b) Change manager
(c) Change supervisor
(d) Change player

56. The provision of training and practice in an increased range of skills is known as
(a) Job enrichment
(b) Job enlargement
(c) Job evaluation
(d) Job analysis

57. A level or standard in a scale against which performance can be evaluated in order to structure similar process is described as
(a) Standardisation
(b) Bench-marking
(c) Structuring
(d) Institutionalizing

58. A wage period shall not exceed
(a) Forty days
(b) One week
(c) One month
(d) Thirty Five days

59. In a factory, where more than __________ women workers are ordinarily employed crèche facility has to be provided
(a) 20 (b) 30
(c) 40 (d) 50

60. ______________ is temporarily closing down the undertaking with the intention of forcing workers to accept the demands of employer
(a) Lay-off
(b) Lockout
(c) Retrenchment
(d) Termination

61. NPV is the sum of
(a) Present value of all the cash flows expected to occur over the life of the project
(b) Present value of the positive cash flows expected to occur over the life of the project
(c) Cumulative values of the positive cash flows expected to occur over the life of the project
(d) Total value of positive and negative cash flows expected to occur over the life of the project

62. Decision tree analysis is a useful tools for analyzing
(a) Sequential decisions in the face of cost
(b) Isolated decisions in the face of risk
(c) Sequential decisions in the face of risk
(d) Quick decisions in the face of returns

63. The division of a company into two separate independent companies is known as
(a) Spin off
(b) Dissolution
(c) Liquidation
(d) Split off

64. The nature of amalgamation as uniting of interests or acquisition is described in which Accounting Standard issued by ICAI
(a) AS-13 (b) AS-14
(c) AS-15 (d) AS-16

65. Marketing Concept emphasizes on
(a) Customer Oriented Products
(b) Value creation
(c) Sales Orientation
(d) Production Orientation

66. Skimming pricing implies
(a) Purchasing with little or no advance planning
(b) Purchasing involving all family members
(c) Purchasing involving all relatives
(d) None of the above

67. The first commandment in marketing is
(a) Know thy customer
(b) Know thy product
(c) Know thy market
(d) Know thy price

68. A brand is a means of
(a) Naming the products
(b) Packaging of the products
(c) Contents of the products
(d) Identifying and differentiating the company's product

69. Consumer's risk is
(a) The lowest allowable probability of accepting a bad lot
(b) The largest allowable probability of accepting a good lot
(c) The largest allowable probability of accepting a bad lot
(d) None of the above

70. Surplus variables are introduced for solving
(a) The transportation problem
(b) The L.P.P problem with constraints are of less than type
(c) The CPM problem
(d) The L.P.P problem with constraints are of greater than type

71. The Economic Order Quantity is that quantity which
(a) The fixed cost is equal to the variable cost
(b) The ordering cost is equal to the carrying cost
(c) The economic cost is less than economic value
(d) The value of the product is equal to the operating cost

72. The ABC analysis in Inventory management is carried out on the basis of
(a) 60/30/10 rule
(b) 80/20 rule
(c) 50/30/20 rule
(d) None of the above

73. The technique used to measure the degree of association between the two variables is known as
(a) Regression analysis
(b) Dispersion
(c) Skewness
(d) Correlation analysis

74. Entry barriers for a company to enter an industry can occur due to
(a) Economies of Scale
(b) Product Differentiation
(c) Switching Costs
(d) All of the Above

75. In growth share matrix 'cash cow' indicates
(a) High business growth-High market share
(b) High business growth-Low market share
(c) Low business growth-High market share
(d) Low business growth-Low market share

76. Principles of entrepreneurship may be applied to
(a) Government
(b) Business
(c) Society
(d) All of the above

77. Which of the following standardization certificate is related to the protection of environment?
(a) ISO 2000
(b) ISO 9000
(c) ISO 14000
(d) None of the above

78. 'Global Compact' is related to
(a) Human rights
(b) Social responsibilities
(c) Environment protection
(d) Women rights

79. Combination of character and behaviour is known as
(a) Technological ethics
(b) Ethico-moral ethics
(c) Value based ethics
(d) None of these

80. The most appropriate goal of Corporate Social Responsibility is centered around
(a) Philanthropic activities
(b) Voluntary activities
(c) Ethical based activities
(d) Legalistic activities

81. Price elasticity of demand for luxury goods will be
(a) Relatively elastic
(b) Relatively inelastic
(c) Infinitely elastic
(d) Infinitely inelastic

82. In the short run, profit maximizing firm produces additional units of production, as long as
(a) Elasticity of demand is infinite
(b) Total revenue is increasing
(c) Price covers at least Average fixed cost
(d) Additional revenue per unit

83. In the typical demand schedule, quantity demanded
(a) Varies directly with price
(b) Varies indirectly with price
(c) Varies inversely with price
(d) Is independent of price

84. Which is not an instrument of monetary policy?
(a) Bank rate
(b) Tax policy
(c) Reserve ratio
(d) Buying/selling securities in open market

85. The assumption that conflict between the organisation (if limited to the economic context only) leads to innovativeness and the new product development, technological advancement and better services at low prices, refers to
(a) Law of diminishing return
(b) Cost reduction strategy
(c) Deflationary economy
(d) Laissez-faire economy

86. Cognitive reorganization and reinforcement takes place in which of the following stages of adoption
(a) Implementation
(b) Proposal
(c) Deliberation
(d) All of the above

87. Those functions in an organisation, which are perceived to be directly contributing to the organization's objectives, are known as
(a) Staff function
(b) Corporate function
(c) Integration function
(d) Line function

88. The process of establishing the value of jobs in a job hierarchy, is known as
(a) Job analysis
(b) Job evaluation
(c) Job requirement
(d) Performance evaluation

89. Standing Labour Conference (SLC) was constituted in line with the recommendations of Royal Commission on Labour in the year
(a) 1942
(b) 1944
(c) 1946
(d) 1948

90. The art of estimating a person's character etc ; from the handwriting, is known as
(a) Physiognomy
(b) Phrenology
(c) Numerology
(d) Graphology

91. The method of management development which purports to develop awareness and sentiments to one's own and other's behaviour patterns, is known as
(a) Structured Insight
(b) The Syndicate system
(c) Business games
(d) Sensitivity training

92. In capital lease depreciation is claimed by
(a) Lessor
(b) Lessee
(c) Both
(d) None of the above

93. Convertible Bond is
(a) Debt + Warrant
(b) Debt + Cash
(c) Debt + Share
(d) None of the above

94. In a merger bargain, the exchange ratio is determined on the basis of
(a) Potential of merging company
(b) Issued capital of the two company
(c) Market share
(d) Future value of the company to the acquirer

95. "Marketing myopia" is based upon
(a) Current products offerings
(b) Forecasting the future product offering
(c) Dropping a product
(d) Adding a product

96. Marketing Information System is the system of information that flows to assist
(a) Promotional programmes
(b) Sales Personnel development
(c) Marketing planning and decision making
(d) Marketing mix

97. SWOT analysis is an analytical tool for having
(a) Comparative advantage
(b) Competitive advantage
(c) Comprehension of current overall situation of a firm
(d) None of the above

98. Consumers are aware of damages caused by packaging, product access, so they prefer
(a) Social marketing concept
(b) Green marketing concept
(c) Direct marketing concept
(d) Organic marketing concept

99. Just-in-time inventory system implies
(a) EOQ in purchase orders
(b) Production lots in manufacturing process
(c) Time lag in new material procurement
(d) Production quantities ideally equal to delivery quantities

100. The Critical Path of a network is
(a) For which earliest start time and latest start time are same
(b) The shortest time path through a network
(c) For which earliest finish time and latest finish time are same
(d) The longest time path through a network

101. AQL (Acceptable Quality Level) is
(a) Quality Level acceptable in the society
(b) Fraction defective (f) that the user considers acceptable
(c) Fraction defective (f) that the producer considers acceptable
(d) Quality Level acceptable in the industry

102. The two lines of regression are $3x + 2y = 26$ and $6x + y = 31$ then the mean of x is
(a) 3
(b) 4
(c) 7
(d) 9

103. A maximum likelihood estimator is always
(a) Consistent
(b) Unbiased
(c) Efficient
(d) None of these

104. Growing popularity of DBMS is due to
(a) User friendliness
(b) Report writing
(c) Better response time
(d) Query processing

105. The dimensions of Group Mapping are
(a) Price versus Product-line breadth
(b) Price versus Volume of Production
(c) Production capacity versus Cost of production
(d) Total demand versus Market share

106. Which of the following is not an element of the Porter's Generic competitive strategy model?
(a) Cost leadership
(b) Focus
(c) Leadership
(d) Product Differentiation

107. Policies of WTO are formulated by
(a) Heads of the member countries
(b) Trading blocks
(c) Ministerial conference
(d) None of the above

108. Which of the following is not a part of incentives to SSI?
(a) Subsidy to labour wages
(b) Lower interest rate
(c) Tax holidays
(d) Interest subsidy

109. The interest rate for bridge finance is
(a) Lower than that of short term loan
(b) Equal to that of short term loan
(c) Higher than that of short term loan
(d) None of the above

110. Which of the following operations provide institutional support to SSI?
(a) Indian Institute of Foreign trade
(b) Tool design institute of SIDO
(c) Nationalised Bnk
(d) Ministry of commerce

111. The process through which people are baptized to accept the tradition and maintain the homogeneity of ethos and behaviours is termed as
(a) Induction
(b) Orientation
(c) Acclimatization
(d) Socialization

112. In demand forecasting important method used in trend projection is
(a) Scientific method
(b) Bax- Jenkin method
(c) Delphi method
(d) Logistics method

113. Under monopoly market structure, the degree of freedom in pricing decision is
(a) Very low (b) Quite good
(c) Very high (d) zero

114. Capital budgeting decisions are generally
(a) Irreversible (b) Reversible
(c) Irrevocable (d) Revocable

115. Which one is not the pre-requisite of a market?
(a) Existence of large number of buyers and sellers in the market
(b) Free flow of market information among the buyers
(c) Free entry and free exist
(d) Intervention of the government

116. Two factors theory was propounded by
(a) Fredrick Herzberg
(b) David Mc Clelland
(c) Mc Cardhy
(d) Phillip Kotler

117. Mutual cooperation and networking
(a) Win-win strategy
(b) Networking strategy
(c) Franchise strategy
(d) Competitive strategy

118. A list of Jobs duties, responsibilities, reporting relationships and working conditions is called
(a) Job enhancement
(b) Job specification
(c) Job description
(d) Job enlargement

119. The use of predesigned set of questions used in an interview process leads to
(a) Behavioral interview
(b) Situational interview
(c) Non-directional interview
(d) Structured interview

120. One of the following is a training provided to enhance decision making skills
(a) On the job training
(b) Management games
(c) Behavioural modelling
(d) Action learning

121. Dramatic reduction of manpower is called as
(a) Termination
(b) Retrenchment
(c) Downsizing
(d) Rightsizing

122. Monitoring employees through a preplanned series of position is called
(a) Promotion
(b) Job reporting
(c) Succession Planning
(d) Job rotation

123. A situation where management is unable to provide employment due to non-availability of work is called as
(a) Lock out
(b) Layoff
(c) Termination
(d) Closure

124. The basic goal of financial management is
(a) Maximizing the profit
(b) Maximizing shareholder's wealth in the long run
(c) Maximising the rate of dividend
(d) Minimising the business risk

125. The principal value of a bond is called as
(a) Coupon rate
(b) Par value
(c) Maturity value
(d) Market value

126. IRR is the rate of interest
(a) Equal to the cost of capital of the proposal
(b) Prevailing rate of interest in the market
(c) Paid on the bonds by the firm
(d) That makes the NPV of capital proposal almost zero

127. The technique used for classifying inventory according to the value of usage is called
(a) XYZ Analysis
(b) DCF Analysis
(c) ABC Analysis
(d) DEF Analysis

128. When a consumer derives more satisfaction from a product/service than his expectation, it is known as
(a) Consumer satisfaction
(b) Consumer dazzling
(c) Consumer delight
(d) Consumer dissatisfaction

129. Classical conditioning is associated with
(a) Repetition
(b) Generalization
(c) Discrimination
(d) All of the above

130. The process of selecting, organizing and interpreting or attaching meaning to events happening in the environment is called
(a) Perception (b) Attitude
(c) Personality (d) Learning

131. Under which concept of marketing customer remains focused?
(a) Marketing (b) Production
(c) Product (d) selling

132. Segmentation of consumers based on factors like climatic conditions and regions are known ass
(a) Demographic segmentation
(b) Cultural segmentation
(c) Geographic segmentation
(d) Administrative segmentation

133. Decision on plant layout and location are concerned with
(a) System design
(b) System operations
(c) System control
(d) None of the above

134. Production system which enables for maximizing the scale advantage is
(a) Flow type
(b) Batch type
(c) Intermittent type
(d) None of the above

135. Which one of the following is the objective of Production scheduling
(a) Where to start the operation
(b) When to start the operation
(c) How to start the operation
(d) When to start as well as stop the operation

136. SQC requires determination of
(a) Quality tolerance limit
(b) Quantity measurement
(c) Production lead time
(d) Inventory carrying cost

137. Chi-square distribution is
(a) Continuous
(b) Multimodal
(c) Symmetrical
(d) None of the above

138 Mode of process in Decision Support System is
(a) Batch (b) Online
(c) Offline (d) Real time

139. Which level of managers translate the corporate strategy in to concrete objectives for their indivisual business?
(a) Corporate level
(b) Functional level
(c) Multi-business level
(d) Business level

140. Which of the following books was not written by Michael E Porter?
(a) Competitive strategy
(b) Competiting for the future
(c) Competitive advantage
(d) Competitive advantages of nations

141. Which of the following provides directions, serves as standard for evaluating performance and motivates members of the organization?
(a) Mission (b) Values
(c) Objectives (d) Goals

142. Sick enterprise is referred to
(a) World Bank (b) RBI
(c) SIDBI (d) BIFR

143. Which of the following is the causes of sickness of the enterprise?
(a) Lack of adequate capital
(b) Lack of demand for the products
(c) Lack of competition
(d) All of the above

144. Consumer's willingness to pay minus (-) actual payment is called
(a) Producer's surplus
(b) Supplier's surplus
(c) Utility cost
(d) Consumer's surplus

145. The Pay Back period refers to
(a) The number of years a project takes to recover its investment cost
(b) The number of installments to make payment
(c) The time lag between investment and commissioning of the project
(d) The time required to attain breakeven point

146. In the case of Veblen goods, demand curve will slope
(a) Upwards (b) Downwards
(c) Horizontal (d) Vertical

147. Monopolist can decide
(a) Quantity
(b) Price
(c) Both quantity and price
(d) Either quantity or price

148. The conflict resulting into anxiety, tension, frustration or hostility due to one's emotional involvement, may be termed as
(a) Perceived conflict
(b) Manifest conflict
(c) Felt conflict
(d) Conflict aftermath

149. The obstruction in communication caused due to differences in indivisual's interpretation of words and symbols, is known as
(a) Psychological barrier
(b) Social barrier
(c) Semantic barrier
(d) Organisational behaviour

150. The usual tendency with negative feedback is to find the reason to explain one's own behaviour. This tendency is termed as
(a) Mutuality
(b) Displacement
(c) Rationalization
(d) Projection

151. In which of the following inter-personal style, people are critical of other's behaviour, develop rules and regulations and impose them on others?
(a) Rescuing style
(b) Prescriptive style
(c) Normative style
(d) Supportive style

152. Sometimes your behaviour results in a congruent response from other person, such behaviour is referred to as
(a) Perceptual defense
(b) Perceptual bias
(c) Misperception
(d) Response evocation

153. 'Employees selected for the work should be as perfectly matched to the demand of the job as possible.' Which of the following theories says this?
(a) Scientific management theory
(b) Bureaucracy theory
(c) Classical theory
(d) Traditional theory

154. The assumption that employees like work, are creative and can exercise self-direction is known as
(a) Theory X
(b) Theory Y
(c) Theory Z
(d) None of the above

155. The horizontal expansion of jobs is termed as
(a) Job enrichment
(b) Job enlargement
(c) Job sharing
(d) Job involvement

156. The communication taking place among members of the same group or among members or work groups at the same level is described as
(a) Downward communication
(b) Upward communication
(c) Lateral communication
(d) Parallel communication

157. An incentive plan, where improvements in group productivity determines the total amount of money that is allocated, is described as
(a) Profit sharing
(b) Gain sharing
(c) Bonus sharing
(d) None of these

158. Which of the following is related with 'change in condition of service' of a worker?
(a) Section 9A of the ID Act
(b) Section 9B of the ID Act
(c) Section 3 of the ID Act
(d) Section 25 of the ID Act

159. The central trade Union AITUC is associated with
(a) Bharatiya Janta Party
(b) Congress Party of India
(c) Communist Party of India
(d) Communist Party of India (Marxist)

160. The weightage average of possible returns, with weights being the possibilities of occurrence is known as
(a) Expected return
(b) Average return
(c) Variability of returns
(d) Variance of returns

161. The amount by which the face value of a bond exceeds its current price is known as
(a) Bond premium
(b) Bond discount
(c) Intrinsic value of bond
(d) Future value of bond

162. The project whose acceptance precludes the acceptance of one or more alternative projects is known as
(a) Independent project
(b) Dependent project
(c) Mutually exclusive project
(d) Contingent project

163. Annual cash dividends divided by annual earnings or alternatively dividend per share divided by earnings per share is known as
(a) Cost of equity
(b) Cash profit
(c) Dividend payout ratio
(d) Retention ratio

164. When merger is in the nature of amalgamation, the accounting approach will be
(a) Purchase method
(b) Combination
(c) Pooling of interest method
(d) None of the above

165. Which of the marketing mix order is correct?
(a) Price, product, place, promotion
(b) Product, price, promotion, place
(c) Product, price, place, promotion
(d) Product, place, price, promotion

166. Which of the following coined the term 4 Ps
(a) Philip Kotler
(b) Stanton
(c) Mc Carthy
(d) Theodore Levitt

167. Which of the following PLC stage is in correct order
(a) Introduction, adolescence, growth, decline
(b) Introduction, growth, maturity, decline
(c) Introduction, maturity, growth, decline
(d) Introduction, growth, saturation, maturity

168. Who authored the book, 'positioning- the battle for the mind'
(a) Philip Kotler and Gary Armstrong
(b) Theodore Levitt
(c) Peter F. Drucker
(d) Alrice and Jack Trout

169. How do you distinguish marketing and selling?
(a) They are one and the same
(b) Marketing is a part of selling
(c) Selling is a part of marketing
(d) No relationship between selling and marketing

170. In case of product/ market, the cell showing New product and existing market is referred as
(a) Market Penetration
(b) Product Development
(c) Market Development
(d) Diversification

171. A transportation problem will be said to be balanced if
(a) No. of rows is equal to the no. of columns
(b) No. of warehouses is equal to the no. of factories
(c) Total supply is equal to the Total demand
(d) Total demand is equal to the estimated demand

172. The level of significance is the probability of committing the
(a) Type I error
(b) Type II error
(c) Standard error
(d) Probable error

173. The type of transmission channel which can transmit data in one direction only is known as
(a) Simplex
(b) Half Duplex
(c) Double Duplex
(d) Half Simplex

174. As per BCG model "HOLD" strategy is used for
(a) Dogs
(b) Stars
(c) Question marks
(d) Cash cows

175. Which one is not part of Ansoff's Growth matrix?
(a) Focused development
(b) Product development
(c) Market development
(d) Diversification

176. The concept of 'Core Competence' is advanced by
(a) Ghoshal and Nohria
(b) Prahallad and Hammel
(c) Evans and Lorange
(d) Friedman and Philips

177. The name of Victor Vroom is associated with which theory of motivation
(a) Two factor theory
(b) Equity theory
(c) Expectancy theory
(d) Contingency theory

178. The pre-conventional, conventional and principled levels of moral development influencing business ethics is given by
(a) J. Keogh
(b) L. Kohlberg
(c) K. Kumar
(d) I. Kant

179. In India the first report on Corporate Governance was given by
(a) Narayana Murty Committee
(b) Cadbury's Committee
(c) Naresh Chandra Committee
(d) Kumarmanglam Birla Committee

180. M. Friedman is associated with which philosophy of Social Responsibility of Business
(a) Profit maximization
(b) Following legal provision
(c) Adopting ethical practices
(d) None of the above

181. Under perfect competition in the long run
(a) All firms earn normal profit
(b) Few firms earn super normal profits
(c) All firms are at break even
(d) None of the above

182. National Income at market prices is equal to
(a) Gross National Product at market prices
(b) Gross domestic product at market prices
(c) Net National Product at factor cost
(d) None of the above

183. A firm having increasing returns to scale means
(a) It is having increasing costs
(b) It is having constant costs
(c) It is having decreasing costs
(d) None of the above

184. Under which of the following method time value of money is not considered
(a) Net present value method
(b) Internal rate of return method
(c) Payback period method
(d) All of the above

185. Which of the following can be measured in terms of production?
(a) Total productivity
(b) Marginal productivity
(c) Average productivity
(d) All of the above

186. Which of the following is an objective of Human Resource Management?
(a) Productivity
(b) Quality of work life
(c) Organisational change
(d) All of the above

187. Communication can be
(a) Oral
(b) Written
(c) Oral and Written
(d) None of the above

188. Which of the following is not considered as a Leadership style?
(a) Task Oriented Leadership style
(b) Society Leadership style
(c) Employee Oriented Leadership style
(d) Autocratic Leadership style

189. Which of the following method has not been given a statutory form in India?
(a) Conciliation
(b) Mediation
(c) Voluntary Arbitration
(d) Adjudication

190. What are the operative functions of HRM?
(a) Organising
(b) Directing
(c) Controlling
(d) All of the above

191. Which one is not a performance appraisal method?
(a) 3600 performance appraisal
(b) Balanced Score Card
(c) Rank method
(d) Interview method

192. Labour welfare facilities were incorporated in which legislation?
(a) MRTP Act
(b) Essential Commodities Act
(c) Consumer Protection Act
(d) Factories Act

193. When a firm proposes to acquire another firm, the merger proposal may be viewed as a
(a) Working capital management proposal
(b) Capital budgeting proposal
(c) Financing decision
(d) Dividend decision

194. At which stage of the new product development, the launching of new product will take place?
(a) Business analysis
(b) Test marketing
(c) Commercialization
(d) Marketing strategy development

195. If a company uses its company name for its entire range of products offered, then it is called
(a) Private branding
(b) Corporate branding
(c) Product- Line branding
(d) Specific Product branding

196. One level channel of distribution consists of
(a) Producer-consumer
(b) Producer-dealer-consumer
(c) Producer-dealer-retailer-consumer
(d) Producer-dealer-retailer-consumer

197. Fixed position layout is used when
(a) The operation requires heavy tools and equipments
(b) The operation requires only hand tools and simple equipments
(c) The continuity of production process is to be maintained
(d) None of the above

198. Critical Path Method (CPM) is concerned with estimation of
(a) Optimistic time
(b) Production lead time
(c) Pessimistic time
(d) None of the above

199. Firms choose a focus strategy to serve the needs of which of the following
(a) Buyers group
(b) Different segment of a product line
(c) Different geographical market
(d) All of the above

200. Which of the following is not the component of work study?
(a) Work measurement
(b) Case study
(c) Time study
(d) Motion study

201. Which of the following is the characteristic of a strategic business unit
(a) It serves a homogenous set of markets
(b) It serves a unique set of products
(c) It owns responsibility for its own profitability
(d) All of the above

202. A retrenchment strategy is designed to reduce
(a) Technical losses
(b) Financial losses
(c) The scale and scope of the business operation
(d) All of the above

203. World Trade Organization provides
(a) Rule based multilateral trading system
(b) Equity and Social justice to the people
(c) Rule based regional trading system
(d) All of the above

204. Value driven companies are
(a) Sure to be successful in the long run
(b) Sure to be unsuccessful as they are not maximizing profits
(c) Sure to become charitable institutions
(d) None of the above

205. Which of the following is not the social responsibility of the business
(a) Promotion of markets
(b) Promotion of green environments
(c) Promotion of education
(d) Promotion of public health

206. Consider a demand curve which takes the form of a straight line cutting both axes. Elasticity at the midpoint of the line would be
(a) 0
(b) 1
(c) 1.5
(d) 2

207. Price taker firms
(a) Advertise to increase the demand for their product
(b) Do not advertise, because most advertising is wasteful
(c) Do not advertise because they can sell as much as they want at the current price
(d) Who advertise will get more profits than those who do not

208. OPEC is an example of the type of producer's organization as a
(a) Marketing board
(b) Producer's cooperative
(c) Trust
(d) Cartel

209. Opportunity costs are also known as
(a) Spill over costs
(b) Money costs
(c) Alternative costs
(d) Sunk costs

210. Employees with high esteem set
(a) Higher goals for others
(b) Higher goals for themselves and lower goals for others
(c) Higher goals for themselves
(d) None of the above

211. Need for achievement goes with
(a) F.W.Taylor
(b) Hertzberg
(c) McGregor
(d) McClelland

212. An organizational memory includes the following except
(a) Information that employees possess
(b) Knowledge embedded in the organization's systems and structures
(c) Documents
(d) Knowledge that is not document

213. Which patterns reflects a pure executive form of management
(a) Functional
(b) Line
(c) Line and Staff
(d) Committee

214. 'a place for everything, everything in its place' is
(a) American management
(b) 5s Principle
(c) Peter Principle
(d) A Chinese practice

215. 'Appreciative Enquiry' is
(a) Pshychological theory
(b) Training method
(c) Business Etiquette
(d) HR quality

216. Robert Keplan is associated with
(a) A benchmark practice
(b) Balance Score Card
(c) Kaizen
(d) Quality of Life

217. Strategic Intent of an Organisation should clearly spell out
(a) Who we are
(b) Where are we going in the long term
(c) How will we measure the performance
(d) All of the above

218. In an endeavor to maximize corporate value, a finance manager has to factor in the following
(a) Optimisation principle
(b) Risk-return principle
(c) Marginal principle
(d) All of the above

219. The term financial engineering relates to
(a) Cost of production
(b) Financial restructuring
(c) Product Planning
(d) Capital Issue

220. Value of the merged firm shall be
(a) Less than the acquired firm
(b) More than the acquirer
(c) Greater than the target and the acquirer
(d) None of the above

221. When borrowing is less than equity, the company is said to be
(a) Trading on equity
(b) Trading on thick equity
(c) Trading on thin equity
(d) None of these

222. It is a set of procedures and sources used by managers to obtain everyday information about developments in the marketing environment. It is known as
(a) Internal Records system
(b) Marketing Intelligence System
(c) Marketing research System
(d) Marketing Decision Support System

223. It has been described as a cluster of complementary goods and services that are closely related in the minds of consumers but are spread across a diverse set of industries. It is known as
(a) Market Place (b) Market Space
(c) Meta Market (d) Marketing mix

224. When work stations are arranged according to the progressive steps by which the product is produced, it is called as
(a) Functional layout
(b) Process layout
(c) Job shop layout
(d) Product layout

225. CRAFT considers exchanges of locations for those departments which
(a) Have the same area
(b) Have a common border
(c) Either (A) or (B)
(d) Have same costs per unit of distance

226. PERT/CPM network analysis can be used for
(a) Project Appraisal
(b) Project implementation
(c) Project monitoring and control
(d) All of the above

227. Eliminate-reduce-raise-create grid is germane to
(a) Competitive strategy
(b) Blue ocean strategy
(c) Red ocean strategy
(d) None of the above

228. Service sector is also called as
(a) Primary sector
(b) Secondary sector
(c) Tertiary sector
(d) None of the above

229. Growth strategy requires that a firm should
(a) Realign its product market environment
(b) Acquire new firm
(c) Go for new product development
(d) Expand the market

230. Michel Porter's five forces model belongs to
(a) Planning school
(b) Positioning school
(c) Cognitive school
(d) Competitive strategy school

231. The theory of business of an organisation is embodied in
(a) Vision
(b) Mission statement
(c) Objectives
(d) Policies

232. The definition of Micro, Small and Medium Enterprises (MSMEs) in India is based on
(a) Total sales of the unit
(b) Investment in machines and Equipments
(c) Market coverage
(d) Export capacity

233. The head quarter of the SIIDBI is in
(a) Mumbai
(b) Lucknow
(c) New Delhi
(d) Hyderabad

234. The study of moral principles or values that determine whether actions are right or wrong and outcomes are good or bad refers to following
(a) Quality (b) Ethics
(c) Opinions (d) Controls

235. An organization's moral obligation towards toward all of its stakeholder is referred as following
(a) Value in workplace
(b) Corporate social responsibility
(c) Ethical behaviour
(d) Professionalism

236. ______________ is the manager's dilemma when he has to decide on the commercial proposal submitted by his son
(a) Favouritism
(b) Moral Dilemma
(c) Conflict of Interest
(d) None of the above

237. 'Make use of paper bags instead of plastic bags'. It is related to
(a) Environmental ethics
(b) Green marketing
(c) Socially responsible marketing
(d) Social marketing

238. A rightward shift in supply curve indicates
(a) A decrease in supply
(b) An increase in quantity demanded
(c) An increase in supply
(d) Law of Variable Proportion

239. Which of the following is the condition of equilibrium for the monopolist
(a) MR = MC
(b) MC = AR
(c) AR = MC = Price
(d) AC = AR

240. If the demand curve is a rectangular hyperbola, elasticity is
(a) 1
(b) 0
(c) Infinity
(d) Less than One

241. If supply and demand increase by the same proportion
(a) Quantity remains constant
(b) Price remains constant
(c). Price increases
(d) Quantity increases

242. What leads to rationing and black marketing under perfect competition
(a) Price support
(b) Excess demand
(c) Decrease in supply
(d) Price control

243. In Maslow's need hierarchy theory, which of the following is not high order need
(a) Safety (b) Social
(c) Esteem (d) Self-Actualization

244. Feeling arising out of inadequate definition of roles is known as
(a) Role ambiguity
(b) Role conflict
(c) Role perception
(d) Role expectation

245. A type of group discussion process in which members of the group are encouraged to generate as many imaginative and creative ideas in relatively short period of time is known as
(a) Brain Storming
(b) Group working
(c) Simulation
(d) Group effect

246. The amount of satisfaction an indivisual anticipates from a particular outcome is known as
(a) Theory Y (b) Theory Z
(c) Valence (d) Stress

247. A personality characteristic indicating one's willingness to do whatever it takes get one's own way is known as
(a) Locus of control
(b) Machiavellism
(c) Job characteristic model
(d) Personality legitimacy

248. A bargaining which is based upon zero-sum game is known as
(a) Integrative bargaining
(b) Principled bargaining
(c) Distributive bargaining
(d) None of these

249. The final link in the communication process that puts the message back into the system as a check against misunderstandings is described as
(a) Decoding
(b) Feedback
(c) Feedback loop
(d) None of these

250. In the "Managerial Grid" 9,1 style is termed as
(a) Team management
(b) Country club management
(c) Authority Obedience
(d) Improvised management

251. Which of the following is not one of the 5s of housekeeping
(a) Sort
(b) Systemize
(c) Specialize
(d) Shine

252. Beta in the context of security analysis indicates
(a) Systematic risk
(b) Unsystematic risk
(c) Total risk
(d) All of the above

253. A high risk, high yield often unsecured bond rated below investment grade is known as
(a) Income Bond
(b) Subordinate debentures
(c) Junk bond
(d) Serial bond

254. Who coined the term 4P's?
(a) Philip Kotler
(b) Stanton
(c) Peter Drucker
(d) Mc Carthy

255. In the buying decision process, what is the term used for a person who first suggests buying the product
(a) Consumer (b) Decider
(c) Initiator (d) Influencer

256. Which of the demographic variable is not used by marketers for demographic segmentation?
(a) Income (b) Gender
(c) Family life cycle (d) Poverty

257. Companies that attack other firms including market leader in an attempt to build market share are called
(a) Market Followers
(b) Market Challengers
(c) Market Penetrators
(d) Market Nichers

258. Tactical production planning includes
(a) Layout Planning
(b) Machine Loading
(c) Production Process
(d) Plant Location

259. Which of the following is not to be considered for plant location decision
(a) Availability of labour
(b) Availability of power
(c) Availability of raw material
(d) Brand image

260. Moving average forecasting is used to
(a) Establish the trend of the past
(b) Forecast the current demand
(c) Forecast market competition
(d) Forecast quasi demand

261. Work measurement is essential for
(a) Labour utilization and efficiency
(b) Determining the delivery dates
(c) Introducing financial incentive schemes
(d) All of the above

262. EOQ refers to the quantity for which
(a) Fixed cost is equal to variable cost
(b) Carrying cost is equal to ordering cost
(c) Average cost is equal to marginal revenue
(d) Total contributions minus cost of goods

263. Judgement sampling to select the sampling units from population is termed as
(a) Probability sampling
(b) Non- Probability sampling
(c) Systematic sampling
(d) None of these

264. Wide area network is
(a) Network of computers spread over a small geographical area
(b) Networks of computer spread over a large geographical area
(c) Network of computer in a room
(d) None of these

265. Which of the following helps a manager to identify the opportunities and threats in the competitive industry environment
(a) Analysis of organisational structure
(b) Analysis of competitive forces
(c) Analysis of operations
(d) Market research

266. Which of the following is a primary activity in a value chain
(a) Operations
(b) Firm infrastructure
(c) Technology development
(d) Procurement

267. Which of the following is an example of external threat
(a) New trade regulations
(b) Global sales potential
(c) Economies of scale
(d) Decreased competition

268. 'Innovation and Entrepreneurship' is abook written by
(a) Philip Kotler
(b) C. K. Prahlad
(c) Pradeep Khandwala
(d) Peter F Drucker

269. Demand curve under perfect competition is
(a) Downward sloping and flat
(b) Downward sloping and steep
(c) Parallel to X-Axis
(d) Parallel to Y- Axis

270. If two commodities are complementary, then a rise in the price of one commodity will induce
(a) An upward shift of demand curve
(b) A rise in the price of the other commodity
(c) No shift in demand for the other commodity
(d) A backward shift in demand for the other commodity

271. Which of the following is a flow variable?
(a) Income
(b) Product
(c) Capital formation
(d) Wealth

272. Functional type of organisation was first developed by
(a) Pavlov (b) Taylor
(c) Henry Fayol (d) Elton Mayo

273. 'Gang Plank' means
(a) Quick communication
(b) Vertical communication
(c) Horizontal communication
(d) None of the above

274. Minimum wages are fixed by
(a) Trade Unions
(b) Employees
(c) State Government
(d) Court of Law

275. Under the Factories Act 1948, safety officer is to be appointed in a factory where
(a) 500 workers are working
(b) 1000 or more workers are working
(c) 3000 or more workers are working
(d) 5000 or more workers are working

276. The amount of Dividend that can be legally distributed is governed by
(a) Company law
(b) Judicial pronouncement in leading cases
(c) Contractual restrictions
(d) All of the above

277. Pre-emptive right of equity shareholders means that they
(a) Can vote to protect their right
(b) Can maintain their proportional ownership by purchasing additional equity shares issued by firm
(c) Can be elected as directors
(d) None of the above

278. Distribution channel aims at moving products from producer to
(a) Ultimate consumer
(b) Retailors
(c) Stockists
(d) Brokers

279. In business situations, personal selling is
(a) Inside selling
(b) Outside selling
(c) Both of the above
(d) None of the above

280. Mass communication tools available to marketers include
(a) Advertisements
(b) Sales Promotion
(c) Public relations
(d) All of the above

281. The financial results of marketing endeavours should be analysed in terms of
(a) Sales volume
(b) Market share
(c) Marketing costs
(d) All of the above

282. Production scheduling is done to achieve
(a) Maximum utilization of the plant and machinery
(b) Maximum utilization of the work force
(c) Minimize inventories of raw materials and work in progress
(d) All of the above

283. Vogel's Approximation method is used to
(a) Study the time and motion problem
(b) Solve the transportation problem
(c) Study the production scheduling problem
(d) Forecast demand

284. From the following which is the causal relationship between two or more variables
(a) Correlation
(b) Regression
(c) Factor analysis
(d) Time series analysis

285. A measure characterizing a population is called a parameter and a similar measure for a sample is called
(a) Standard error
(b) Non parameter
(c) Sampling error
(d) Statistic

286. Which of the following act, regulates sick industries specially in India
(a) ID Act (b) Companies Act
(c) SICA (d) None of the above

287. Global compact is an initiative of
(a) UN (b) WTO
(c) IMF (d) ILO

288. Agenda 21 relates to
(a) Child labour
(b) Human rights
(c) Environment Protection
(d) None of these

289. Which of the following convention of ILO provides for gender equity?
(a) 87 (b) 98
(c) 100 (d) 105

290. Disclosure of Information is a part of
(a) Global compact
(b) Human rights
(c) Social clause
(d) Corporate Governance

291. Corporate governance revolves around
(a) Stakeholders
(b) Organisation
(c) Society
(d) All of these

292. A monopoly always charges a price which is
(a) Greater than marginal cost
(b) Equal to marginal cost
(c) Equal to average cost
(d) Equal to average variable cost

293. The kinked demand curve hypothesis predict
(a) Stable cost for the firm
(b) Sticky prices in the market
(c) More advertisement by the firms
(d) None of the above

294. Organisational behaviour was popularized by
(a) F.W. Taylor (b) Keith Davis
(c) Elton Mayo (d) David Cock

295. The Neo-classical theory of organisational structure was propounded by
(a) Henry Fayol
(b) Elton Mayo
(c) Peter Drucker
(d) J.T. Dunlop

296. The conditioned - learning theory was propounded by
(a) Thoradike (b) Pevlov
(c) Maslow (d) Luther Gullick

297. Two factor theory of motivation was propounded by
(a) Vroom (b) Maslow
(c) Peterson (d) Hertzberg

298. The book 'Human Behaviour at Work' was written by
(a) McGregor
(b) Abraham Maslow
(c) Hertzberg
d) Keith Davis

299. Human Resource Planning is the ________________step in the management function
(a) Primary (b) Middle
(c) Final (d) None of the above

300. Giving general information regarding organisation to the newly appointed personnel is called
(a) Training (b) Development
(c) Induction (d) Promotion

301. Job evaluation is the evaluation of
(a) Work
(b) Worker
(c) Working condition
(d) All of the above

302. Industrial relations consists of the relation between
(a) Employer and employees
(b) Union and management
(c) State and management
(d) All of the above

303. The father of the Employees State Insurance Scheme in India is
(a) B.P. Wadia (b) B.P. Adarkar
(c) V.V.Giri (d) N.M. Lokhande

304. Dividend is paid only out of
(a) Capital profit
(b) Current or Accumulated Profit
(c) Reserves
(d) Cash

305. Net Operating income approach is very close to
(a) Modigilliani - Miller approach
(b) Net Income approach
(c) Optimum capital structure
(d) None of the above

306. Point out which one does not affect the capital of a firm
(a) Cash
(b) Corporate tax
(c) Current liabilities
(d) None of the above

307. The Psychological factors in Market Segmentation will include
(a) Income and Occupation
(b) Sex and Age
(c) Hobbies and Activities
(d) Living standard and Income

308. In channels of distribution of Industrial products which level is popular
(a) Zero level and One level
(b) One level and two level
(c) Two levels only
(d) Multi-level

309. Product management do not cover the following aspects
(a) Product Line
(b) Product mix
(c) Product Coordination
(d) Product Life cycle

310. Which one of the following is not the type of Plant Layout
(a) Process layout
(b) Product layout
(c) Fixed layout
(d) Selection layout

311. In the EOQ model
(a) Demand is known and occurs at a constant rate
(b) Shortages are allowed
(c) Inventory replenishment time is finite
(d) Both (b) and (c)

312. Duality problem relates to
(a) Markov Analysis
(b) Inventory problem
(c) Linear programming problem
(d) Queuing Problem

313. Which 'S' is not included in the McKinsey's 7 S frame work
(a) Style (b) System
(c) Staff (d) Selection

314. Five forces model is used
(a) Competitor analysis
(b) Organisational analysis
(c) Industry analysis
(d) Corporate analysis

315. Which of the following book is not written by Michel E. Porter?
(a) Fortune at the bottom of the Pyramid
(b) Competitive Strategy
(c) Competitive advantage
(d) Competitive advantage of the nation

316. ISO cost line is also called as
(a) Profit line
(b) Budget line
(c) Production line
(d) Return line

317. Quadratic cost function is written as
(a) TC = a + bQ
(b) TC = a + bQ + Q2
(c) TC = a + bQ + cQ
(d) TC = a + b + c

318. Transfer Pricing is referred to as
(a) Price charged to a customer
(b) Price charged to different firms
(c) Price charged under intra firm transfer of product
(d) Inter firm transfer of price

319. Margin of safety is
(a) Above the breakeven point
(b) Below the breakeven point
(c) Equals to breakeven point
(d) Not related to breakeven point

320. Profit is maximized when
(a) MC =MR
(b) MC > MR
(c) MC < MR
(d) MC ≤ MR

321. What is the shape of short run average cost curve
(a) 'L' shape
(b) 'U' shape
(c) Horizontal
(d) Vertical

322. Emotional Intelligence includes
(a) Self-awareness, self-knowledge and self-management
(b) Self-awareness, self-knowledge and self-understanding
(c) Knowledge, belief, culture and perception
(d) All of the above

323. What is the most important attribute of an effective leader
(a) Behaviour
(b) Character
(c) Drive
(d) Influence

324. Graphical representation of manpower in the Organisation is known as
(a) Replacement chart
(b) Placement chart
(c) Organisational structure
(d) Manpower inventory

325. The process of retaining suitable successor is called as
(a) Alternative staffing
(b) Succession planning
(c) Position replacement
(d) Promotion

326. A systematic comparision done in order to determine the value of Job is called
(a) Job upgradation
(b) Job enrichment
(c) Job evaluation
(d) Job analysis

327. A wilful disobedience of authority is called as
(a) Insubordination
(b) Indiscipline
(c) Wilful disobedience
(d) Misconduct

328. Interviews conducted for knowing why an employee is leaving the organisation is called
(a) Exit interview
(b) Separation interview
(c) Termination interview
(d) Retrenchment interview

329. Under Profit-maximisation goal, which of the following decisions does not fall under the purview of financial management
(a) Investment decision
(b) Debt-equity-mix decision
(c) Dividend decision
(d) Capitalization decision

330. Which of the following is not a determinant of dividend policy of the firm
(a) The employee efficiency
(b) The future financial need
(c) The market rate of dividend
(d) The future profit projection of the firm

331. Capital budgeting is concerned with budgeting for
(a) Capital income
(b) Capital expenditure
(c) Capital gains
(d) Revenue expenditure

332. Who is associated with the statement that marketing is the process of creation and delivery of Standard of Living?
(a) Mc Nair (b) Mazur
(c) Kotler (d) Peter Drucker

333. Which concept of marketing maintains that the company aggressively promotes and pushes its products?
(a) Marketing
(b) Product
(c) Production
(d) Selling

334. Which of the following is the sole vehicle for creating and delivering customer value?
(a) Marketing channel
(b) Marketing mix
(c) Marketing communication
(d) Marketing research

335. Who is associated with concept of marketing myopia?
(a) Levitt
(b) Drucker
(c) Mc carthy
(d) Kotler

336. Who introduced the concept of time and motion study?
(a) Hanry Fayol (b) Peter Drucker
(c) R Likert (d) F W Taylor

337. The objective of Inventory control is
(a) To produce required quality of produce
(b) To produce required quantity of produce
(c) To minimize the cost
(d) All of the above

338. The pessimistic time estimate is
(a) The maximum time to be taken in the accomplishment of goal
(b) The minimum time to be taken in the accomplishment of goal
(c) The total time estimate of the critical path
(d) None of the above

339. Under which stage of the industry life cycle, entry barriers increase and threat of potential competitors decrease
(a) Beginning stage
(b) Growing stage
(c) Mature stage
(d) Declining stage

340. Which of the following is short term strategy for a firm?
(a) Business strategy
(b) Annual objectives
(c) Functional strategy
(d) None of the above

341. Which of the following terms best explains the partnership between two or more firms to carry out a specific project in a selected area of business?
(a) Joint venture
(b) Merger and acquision
(c) Cartel
(d) None of the above

342. "SEZ" stands for
(a) Small Entrepreneurship Zone
(b) Special Entrepreneurship Zone
(c) Sick Economic Zone
(d) Special Economic Zone

343. Which one of the following is not the purpose of the promotion of small business?
(a) Generation of employment
(b) Balanced regional development
(c) Promotional of communal harmony
(d) Reduction in rural poverty

344. A decision maker has to remember the proverb "a bird in-hand is worth two in the bush", while he examines
(a) Opportunity Cost Principle
(b) Discounting and Compounding Principle
(c) Marginal or Incremental Principle
(d) Equi-Marginal Principle

345. Market with one buyer and one seller is called
(a) Monopsony (b) Bilateral monopoly
(c) Monopoly (d) Duopoly

346. Cardinal measure of utility is required in
(a) Utility theory
(b) Indifference curve analysis
(c) Revealed preference
(d) Inferior goods

347. In case of Giffen goods, price effect is
(a) Negative (b) Zero
(c) Positive (d) -1

348. Which of the following theories state that, 'employees make comparision of their efforts and rewards with those of others in similar work situation?
(a) Vroom's Expectancy Theory
(b) Adam's Equity Theory
(c) Alderfer's ERG Theory
(d) Hertzberg's two factor Theory

349. Which of the following is a method of indicating the feeling of acceptance or rejection among members of a group?
(a) Mutuality (b) Simulation
(c) T-group (d) Sociometry

350. In which of the following processes a person rejects his own feelings about the other person?
(a) Self-analysis (b) Projection
(c) Empathy (d) Denial

351. Which of the following methods avoid a win-lose attitude?
(a) Integration
(b) Compromise
(c) Domination
(d) All of the above

352. In Transactional analysis the "I am OK, you are not OK" life position is also referred to as
(a) Bossing (b) Avoidant
(c) Diffident (d) Bohemian

353. At which level of PCMM the concept of competency management is brought into workforce practices
(a) One (b) Two
(c) Three (d) Five

354. The total dedication of continuous improvement so that customers' needs are met is known as
(a) Theory X
(b) Theory Y
(c) Total Quality Management
(d) Change Management

355. The Employee Stock Ownership Plan (ESOP) was developed by
(a) Michael Porter (b) Kaplan
(c) P.F.Drucker (d) Louis Kelso

356. The book "HR Champion" is authored by
(a) Mintzberg
(b) J. Pfeffer
(c) M. Porter
(d) David Ulrich

357. The Central Trade Union CITU is associated with
(a) Congress Party of India
(b) Communist Party of India
(c) Communist party of India Marxist
(d) Shiv Sena

358. If bond sells at discount, the price is less than par value and
(a) YTM = Coupon rate
(b) YTM > Coupon rate
(c) YTM < Coupon rate
(d) None of the above

359. The discount rate that equates the present value of the future net cashflows from an investment project with the project's initial cash outflow is known as
(a) Average rate of return
(b) Internal rate of return
(c) Cost of capital
(d) Hurdle rate

360. As per Pecking order theory of capital structure
(a) Internal equity is preferred over external debt
(b) External debt is preferred over internal equity
(c) External equity is preferred over external debt
(d) None of the above

361. A formal legal commitment to extend credit up to some maximum amount over a stated period of time is known as a
(a) Line of credit
(b) Revolving credit agreement
(c) Commitment charge
(d) Secured term loan

362. Economics realized in a merger where the performance of the combined firms exceed that of its previously separate parts is known as
(a) Economies of Scale
(b) Synergy
(c) Horizontal integration
(d) Leveraged Buyout

363. Macro marketing environment refers to
(a) Internal to the company
(b) Employees and stakeholders of a company
(c) External to the company
(d) Stakeholers

364. Environmental scanning involves
(a) Weather forecasting
(b) Studying depression in a sea
(c) Identifying threats and opportunities
(d) Identifying strength and weakness

365. Industrial marketing involves
(a) Business to Business
(b) Customers to customers
(c) Online marketing
(d) Customers to Business

366. News Boy Problem is related to
(a) Queuing problem
(b) Inventory problem
(c) Replacement problem
(d) Linear Programming problem

367. Mini max method of decision making is done when
(a) The problem is under risk
(b) The problem is under uncertainty
(c) The problem is of deterministic type
(d) The problem is under linear programming problem

368. Cost leadership strategy is related to
(a) Corporate strategy
(b) "SBU" strategy
(c) Functional strategy
(d) Global strategy

369. Who is associated with the concept that competence is nothing but 'resourcefulness' meaning 'identifying resources when there are no resources'
(a) Gopal C Pati and Rabi K Kanungo
(b) M Maurice and F Sellier
(c) Evans and Lorange
(d) A Martin and G Ross

370. Who has advanced the Motivational Analysis of Organisation?
(a) D M Pestonjee
(b) T V Rao
(c) Udai Pareek
(d) None of the above

371. Who had advanced the Stewardship theory of Corporate Governance
(a) J H Davis et.al
(b) D R Dalton et.al
(c) D C Hambrick et.al
(d) None of the above

372. In the four classification of Social Responsibility of Business proposed by Archie Carroll, which one of the following is not included
(a) Economic
(b) Legal
(c) Ethical
(d) Environmental

373. In India, which method is used to measure National Income in service sector
(a) Expenditure method
(b) Income method
(c) Output method
(d) Value added method

374. Where anchors are illustrations that help to define as superior, average or below average, the behaviour of the employee, the rating scale is known as
(a) Behavioural expectation scale
(b) Behavioural observation scale
(c) Forced choice description
(d) Grading scale

375. In the decision tree model of Vroom and Yetton the style when the leader shares the problem with subordinates as a group, obtains collective ideas, and then makes the decision is identified as
(a) A I (b) A II
(c) C I (d) C II

376. Leaders who provide indivisual consideration and intellectual stimulation and who possess charisma is known as
(a) Transactional leader
(b) Transformational leader
(c) Situational leader
(d) Visionary leader

377. Which of the following is not one of the career anchors
(a) Managerial competence
(b) Technical-functional competence
(c) Personal competence
(d) Security

378. Theory 'Z' has been propounded by
(a) Mc Gregor (b) Ouchi
(c) Maslow (d) Skinner

379. Outplacement services are provided to employees who have been
(a) Dismissed
(b) Retired
(c) Rendered surplus
(d) Given voluntary retirement

380. What notice period is required for going on strike/lockout in public utility services?
(a) One weeks (b) Six weeks
(c) Three weeks (d) Nine weeks

381. Modifying methods of performing a job to reduce the levels of skills or skills demanded is known as
(a) Multi-tasking
(b) Multi-skilling
(c) De-skilling
(d) None of the above

382. For determination of cost of equity the formula used is
(a) Rate of interest (1-t)
(b) Rate of dividend (1-t)
(c) Rate of dividend (Face value – Issue Price)/n
(d) D1 --- + g Po

383. A stable dividend policy indicates
(a) Stable earning/price ratio
(b) Stable price/earnings ratio
(c) Stable payout ratio
(d) Stable earnings per share

384. The vertical merger represents merger of firms engaged in
(a) The same line of business
(b) Different stages of production in an industry
(c) Unrelated lines of business
(d) All of the above

385. Which of the following factors does not influence the pricing strategy of the firm
(a) Market (b) Competition
(c) Cost (d) R & D

386. In the diffusion of innovation who are the people not willing to try out the product
(a) Innovators (b) Progressive people
(c) Early adopters (d) Early majority

387. Who divided all activities of the group of organisations into six groups: Technical, Commercial, Financial, Security, Accounting and Managerial
(a) Peter Drucker
(b) D.E McForland
(c) Henry Fayol
(d) George R Terry

388. The policy of introducing "New Blood" is
(a) Stereotype
(b) Bureaucratic as well as red tapic
(c) Innovative and challenging, making the corporate dynamic
(d) Innovative challenging, but often resisted

389. An 'Assessment Centre' is
(a) A Place
(b) A Technique
(c) Both (a) and (b)
(d) None of the above

390. Which of the following is an important money market instrument
(a) Debentures
(b) Commercial paper
(c) Public deposit
(d) None of the above

391. Globalisation involves
(a) Free flow of technology from one country to another
(b) Free flow of investment from one country to another
(c) Free flow of people from one country to another
(d) All the above

392. Whistle blowers are
(a) To be removed from the job
(b) To be promoted
(c) To be rewarded for the organisational enrichment
(d) To be protected since they intend to bring out truth

393. Hawthorne experiment is related to the
(a) Systems theory
(b) Classical theory
(c) Human relations theory
(d) Scientific management theory

394. Cost incurred in the past and is not affected by a current decisions referred to as
(a) Sunk cost
(b) Marginal cost
(c) Incremental cost
(d) Replacement cost

395. The principle of organisation that no employee should report to more than one superior is called
(a) Scalar principle
(b) Span of control
(c) Unity of command
(d) Unity of direction

396. 'Rage Committee' is related to
(a) Working conditions in Indian Industry
(b) Social security in Indian Industry
(c) Indivisual differences in Indian Industry
(d) None of the above

397. Balance of Payment is
(a) Balance of trade + Net earnings on invisibles
(b) Foreign exchange inflow – foreign exchange outflow
(c) Balance of current account + Balance of capital account + Statistical discrepancy
(d) Exports of goods – Import of goods

398. Strategies formulated to convert a sick unit to healthy is referred to
(a) Turn around (b) Expansion
(c) Diversification (d) Stabilisation

399. Which one is an international credit rating agency
(a) Mc Kinsey
(b) B C G
(c) Standard and Poor
(d) I M F

400. "The optimal payout ratio for growth firm is nil and declining firm is 100%" is established by
(a) Gordon Model
(b) Walter Model
(c) Both (a) and (b)
(d) None of the above

401. The time that elapses between the purchase of raw materials and the collection of cash for sales is referred as
(a) Production cycle
(b) Operating cycle
(c) Cash cycle
(d) Organisation cycle

402. Taylor differential piece rate system is related to
(a) Time wage system
(b) Piece wage system
(c) Incentive wage system
(d) None of the above

403. A company using high price and high promotion policy is adopting the following strategy
(a) Slow skimming
(b) Rapid penetration
(c) Slow penetration
(d) Rapid skimming

404. The number of product lines a company carries is called
(a) Product range
(b) Product mix depth
(c) Product mix width
(d) Product line length

405. Sensitivity analysis may be used in
(a) Linear Programming
(b) Transportation Programming
(c) PERT
(d) CPM

406. Sensitivity training method is part of
(a) On the job training method
(b) Off the job training method
(c) Vestibule training method
(d) None of the above

407. The full form of GATT is
(a) General Agreement of Trade and Tariff
(b) General Agreement of Trade or Tariff
(c) General Agreement of Tariff and Trade
(d) General Agreement of Tariff or Trade

408. GATT came in the year
(a) 1947 (b) 1948
(c) 1949 (d) 1995

409. WTO was established in the year
(a) January 1, 1995
(b) June 1, 1995
(c) July 1, 1995
(d) Dec 1, 1995

410. Head quarter of WTO
(a) Washington DC
(b) Geneva, Switzerland
(c) New York, USA
(d) Nairobi, Kenya

411. Full form of WTO
(a) World Trust Organization
(b) World Tariff Organization
(c) World Trade Organization
(d) World Treaty Office

412. Current Membership of WTO till date 30/11/2015 is
(a) 150 (b) 143
(c) 162 (d) 192

413. Director General of WTO is
(a) Knut Hove
(b) Roberto Azevedo
(c) Christine Lagarde
(d) José Graziano da Silva

414. Market access commitment under Agreement on Agriculture in WTO was fixed at
(a) 2% (b) 3%
(c) 4% (d) 5%

415. Domestic support under AOA for developed countries was fixed at
(a) 2% (b) 3%
(c) 4% (d) 5%

416. Domestic support under AOA for developing countries was fixed at
(a) 5% (b) 10%
(c) 20% (d) 25%

417. The support given in the form of research, extension, disaster payment, control of pests and diseases, structural adjustment programmes etc fall under
(a) Blue Box
(b) Green Box
(c) Amber Box
(d) S&D Measures

418. The support given in the form of direct payments, decoupled income support, deficiency payments, single payments etc fall under
(a) Blue Box
(b) Green Box
(c) Amber
(d) S&D Measures

419. The support given in the form of export subsidies fall under
(a) Blue Box
(b) Green Box
(c) Amber Box
(d) S&D Measures

420. Protection of Plant Varieties and Farmers' Right Act came in the year
(a) 1999 (b) 2001
(c) 2003 (d) 2005

421. Geographical Indications of Goods (Registration and Protection) Act came in the year
(a) 1999 (b) 2001
(c) 2003 (d) 2005

422. Biological Diversity Act came in the year
(a) 1999 (b) 2001
(c) 2003 (d) 2005

423. Trade Marks Act came in the year
(a) 1999 (b) 2001
(c) 2003 (d) 2005

424. Patents Act came in the year
(a) 1960 (b) 1968
(c) 1970 (d) 2005

425. Copy right act came in the year
(a) 1948 (b) 1957
(c) 1999 (d) 2011

426. Layout Design Act came in the year
(a) 1999 (b) 2000
(c) 2003 (d) 2005

427. Layout Design comes under
(a) Ministry of Agriculture
(b) Ministry of Communication and IT
(c) Ministry of Human Resource Development
(d) Ministry of Commerce and Industries

428. Patent, Industrial design, trademark, Geographical indications comes under
(a) Ministry of Agriculture
(b) Ministry of Communication and IT
(c) Ministry of Human Resource Development
(d) Ministry of Commerce and Industries

429. Copy right comes under
(a) Ministry of Agriculture
(b) Ministry of Communication and IT
(c) Ministry of Human Resource Development
(d) Ministry of Commerce and Industries

430. Literary, musical, dramatic and artistic works and cinematograph films and sound recordings mainly protected through
(a) Patent
(b) Copyright
(c) Trademark
(d) Geographical Origin

431. Patent work must contains
(a) Novelty
(b) Non-Obviousness
(c) Usefulness
(d) All of the above

432. Patent protect must be available for inventions for
(a) 5 year
(b) At least 10 years
(c) At least 20 years
(d) No time limit

433. Darjeeling has the Geographical Indications for
(a) Coffee (b) Tea
(c) Saree (d) Petha

434. Ponchampali and Chanderi has the Geographical Indications for
(a) Cotton (b) Silk
(c) Nylon (d) Polyster

435. Nagpur has the Geographical Indications for
(a) Apple (b) Orange
(c) Peas (d) Pomegranate

436. Bikaner has the Geographical Indications for
(a) Petha (b) Rasgoola
(c) Bhujia (d) Mustard Oil

437. The registration of a GI shall for a period of
(a) 5 years (b) 10 years
(c) 20 years (d) 50 years

438. Industrial design must be protected for at least
(a) 5 years
(b) 10 years
(c) 20 years
(d) 50 years

439. According to quantification, tariff that has been imposed on different products according to their physical weight is called as
(a) Ad-valorem duties
(b) Specific duties
(c) Both
(d) None

440. According to quantification, tariff that been imposed on different products according to their money price is called as
(a) Ad-valorem duties
(b) Specific duties
(c) Combined duty
(d) Tariff Quota

441. Duties may be imposed on certain imports when they have been subsidized by foreign governments is called as
(a) Countervailing duties
(b) Specific duties
(c) Combined duty
(d) Anti-dumping duty

442. Duties are applied to imports which are dumped in the domestic market at prices below their cost of production or substantially lower than their domestic prices is called as
(a) Countervailing duties
(b) Specific duties
(c) Combined duty
(d) Anti-dumping duty

443. Value Added tax is a
(a) Single stage, Non-cumulative tax on consumption
(b) Multistage, Non-Cumulative tax on consumption
(c) Multistage, Cumulative tax on consumption
(d) Single stage, Cumulative tax on consumption

444. Non-tariff barrier where imports of a commodity up to a specified volume are allowed duty free or at a special low rate; but any imports in excess of this limit are subject to duty or higher rate of duty is called as
(a) Unilateral Quota
(b) Tariff Quota
(c) Bilateral Quota
(d) Mixing Quota

445. Antidumping action is not applicable if the margin of dumping is
(a) Less than 2% of the import price
(b) Volume from one country is less than 3% of the total imports of that product
(c) Aggregate imports from such countries do not account for more than 7% of total imports
(d) All of the above cases

446. Pradhan Mantri Fasal Bima Yojana came in the year
(a) 2013 (b) 2014
(c) 2015 (d) 2016

447. According to Agricultural Census report, 2010-11, Operational Land Holding Capacity of small and marginal farmers is
(a) 85.01% (b) 80%
(c) 90% (d) 75%

448. According to Agricultural Census report, 2010-11, how many cultivators are in the country
(a) 110 million
(b) 118.6 million
(c) 150 million
(d) 200 million

449. Under Pradhan Mantri Fasal Bima Yojana, a farmer has to pay a uniform premium of
(a) 2% of the total crops for all Kharif Crop
(b) 1.5% of the value of Rabi Crop
(c) 5% on all commercial crops like Cocoa, Coffee, Cotton, Tea and Tobacco
(d) All of the Above

450. Operational Boundary of all dairy Cooperative is
(a) A block (b) A District
(c) A State (d) A Country

451. Part IX of the Companies Act, 1956 deals with
(a) Cooperative (b) Producer Company
(c) SHG (d) LIC

452. Board Strength in Producer Company is up to
(a) 2 (b) 5
(c) 7 (d) 15

453. Minimum member of director in Private Limited Company
(a) 2 (b) 3
(c) 5 (d) 7

454. Minimum number of director in Public Limited Company is
(a) 2 (b) 3
(c) 5 (d) 7

455. Minimum and Maximum number of members in Private Limited Company is respectively
(a) 2,20 (b) 5,40
(c) 2,50 (d) 10,100

456. Minimum number of members in public limited company
(a) 2 (b) 3
(c) 5 (d) 7

457. Basic requirements for formation of producer company is
(a) Minimum 10 Primary Producers members
(b) Two Producer institutional member
(c) Both of the above
(d) None

458. Types of shares circulate in producer company
(a) Equity share
(b) Preferential share
(c) Both
(d) None

459. The concept of producer company was introduced in the year
(a) 2000 (b) 2001
(c) 2002 (d) 2004

460. Producer Company has been recommended by
(a) Sivaraman Committee
(b) Narasimham Committee
(c) Y. K. Alagh Committee
(d) Kelko Committee

461. Full form of PES
(a) Payment for Ecological System
(b) Payment for Ecosystem Services
(c) Provision for Environmental Substances
(d) None

462. Payment for Ecosystem Services (PES) scheme revolve around
(a) Water and Soil related Services
(b) Climate Stabilisation
(c) Biodiversity Conservation
(d) All of the above

463. Climate Smart Agriculture deals with
(a) Sustainable increasing agricultural productivity
(b) Adopting and building resilience of agricultural and food security
(c) Reducing greenhouse gas emission from agriculture
(d) All of the above

464. Second largest consumer and largest importer of fertiliser
(a) India (b) China
(c) USA (d) Russia

465. How much emission of GHG has been contributed by agriculture
(a) 10% (b) 14%
(c) 20% (d) 22%

466. How much emission of GHG has been contributed by agriculture and forestry
(a) 20% (b) 31%
(c) 41% (d) 50%

467. Contribution of GHG from agricultural activities in Asia is
(a) 40% (b) 44%
(c) 50% (d) 85%

468. Rice cultivation contributes, how much percentage of GHG
(a) 10% (b) 16%
(c) 20% (d) 22%

469. Agenda 21 deals with
(a) Biological Diversity
(b) Climate Change
(c) Both
(d) None

470. The first land mark of environmental at Stockholm has been enunciated in
(a) 1972 (b) 1992
(c) 2012 (d) 2016

471. The Union Government has launched a toll free helpline number to provide assistance to refrigerated transport truck in the middle of highways in a bid to bring down losses of perishable food items in transit. The toll free number is
(a) 1800-2676-223
(b) 1800-180-1551
(c) 1800-180-1503
(d) 0124-614-9200

472. Marketing research and Information Network (MRIN) provides electronic connectivity to important wholesale markets in the country for collection and dissemination of price and market related information. This scheme has been launched in the year
(a) 1990 (b) 2000
(c) 2002 (d) 2007

473. Kisan Call Centers (KCC) have been functioning since
(a) 20 February, 2002
(b) 18 October, 2004
(c) 21 January, 2004
(d) 1 January, 2005

474. MS Swaminathan Research Foundation launched the village knowledge enters in 1998 in
(a) Chennai (b) Pondicherry
(c) Bangalore (d) Hyderabad

475. Bhoomi project deals with computerization of land records is an innovative concept of
(a) Maharashtra
(b) Karnataka
(c) Gujarat
(d) Tamil Nadu

476. Akashganga project launched by Dairy Cooperative Society of
(a) Maharashtra
(b) Karnataka
(c) Gujarat
(d) Tamil Nadu

477. Pardarshi Kisan Sewa Yojana was launched in September, 2004 for distribution of hybrid seeds through DBT (Direct Benefit Transfer) by
(a) Uttar Pradesh
(b) Madhya Pradesh
(c) Andhra Pradesh
(d) Gujarat

478. National Agriculture Market (NAM) through Agritech Infrastructure Fund (ATIF) was approved by the Cabinet Committee on Economic Affairs on
(a) 1st June, 2014
(b) 1st July, 2015
(c) 1st February, 2013
(d) 1st August, 2012

479. Total number of mandis covered in NAM for the year 2015-16, 2016-17 and 2017-18 are respectively
(a) 250,200,135 (b) 135,200,250
(c) 205,135,200 (d) 135,250,200

480. Tenth ministerial conference of the WTO held in
(a) Geneva
(b) Nairobi
(c) Philippines
(d) Singapore

481. The average dietary energy intake per person per day in rural India is
(a) 2000 Kcal (b) 2147 Kcal
(c) 2123 Kcal (d) 2500 Kcal

482. The average dietary energy intake per person per day in urban India is
(a) 2000 Kcal
(b) 2147 Kcal
(c) 2123 Kcal
(d) 2500 Kcal

483. Special Economic Zones in India are
(a) Kerala Industrial Infrastructure Development Corporation
(b) Karnataka Industrial Area Development Board
(c) Parry Infrastructure Company Private Limited
(d) All of the above

484. Per capita availability of Pulses
(a) 30g (b) 42g
(c) 70g (d) 110g

485. Retention Price Scheme for nitrogen first came in
(a) 1977 (b) 1979
(c) 1982 (d) 1985

486. Retention Price Scheme for complex fertiliser came in the year
(a) 1977 (b) 1979
(c) 1982 (d) 1985

487. Retention Price Scheme for single super phosphate came in the year
(a) 1977 (b) 1979
(c) 1982 (d) 1985

488. Micro nutrients that included in Nutrient Based Subsidy Scheme in April, 2010 were
(a) Sulphur (b) Zinc
(c) Boron (d) All

489. In Nutrient Based Subsidy Scheme, price decided per kg of Nitrogen, Phosphorus, Potash are respectively
(a) Rs 20.87, Rs18.69, Rs 15.50
(b) Rs 15.50, Rs 18.69, Rs 20.87
(c) Rs 20.87, Rs 15.50, Rs 18.69
(d) Rs 18.69, Rs 20.87, Rs 15.50

490. How much contribution of Indian seed market in Global sphere
(a) 4.4% (b) 6%
(c) 20% (d) 27%

491. Country that is first in seed production
(a) India (b) China
(c) USA (d) Brazil

492. Seed production in India ranks
(a) First (b) Third
(c) Fifth (d) Seven

493. In FY 2013, how much percentage has been contributed by non-vegetable seeds in India
(a) 50% (b) 62.2%
(c) 82.2% (d) 95%

494. Which crop contributes maximum in non-vegetable seed categories in India
(a) Paddy (b) Wheat
(c) Maize (d) Cotton

495. Total number of seed corporations in India are
(a) 13 (b) 15
(c) 17 (d) 19

496. FDI in seed sector came in the year
(a) 1975 (b) 1988
(c) 1991 (d) 2005

497. Protection of Plant Varieties and Farmers' Right Act came in the year
(a) 2000 (b) 2001
(c) 2002 (d) 2003

498. National Seed Research and Training Center is at
(a) Mumbai (b) Ahmedabad
(c) Varanasi (d) Jhansi

499. The farm machinery training and testing institute is at
(a) Budni (Madhya Pradesh)
(b) Hissar (Haryana)
(c) Garladdine (Andhra Pradesh)
(d) All of the Above

500. The ratio of export to import for virtual water in India is
(a) 1 (b) 2
(c) 3 (d) 4

501. The ratio of export to import for virtual water in China is
(a) 0.1 (b) 1
(c) 5 (d) 7

502. According to NASA, India's water tables are declining at a rate of
(a) 0.1 meter per year
(b) 0.2 meter per year
(c) 0.3 meter per year
(d) 0.5 meter per year

503. How much percentage of total subsidy of fertilizer reaches to ultimate beneficiary?
(a) 10% (b) 25%
(c) 35% (d) 50%

504. How many recognized agencies are there in India for importing fertilizer?
(a) 1 (b) 2
(c) 3 (d) 4

505. How many domestic urea producers are there in India?
(a) 10 (b) 20
(c) 30 (d) 40

506. The fertilizer that is most produced, most consumed and the most imported in India
(a) Urea (b) DAP
(c) MOP (d) CAN

507. According to NSSO, the share of agriculture in employment is (as on 2011-12)
(a) 52% (b) 48.9%
(c) 60% (d) 45%

508. The food grain production in India as on 2015-16
(a) 253.16 million tonne
(b) 264.16 million tonne
(c) 252.25 million tonne
(d) 250 million tonne

509. Total tractor penetration among large farmers in India
(a) 25% (b) 38%
(c) 45% (d) 50%

510. Total tractor penetration among medium farmers in India
(a) 15% (b) 18%
(c) 20% (d) 25%

511. Total pesticide consumption in India per kg is
(a) 0.2 Kg (b) 0.25Kg
(c) 0.5 Kg (d) 0.1 kg

512. Indian famers can avail loan up to 3 lakhs at
(a) 3% interest (b) 5% interest
(c) 4% interest (d) 7% interest

513. Gross capital formation in agriculture as on 2013-14 in India is
(a) 5.5% (b) 8.6%
(c) 7.4% (d) 9%

514. According to WTO agreement on agriculture, market access that has been decided is
(a) 3% (b) 5%
(c) 7% (d) 10%

515. How much percentage of Aggregate measurement of support in value was fixed for developed country?
(a) 5% (b) 10%
(c) 15% (d) 20%

516. How much percentage of Aggregate measurement of support in value was fixed for developing country?
(a) 5% (b) 10%
(c) 15% (d) 20%

517. Safal Market is the innovative concept of
(a) Food Corporation of India
(b) National Dairy Development Board
(c) Walmart
(d) NAFED

518. Central Storage Institute is at
(a) Pune (b) Indore
(c) Hapur (d) Chandigarh

519. Indian Institute of Packaging is at
(a) Delhi (b) Mumbai
(c) Chennai (d) Kolkata

520. Central AGMARK laboratory is at
(a) Hyderabad (b) Nagpur
(c) Luckow (d) Bhubaneswar

521. The grading and marking of goods was first started in
(a) 1935 (b) 1938
(c) 1941 (d) 1955

522. Head quarter of DMI
(a) Faridabad (b) Nagpur
(c) Delhi (d) Asunsol

523. ECOMARK scheme is administered by
(a) WTO (b) BIS
(c) UN (d) FAO

524. World Standard day is celebrated annually is on
(a) 14th October (b) 15th October
(c) 16th October (d) 17th October

525. Ministry of Food Processing was set up in
(a) 1986 (b) 1988
(c) 1990 (d) 2000

526. Total number of commodities are included under MSP
(a) 20 (b) 24
(c) 25 (d) 30

527. Jute Corporation of India was set up in
(a) 1965 (b) 1970
(c) 1971 (d) 1972

528. Cotton Corporation of India is responsible for price support operation in all states, except
(a) West Bengal (b) Maharashtra
(c) Gujarat (d) Haryanan

529. APEDA came in the year
(a) 1986 (b) 1988
(c) 1995 (d) 2005

530. National Dairy Development Board was set up in
(a) 1960 (b) 1965
(c) 1970 (d) 1971

531. MPEDA was set up in the year
(a) 1971 (b) 1972
(c) 1991 (d) 2001

532. Bureau of Indian Standard came in the year
(a) 1986 (b) 1998
(c) 2005 (d) 2011

533. Standards of weights and measures (packed goods) Act came in the year
(a) 1957 (b) 1991
(c) 1997 (d) 2001

534. Essential Commodity Act came in the year
(a) 1955 (b) 1965
(c) 1971 (d) 1975

535. MSP is announced on the recommendation of
(a) FCI
(b) Government of India
(c) CACP
(d) Ministry of Commerce

536. Head Quarter of Cotton Corporation of India is at
(a) Maharashtra (b) Gujarat
(c) Haryana (d) Karnataka

537. Big retailer or small wholesaler is called as
(a) Itinerant trader (b) Mashakhore
(c) A Class trader (d) Vendor

538. Petty merchant who moves from village to village is called as
(a) Itinerant trader (b) Wholesaler
(c) Broker (d) Agent

539. Pacca Arhatiyas in the unregulated market deals with
(a) Farmer (b) Trader
(c) Agent (d) Broker

540. Processing of products add ______ utility
(a) Time (b) Space
(c) Form (d) Possession

541. The head quarter of NAFED is at
(a) New Delhi (b) Mumbai
(c) Kolkata (d) Chennai

542. Sneha is the milk cooperative of ___ state
(a) Haryana
(b) Karnataka
(c) Madhya Pradesh
(d) Andhra Pradesh

543. Indian Society of Agricultural Marketing came in the year
(a) 1986 (b) 1988
(c) 1990 (d) 1991

544. Head Quarter of National Institute of Agricultural Marketing is at
(a) Hyderabad (b) Bhubaneswar
(c) Lucknow (d) Jaipur

545. According to CIPHET, The post-harvest losses in fruits and Vegetables is around
(a) 10-12% (b) 12-14%
(c) 16-18% (d) 20-22%

546. Average land holding capacity of Indian farmers' is approximately
(a) 2.25 ha (b) 1.15 ha
(c) 3.35 ha (d) 0.75 ha

547. Number of regulated market in India in 2014 is
(a) 7000 (b) 7114
(c) 7333 (d) 7426

548. Percentage of marginal farmers' in India is approximately
(a) 55% (b) 67%
(c) 80% (d) 90%

549. Open Auction method in which one bid is given for all the lots in a particular shop is called as
(a) Phar system of Open Auction
(b) Close Bid method of Open Auction
(c) Random bid system of open Auction
(d) Roster Bid System of Open Auction

550. Market information that gathers near future is called as
(a) Market News
(b) Market Intelligence
(c) Market Outlook
(d) Market Research

551. The monthly situation and outlook reports of agricultural price are published by
(a) FCI (b) NAFED
(c) CCI (d) DMI

552. The institute that has taken the initiatives to prepare National Atlas of Indian Agricultural market is
(a) MANAGE (b) IRMA
(c) NIAM (d) SYMBIOSIS

553. A calculated risk in business for transaction of agricultural commodity is called as
(a) Speculation (b) Hedging
(c) Both (d) None of the Above

554. Future trading includes both speculation and
(a) Hedging
(b) Arbitrage
(c) Option
(d) Swap

555. Indivisuals who take the title to the goods they handle are called as
(a) Facilitative Middleman
(b) Merchant Middleman
(c) Agent Middleman
(d) Speculative Middleman

556. In regulated market, only one category of commission agent exists is called as
(a) Kachha Arhatiyas
(b) Pacca Aarhatiyas
(c) A Class Trader
(d) Broker

557. In Apni Mandi, number of channels members between producer and consumer is
(a) 0 (b) 1
(c) 2 (d) 3

558. The first Apni Mandi was started in
(a) Punjab (b) Haryana
(c) Gujarat (d) Rajasthan

559. Vegetable and Fruit Promotion Council is seen in
(a) Andhra Pradesh (b) Karnataka
(c) Tamil Nadu (d) Kerala

560. Pepsi started contract farming for which commodity
(a) Potato (b) Tomato
(c) Onion (d) Tobacco

561. Market charges that have taken due to impurity in agricultural commodity is called as
(a) Karda (b) Muddat
(c) Dhalta (d) Darmada

562. Market charges that have taken due to excessive moisture content in the agricultural commodity is called as
(a) Karda (b) Muddat
(c) Dhalta (d) Darmada

563. In India, complete wholesale trade was taken over by Government in 1973 for
(a) Paddy (b) Wheat
(c) Maize (d) Jowar

564. Procurement strategy for food grains in which Government assumes the first attempt right to purchase the grain is called as
(a) Open Market Operation
(b) Monopoly Procurement
(c) Levy On Producer
(d) Preemptive Procurement

565. Unza market deals in
(a) Cumin (b) Sugar
(c) Potato (d) Onion

566. In Input markets farmers mainly face ______________- competition
(a) Perfect Competition
(b) Monopoly Competition
(c) Monopolistic Competition
(d) Oligopolistic Competition

567. The state that abolish commission agent for agro commodity is
(a) Madhya Pradesh
(b) Maharashtra
(c) Karnataka
(d) Kerala

568. The state that has exempted 30% of market fee on purchase under contract agreement is
(a) Maharashtra
(b) Karnataka
(c) Kerala
(d) Gujarat

569. The commission charges for fruits and vegetables is
(a) 2-5% (b) 1-2%
(c) 4-8% (d) >10%

570. Average area served by a regulated market in India in 2011 is
(a) 250 sq km
(b) 454 sq km
(c) 11,200 sq km
(d) None of the above

571. Hapur and Khanna commodity market deals in
(a) Sugar (b) Wheat
(c) Onion (d) Leather

572. Market comes from the word
(a) Murcatus (b) Marcatus
(c) Markat (d) Marucat

573. In marketing, third 'p' represents as
(a) Product (b) Price
(c) Place (d) Promotion

574. Spontaneous market in developed country simulates ______________ market of developing country
(a) Periodic Market
(b) Seasonal Market
(c) Secular Market
(d) Long Period Market

575. Revamped Public Distribution System was launched in
(a) 1961 (b) 1965
(c) 1992 (d) 1997

576. Grading supervisor training centre is located in
(a) Nagpur (b) Faridabad
(c) Pune (d) Hyderabad

577. First cold structure is located at
(a) Bihar (b) Kolkata
(c) Dehradun (d) Delhi

578. Which is an adhoc price support scheme for horticultural products
(a) MSP
(b) MIS
(c) Statutory Price
(d) Procurement Price

579. CAP storage structure is developed by
(a) FCI
(b) CCI
(c) NAFED
(d) NCDC

580. Sale under cover of cloth is called
(a) Dara Sale Method
(b) Hatha Sale Method
(c) Moghum Sale Method
(d) Auction Method

581. The first grading station was establishes in
(a) Andhra Pradesh
(b) Tamil Nadu
(c) Kerala
(d) Madhya Pradesh

582. Sale in which selling price is less than the cost of production is called as
(a) Bumper Price (b) Distress sale
(c) Floor Price (d) Merchant sale

583. The type of marketing middlemen who assist in the process of marketing, but do not involve in buying or selling activities is called as
(a) Merchant Middleman
(b) Agent Middleman
(c) Speculative Middleman
(d) Facilitative Middleman

584. The price below which farmer would find it uneconomical to sell the produce is
(a) Wholesale Price
(b) Minimum Price
(c) Floor Price
(d) Harvesting Price

585. The largest food grain procurement agency at National level is
(a) FCI (b) CCI
(c) NAFED (d) NDDB

586. The market that deals with durable goods is
(a) Secular Market
(b) Periodic Market
(c) Spontaneous market
(d) Long run Market

587. According to the recommendation of National Farmers' commission, availability of market is __________ km from his premises
(a) 5 km (b) 10 km
(c) 12 km (d) No such boundaries

588. According to the recommendation of National Farmers' commission, on an average farmers' need to travel how many kilometer to reach at Mandi
(a) 5 km (b) 12 km
(c) 50 km (d) 65 km

ANSWER KEY (Practice Set 1)

1. (a)	**2.** (b)	**3.** (d)	**4.** (c)	**5.** (d)	**6.** (b)	**7.** (c)	**8.** (c)	**9.** (a)	**10.** (c)
11. (c)	**12.** (d)	**13.** (b)	**14.** (b)	**15.** (d)	**16.** (a)	**17.** (a)	**18.** (b)	**19.** (d)	**20.** (c)
21. (b)	**22.** (d)	**23.** (b)	**24.** (d)	**25.** (c)	**26.** (d)	**27.** (d)	**28.** (d)	**29.** (a)	**30.** (a)
31. (a)	**32.** (a)	**33.** (b)	**34.** (c)	**35.** (b)	**36.** (a)	**37.** (a)	**38.** (b)	**39.** (a)	**40.** (b)
41. (a)	**42.** (a)	**43.** (a)	**44.** (c)	**45.** (c)	**46.** (c)	**47.** (b)	**48.** (b)	**49.** (b)	**50.** (c)
51. (d)	**52.** (b)	**53.** (b)	**54.** (a)	**55.** (a)	**56.** (a)	**57.** (b)	**58.** (c)	**59.** (b)	**60.** (b)
61. (a)	**62.** (c)	**63.** (d)	**64.** (b)	**65.** (a)	**66.** (d)	**67.** (a)	**68.** (a)	**69.** (c)	**70.** (d)
71. (b)	**72.** (b)	**73.** (d)	**74.** (d)	**75.** (c)	**76.** (d)	**77.** (c)	**78.** (a)	**79.** (c)	**80.** (c)
81. (b)	**82.** (b)	**83.** (c)	**84.** (b)	**85.** (d)	**86.** (c)	**87.** (d)	**88.** (b)	**89.** (a)	**90.** (c)
91. (d)	**92.** (b)	**93.** (c)	**94.** (c)	**95.** (a)	**96.** (c)	**97.** (c)	**98.** (b)	**99.** (d)	**100.** (d)
101. (c)	**102.** (a)	**103.** (b)	**104.** (a)	**105.** (a)	**106.** (c)	**107.** (c)	**108.** (a)	**109.** (c)	**110.** (b)
111. (d)	**112.** (b)	**113.** (b)	**114.** (a)	**115.** (d)	**116.** (a)	**117.** (a)	**118.** (c)	**119.** (d)	**120.** (d)
121. (b)	**122.** (c)	**123.** (b)	**124.** (b)	**125.** (b)	**126.** (d)	**127.** (c)	**128.** (c)	**129.** (b)	**130.** (a)
131. (a)	**132.** (c)	**133.** (a)	**134.** (a)	**135.** (d)	**136.** (a)	**137.** (a)	**138.** (d)	**139.** (a)	**140.** (b)
141. (a)	**142.** (d)	**143.** (d)	**144.** (d)	**145.** (b)	**146.** (a)	**147.** (d)	**148.** (b)	**149.** (c)	**150.** (c)
151. (a)	**152.** (b)	**153.** (c)	**154.** (b)	**155.** (b)	**156.** (d)	**157.** (b)	**158.** (a)	**159.** (c)	**160.** (a)
161. (b)	**162.** (c)	**163.** (c)	**164.** (c)	**165.** (c)	**166.** (c)	**167.** (b)	**168.** (d)	**169.** (c)	**170.** (a)
171. (c)	**172.** (a)	**173.** (a)	**174.** (b)	**175.** (a)	**176.** (b)	**177.** (c)	**178.** (b)	**179.** (d)	**180.** (a)
181. (a)	**182.** (d)	**183.** (d)	**184.** (c)	**185.** (a)	**186.** (d)	**187.** (c)	**188.** (b)	**189.** (b)	**190.** (d)
191. (d)	**192.** (d)	**193.** (b)	**194.** (c)	**195.** (b)	**196.** (b)	**197.** (a)	**198.** (d)	**199.** (b)	**200.** (b)
201. (d)	**202.** (b)	**203.** (a)	**204.** (d)	**205.** (a)	**206.** (b)	**207.** (a)	**208.** (d)	**209.** (c)	**210.** (c)
211. (d)	**212.** (d)	**213.** (a)	**214.** (b)	**215.** (a)	**216.** (b)	**217.** (b)	**218.** (d)	**219.** (b)	**220.** (a)
221. (b)	**222.** (b)	**223.** (c)	**224.** (d)	**225.** (c)	**226.** (c)	**227.** (b)	**228.** (c)	**229.** (d)	**230.** (b)
231. (b)	**232.** (b)	**233.** (b)	**234.** (b)	**235.** (b)	**236.** (b)	**237.** (b)	**238.** (c)	**239.** (a)	**240.** (a)
241. (b)	**242.** (b)	**243.** (a)	**244.** (a)	**245.** (a)	**246.** (c)	**247.** (b)	**248.** (c)	**249.** (b)	**250.** (c)
251. (c)	**252.** (a)	**253.** (c)	**254.** (a)	**255.** (c)	**256.** (d)	**257.** (b)	**258.** (b)	**259.** (d)	**260.** (a)
261. (a)	**262.** (b)	**263.** (b)	**264.** (b)	**265.** (d)	**266.** (a)	**267.** (a)	**268.** (d)	**269.** (c)	**270.** (d)
271. (a)	**272.** (b)	**273.** (c)	**274.** (c)	**275.** (a)	**276.** (b)	**277.** (b)	**278.** (a)	**279.** (c)	**280.** (d)
281. (d)	**282.** (d)	**283.** (b)	**284.** (b)	**285.** (d)	**286.** (c)	**287.** (a)	**288.** (c)	**289.** (c)	**290.** (d)
291. (d)	**292.** (a)	**293.** (b)	**294.** (c)	**295.** (b)	**296.** (b)	**297.** (d)	**298.** (d)	**299.** (a)	**300.** (c)
301. (a)	**302.** (a)	**303.** (b)	**304.** (b)	**305.** (a)	**306.** (d)	**307.** (c)	**308.** (d)	**309.** (c)	**310.** (d)
311. (a)	**312.** (c)	**313.** (d)	**314.** (c)	**315.** (a)	**316.** (b)	**317.** (b)	**318.** (c)	**319.** (a)	**320.** (a)

321. (b) **322.** (d) **323.** (d) **324.** (c) **325.** (b) **326.** (c) **327.** (a) **328.** (a) **329.** (d) **330.** (a)
331. (b) **332.** (c) **333.** (d) **334.** (b) **335.** (a) **336.** (d) **337.** (c) **338.** (a) **339.** (c) **340.** (b)
351. (b) **352.** (a) **353.** (c) **354.** (c) **355.** (d) **356.** (d) **357.** (c) **358.** (b) **359.** (b) **360.** (a)
361. (b) **362.** (b) **363.** (c) **364.** (c) **365.** (a) **366.** (b) **367.** (a) **368.** (a) **369.** (a) **370.** (c)
371. (a) **372.** (d) **373.** (b) **374.** (d) **375.** (c) **376.** (b) **377.** (c) **378.** (b) **379.** (c) **380.** (b)
381. (c) **382.** (d) **383.** (c) **384.** (b) **385.** (d) **386.** (b) **387.** (c) **388.** (d) **389.** (c) **390.** (b)
391. (d) **392.** (d) **393.** (c) **394.** (a) **395.** (c) **396.** (b) **397.** (c) **398.** (a) **399.** (c) **400.** (b)
401. (b) **402.** (c) **403.** (d) **404.** (b) **405.** (a) **406.** (b) **407.** (c) **408.** (b) **409.** (a) **410.** (b)
411. (c) **412.** (c) **413.** (b) **414.** (d) **415.** (d) **416.** (b) **417.** (b) **418.** (a) **419.** (c) **420.** (b)
421. (a) **422.** (c) **423.** (a) **424.** (c) **425.** (b) **426.** (b) **427.** (b) **428.** (d) **429.** (c) **430.** (b)
431. (d) **432.** (c) **433.** (b) **434.** (b) **435.** (b) **436.** (c) **437.** (b) **438.** (b) **439.** (b) **440.** (a)
441. (a) **442.** (d) **443.** (b) **444.** (b) **445.** (d) **446.** (d) **447.** (a) **448.** (b) **449.** (d) **450.** (c)
451. (b) **452.** (d) **453.** (a) **454.** (b) **455.** (c) **456.** (d) **457.** (c) **458.** (a) **459.** (c) **460.** (c)
461. (b) **462.** (d) **463.** (d) **464.** (a) **465.** (b) **466.** (b) **467.** (b) **468.** (b) **469.** (c) **470.** (a)
471. (a) **472.** (b) **473.** (c) **474.** (b) **475.** (b) **476.** (c) **477.** (a) **478.** (b) **479.** (a) **480.** (b)
481. (b) **482.** (c) **483.** (d) **484.** (b) **485.** (a) **486.** (b) **487.** (c) **488.** (d) **489.** (a) **490.** (a)
491. (c) **492.** (c) **493.** (b) **494.** (d) **495.** (b) **496.** (b) **497.** (b) **498.** (c) **499.** (d) **500.** (d)
501. (a) **502.** (c) **503.** (c) **504.** (c) **505.** (c) **506.** (a) **507.** (b) **508.** (c) **509.** (b) **510.** (b)
511. (c) **512.** (d) **513.** (c) **514.** (b) **515.** (a) **516.** (b) **517.** (b) **518.** (d) **519.** (b) **520.** (b)
521. (b) **522.** (a) **523.** (b) **524.** (a) **525.** (b) **526.** (c) **527.** (c) **528.** (b) **529.** (a) **530.** (b)
531. (b) **532.** (a) **533.** (c) **534.** (a) **535.** (c) **536.** (a) **537.** (b) **538.** (a) **539.** (b) **540.** (c)
541. (a) **542.** (c) **543.** (a) **544.** (d) **545.** (c) **546.** (b) **547.** (b) **548.** (b) **549.** (a) **550.** (c)
551. (d) **552.** (c) **553.** (b) **554.** (a) **555.** (b) **556.** (c) **557.** (a) **558.** (a) **559.** (d) **560.** (b)
561. (a) **562.** (c) **563.** (b) **564.** (d) **565.** (a) **566.** (c) **567.** (a) **568.** (b) **569.** (c) **570.** (b)
571. (b) **572.** (b) **573.** (c) **574.** (a) **575.** (c) **576.** (a) **577.** (b) **578.** (b) **579.** (a) **580.** (b)
581. (b) **582.** (b) **583.** (d) **584.** (c) **585.** (a) **586.** (a) **587.** (a) **588.** (b)

Practice Set 2

1. ______________ is the process of dividing a market into smaller groups of buyers with distinct needs, characteristics, or behaviors who might require separate products or marketing mixes.
(a) Mass Marketing
(b) Market Segmentation
(c) Target Marketing
(d) Market Positioning

2. ______________ is the process of evaluating each market segment's attractiveness and selecting one or more segments to enter
(a) Mass Marketing
(b) Market Segmentation
(c) Market Targeting
(d) Market Positioning

3. When one buyer and one sellers are found in a market, is called as
(a) Bilateral Monopoly
(b) Oligopoly
(c) Monopoly
(d) Cartel

4. What is the commission charge for food grains?
(a) 1-2% (b) 2-5%
(c) 4-8% (d) 8-10%

5. What is the direct marketing model of Maharashtra?
(a) Apni Mandi
(b) Shetkari Bazar
(c) Krushak Bazar
(d) Kisan Mandi

6. What is the other meaning of Monopsony?
(a) One Buyer (b) One Seller
(c) Two Buyer (d) Multiple Buyers

7. When has TRIFED been established?
(a) 1963 (b) 1987
(c) 1978 (d) 1965

8. What is the other name of complete segmentation?
(a) Macro Marketing
(b) Micro Marketing
(c) Niche Marketing
(d) Mass Marketing

9. Sneha is the milk cooperative of which state?
(a) Maharashtra
(b) Kerala
(c) Andhra Pradesh
(d) Madhya Pradesh

10. When did DMI come?
(a) 1955 (b) 1965
(c) 1935 (d) 1945

11. Which of the following forms of Marketing has been seen in 20th Century?
(a) Mass Marketing
(b) Micro Marketing
(c) Niche Marketing
(d) Segment Marketing

12. Which of the following marketers epitomized the mass marketing strategy?
(a) Henry Ford
(b) Bill Gates
(c) F.W.Woolworth
(d) Thomas A.Edison

13. When General Motors designs specific models for different income and age groups, it is practicing which of the following marketing formats?
(a) Mass Marketing
(b) Micro Marketing
(c) Niche Marketing
(d) Segment Marketing

14. All of the following would be ways to segment within the category of psychographic segmentation EXCEPT:
(a) Values
(b) Occupation
(c) Lifestyle
(d) Personality

15. When did Food Safety and Standard (FSSA) Act come?
(a) 2006 (b) 2009
(c) 2011 (d) 2014

16. "Coke in the morning" is an attempt to segment according to which of the following?
(a) Gender Segmentation
(b) Psychographic Segmentation
(c) Benefit Segmentation
(d) Occasion Segmentation

17. Nasik Market is famous for which commodity?
(a) Cotton (b) Cumin
(c) Potato (d) Onion

18. In market targeting, a firm looks at all of the following factors EXCEPT:
(a) Segment Size and Growth
(b) Segment Structural Attractiveness
(c) Company Objectives and Resources
(d) Segment Public relations value

19. When was Revamped Public Distribution System (RPDS) launched?
(a) 1965 (b) 1944
(c) 1997 (d) 1992

20. Which kind of differentiation criterion are there for choosing a product that different than your competitors and cannot easily be copied?
(a) Important (b) Distinctive
(c) Superior (d) Preemptive

21. What is the name of that sampling when an enumerator segmented every item in different groups on the basis of a particular variable though forming sampling frame?
(a) Stratified Sampling
(b) Cluster Sampling
(c) Systematic Sampling
(d) Quota Sampling

22. Which scaling technique embodies Garrett Ranking technique?
(a) Nominal Scale
(b) Ordinal Scale
(c) Interval Scale
(d) Ratio Scale

23. What is the name of that list for depicting samples of a population?
(a) Sample Design
(b) Sample frame
(c) Sampling Analysis
(d) Census

24. What is basic unit of a population that describes its characteristics?
(a) Sample (b) Statistics
(c) Parameter (d) Characteristics

25. What is basic unit of a sample that describes its characteristics?
(a) Sample
(b) Statistics
(c) Parameter
(d) Characteristics

26. When more will be the size of the sample, then more will be?
(a) Sampling Error
(b) Standard Error
(c) Non Sampling Error
(d) Sampling Frame Error

27. If ten items have to be chosen from a list of two hundred sample units, then according to systematic sampling technique , first unit should be in between
(a) 1-10
(b) 1-20
(c) 1-30
(d) 1-40

28. According to researcher, What is the minimum size of a strata?
(a) Within 5
(b) Within 6
(c) Within 10
(d) No Limit

29. What is the value R1.23 in statistics?
(a) 0 to 1 (b) -1 to +1
(c) $-\infty$ to $+\infty$ (d) 0 to -1

30. A non-probability sampling technique in which samples are collected in waves by obtaining referrals from referrals are
(a) Quota Sampling
(b) Judgmental Sampling
(c) Purposive Sampling
(d) Snowball Sampling

31. Multiplication and division of the variables can only be managed through
(a) Nominal Scale (b) Ordinal Scale
(c) Interval Scale (d) Ratio Scale

32. A comparative scaling technique in which respondents are presented with several objects simultaneously and asked to order or rank them according to some criteria
(a) Paired Comparison Scaling
(b) Rank Order Scaling
(c) Constant Sum Scaling
(d) Q-sort Scaling

33. Who discovered Regression concept?
(a) Karl Pearson (b) Francis Galton
(c) Robert Fisher(d) Mendel

34. What is the other name of Graphic rating scale?
(a) Continuous Rating Scale
(b) Likert Scale
(c) Semantic Differential Scale
(d) Stapel Scale

35. Which is the indicator of BOX-M?
(a) Discriminant Analysis
(b) Factor Analysis
(c) Conjoint Analysis
(d) Cluster Analysis

36. Kaizen is the philosophy of which country?
(a) Japan (b) USA
(c) Korea (d) India

37. In 5S' principle, sweep is otherwise called as__________
(a) Sustain (b) Standardise
(c) Shine (d) Sort

38. Who is associated with the statement that marketing is the process of creation and delivery of standard of living?
(a) Mc Nair (b) Philip Kotler
(c) Peter Drucker (d) Mazur

39. What is Marketing myopia?
(a) Broad View of the Market
(b) Narrow view of MIS
(c) Narrow View of the Market
(d) Broad view of MIS

40. "Stop Smoking", is a type of____________
(a) Green Marketing Concept
(b) Cause related Marketing Concept
(c) Social Marketing Concept
(d) Humanistic Marketing Concept

41. Who stated reality shock syndrome?
(a) Hall
(b) Harrison
(c) Hamburger
(d) Freidman

42. Labour welfare facilities were incorporated in which legislation?
(a) MRTP Act
(b) Essential Commodity Act
(c) Consumer Protection Act
(d) Factories Act

43. According to Plantation Labour Act, 1951, what should be the minimum number of women in a factory for getting crèche facility?
(a) 50 (b) 30
(c) 25 (d) 20

44. Who established administrative management?
(a) Henry Fayol
(b) Max Weber
(c) Henry Gantt
(d) Chester Bernard

45. According to factories Act, 1948, what should be the minimum number of women in a factory for getting crèche facility?
(a) 50 (b) 30
(c) 25 (d) 20

46. What is not included as primary activities in Porter's Value chain Model?
(a) Inbound Logistics
(b) Operations
(c) Marketing and Sales
(d) Procurement

47. What is the listing of employee skills, abilities and qualification for a job?
(a) Job Description
(b) Job Specification
(c) Job Analysis
(d) Job Evaluation

48. What is the logical sequence of job analysis?
(a) Job Enrichment
(b) Job design
(c) Job Engineering
(d) Departmentation

49. A situation where management is unable to provide employment due to non-availability of work is called as ___________.
(a) Lock Out (b) Lay Off
(c) Termination (d) Closure

50. When did Indian Contract Act come in?
(a) 1955 (b) 1872
(c) 1972 (d) 1855

51. Methylisocyanide gas leak tragedy is related with which city?
(a) New Delhi (b) Bhopal
(c) Rourkela (d) Panipat

52. Taylor's differential piece rate scheme is based on
(a) Earning vary in the same proportion as output
(b) Earnings vary less proportionately than output
(c) Earnings vary proportionately more than output
(d) Earnings differ at different levels of output

53. What is Appreciative Enquiry?
(a) Psychological theory
(b) Training Method
(c) HR Quality
(d) Business Etiquette

54. Who has advanced the motivational analysis of an organization?
(a) D M Pestonjee
(b) T V Rao
(c) Udai Pareek
(d) Kelso

55. Which committee recommended ASRB as an independent recruitment agency?
(a) Randhawa Committee
(b) Gajendra Gadkar Committe
(c) Mohan Singh Mehta Committee
(d) Khusro Committe

56. Which committee recommended KVK (Krishi Vigyan Kendra) in India?
(a) Randhawa Committee
(b) Gajendra Gadkar Committe
(c) Mohan Singh Mehta Committee
(d) Khusro Committe

57. Where is Central Institute of Freshwater Aquaculture (CIFA) located?
(a) Bhubaneswar
(b) Mumbai
(c) Kolkatta
(d) Chennai

58. When World Sports Day is celebrated?
(a) 15th October (b) 19th August
(c) 29th August (d) 14th October

59. What is the full form of CeRA?
(a) Consortium for electronic Reply in Agriculture
(b) Consortium for e-Resources in Agriculture
(c) Centre for Resource allocation
(d) Centre for Resilient Agriculture

60. What is the nodal Centre of Krishi Prabha?
(a) NDRI, Karnal
(b) IARI, New Delhi
(c) IVRI, Izzatnagar
(d) HAU, Hisar

61. The theory that reveals the quantitative approach to the study of personality is
(a) Psychoanalytic Theory
(b) Socio-psychological Theory
(c) Trait Theory
(d) Self-Theory

62. Who has not come under the contributor of socio-psychological theory
(a) Adlerb
(b) Cattell
(c) Horneyd
(d) Sullivan

63. Which kind of people are Easy going and donot have time urgency?
(a) Type A Personality
(b) Type B Personality
(c) External Locus Personality
(d) Internal Locus Personality

64. When faced with incomplete information, people fill up the gaps themselves to make the information meaningful is called as_____.
(a) Figure-Ground
(b) Grouping
(c) Simplification
(d) Closure

65. Sometimes your behavior result in a congruent response from other person, such beheviour is referred to as_________
(a) Perceptual defense
(b) Misperception
(c) Perceptual bias
(d) Response evocation

66. Which law describes the concept of supply that creates its own demand?
(a) Keynesian Law
(b) Say's Law
(c) Veblen Law
(d) Ratchet Effect

67. Which tenets of management principles that cover ABC analysis in Inventory of an organization?
(a) Planning Measure
(b) Organizing Measure
(c) Directing Measure
(d) Controlling Measure

68. In growth share matrix 'Dog' indicates
(a) High business growth-High market share
(b) High business growth-Low market share
(c) Low business growth-High market share
(d) Low business growth-Low market share

69. Who established Managerial grid type of leadership determination model?
(a) Rennis Likert
(b) Reddin
(c) Fiedler
(d) Blake and Mouton

70. When Standing Labour Conference (SLC) was constituted?
(a) 1942 (b) 1946
(c) 1944 (d) 1948

71. Who has very little or no conflict with other members of the organization?
(a) Liasion (b) Gatekeeper
(c) Isolates (d) Cosmopolites

72. The tag line, 'An additional idea accompanying every idea that is expressed' relates to___
(a) Gossip
(b) Meta Communication
(c) Jargon
(d) Rumour

73. Who postulates 5 traits theory for personality judgment?
(a) Eyesenck
(b) Raymond Cattel
(c) Murray
(d) Costa and Murree

74. When do individuals with substantially different levels of experience and performance abilities are being paid wages or salaries that relatively equal?
(a) Pay Secrecy
(b) Maturity curve
(c) Pay compression
(d) Point Manual

75. Which organization is referred for Sick enterprise?
(a) World Bank (b) NABARD
(c) SIDBI (d) BIFR

76. When do we find a sustained rise of less than 3 per cent in prices per annum?
(a) Creeping inflation
(b) Running inflation
(c) Walking inflation
(d) Galloping inflation

77. What is the other name of Demonstration effect?
(a) Ratchet effect
(b) Bandwagon effect
(c) Snob effect
(d) Pigou effect

78. When 'Poor people imitate the living style of the rich', this pattern of consumption is
(a) Ratchet effect
(b) Demonstration effect
(c) Snob effect
(d) Pigou effect

79. If the marginal propensity to consume is ¾ then the value of the multiplier is
(a) 1 (b) 3
(c) 2 (d) 4

80. What does Say's law of employment speak about?
(a) The government always interfere in the decision making
(b) The size of the market is unlimited
(c) Savings are equal to investment
(d) There is always a hindrance in the free flow of the money

81. Frictional unemployment occur when one changes himself from one job to another. This was the statement of one of the classical theorists of employment. He was
(a) J B Say (b) James Mill
(c) Pigou (d) David Ricardo

82. Who written the Book, 'General Theory of Employment, Interest and Money'?
(a) J B Say
(b) James Mill
(c) J M Keynes
(d) David Ricardo

83. Which philosophy speaks about 'Strike is a measure weapon for economic warfare'?
(a) Guild Socialism
(b) Syndicalism
(c) Communism
(d) Anarchism

84. What is the name of periodic boom and slumps in economic activity?
(a) Manufacturing Cycle
(b) Production Cycle
(c) Business Cycle
(d) Operation Cycle

85. A cause of variation in a process probably wouldn't be due to
(a) People (b) Standards
(c) Methods (d) Environment

86. What type of control chart would be used to monitor the number of defects in the output of a process?
(a) x-bar Chart
(b) P chart
(c) c chart
(d) R Chart

87. What is the ideal inventory level in JIT management system?
(a) The Economic Order Quantity
(b) The Economic run Size
(c) One Unit
(d) The Capacity of the Standard Container

88. 'Labour markets become the seller's market' in which phases of business cycle?
(a) In the initial phase of Prosperity
(b) In the beginning phase of recovery
(c) In the starting phase of recession
(d) In the later phase of prosperity

89. Where we feel the first impact of recession?
(a) Automobile sector
(b) Household Appliance Sector
(c) Real Estate Sector
(d) Service Sector

90. Market value of all final goods and services minus income earned locally by foreigners plus income earned abroad by the nationals is called as
(a) Gross National Product
(b) Gross Domestic Product
(c) Net National Product
(d) Net National Income at Factor Cost

91. What is the other name of Continuous costing?
(a) Operation Costing
(b) Batch Costing
(c) Process Costing
(d) Contract Costing

92. What is the objective of cost accounting?
(a) Approximation
(b) Historical
(c) Estimation
(d) Empirical

93. What is the total of all direct expenses in accounting?
(a) Total Cost
(b) Work cost
(c) Overhead Cost
(d) Prime Cost

94. DU PONT Analysis deals with___________.
(a) Analysis of Current Assets
(b) Capital Budgeting
(c) Analysis of Fixed Assets
(d) Analysis of Profit

95. In Net Profit Ratio, the denominator is
(a) Net Purchases
(b) Net Sales
(c) Credit Sales
(d) Cost of Goods sold

96. Where we get Ledger Folio from the following list?
(a) Ledger
(b) Journal
(c) Trial Balance
(d) Balance sheet

97. Where we find Folio from the following list?
(a) Ledger
(b) Journal
(c) Trial Balance
(d) Balance sheet

98. What is the name of transferring entries from the journal to the ledger?
(a) Journalizing
(b) Ledger Folio
(c) Ledger Posting
(d) Folio

99. A statement that summarizes the date recorded in the journal as well as in the ledger is called as
(a) Ledger
(b) Journal
(c) Trial Balance
(d) Balance sheet

100. Where we find Direct wages in financial account?
(a) Debit side of trading account
(b) Credit side of trading account
(c) Debit side of Profit and Loss account
(d) Credit side of Profit and Loss account

101. When WTO came in to picture?
(a) 1947 (b) 1991
(c) 1995 (d) 2000

102. What is the full form of GATT?
(a) General Agreement on Trade and Tariffs
(b) Good Agreement on Tariffs and Trade
(c) Good Agreement on Trade and Tariffs
(d) General Agreement on Tariffs and Trade

103. What is called as King of fruit?
(a) Jackfruit (b) Mango
(c) Banana (d) Apple

104. What is the position of India in milk production in world?
(a) First (b) Second
(c) Third (d) Fourth

105. What is the position of India in beef export in world?
(a) First (b) Second
(c) Third (d) Fourth

106. What is the position of India in fruits and vegetables production in world?
(a) First (b) Second
(c) Third (d) Fourth

107. Where is the head quarter of SIDBI located?
(a) Pune (b) Hyderabad
(c) Mumbai (d) Lucknow

108. Which state is known for Sonepur Mela?
(a) Odisha
(b) Bihar
(c) Madhya Pradesh
(d) Kerala

109. Which Edible oil that is produced maximum in world?
(a) Soya Oil (b) Palm Oil
(c) Rape Oil (d) Coconut Oil

110. Which organization relates to E-choupal?
(a) HUL (b) ITC
(c) IOC (d) DSCL

111. Who says "Everything else can wait, but not agriculture"?
(a) Sri J.L.Nehru
(b) Mahatma Gandhi
(c) Dr. (a) P. J.Abdulkalam
(d) Dr. H. Rembold

112. Where we find Lycopene pigment from the following list?
(a) Castor (b) Tobacco
(c) Tomato (d) Onion

113. The nature of amalgamation as uniting of interests or acquisition is described in which Accounting Standard issued by ICAI
(a) AS-13 (b) AS-14
(c) AS-15 (d) AS-16

114. SQC (Statistical Quality Control) requires determination of__________.
(a) Quality Tolerance Limit
(b) Quantity Measurement
(c) Production Lead time
(d) Inventory Carrying cost

115. Under which stage of the industry life cycle entry barriers increase and threat of potential competitors decrease?
(a) Beginning Stage
(b) Growing Stage
(c) Mature Stage
(d) Declining Stage

116. The symbol that denotes direction in the operation site
(a) Δ (b) □
(c) ○ (d) →

117. What is the major driving force of Agro-processing industry presence in rural areas?
(a) Labour Facility
(b) Infrastructure facility
(c) Raw Material Facility
(d) Electricity Facility

118. What is the core focus area of COPC standard in quality management?
(a) Quality
(b) Production
(c) Design
(d) Aftersales Service

119. What is the other name of Stationary Layout?
(a) Process Layout
(b) Product Layout
(c) Cellular Layout
(d) Fixed Layout

120. What is the price effect in Giffen goods?
(a) Negative (b) Zero
(c) Positive (d) -1

121. Which committee deals with Amalgamation of sick unit with healthy unit?
(a) Tiwari Committee
(b) Choksi Committee
(c) Khusro Committee
(d) Narsimhan Committee

122. Which is an international credit rating agency from the following list?
(a) Mc Kinsey
(b) Standard and Poor
(c) IMF
(d) BCG

123. Who has given the term Supply Chain Management?
(a) Peter Drucker
(b) Keith Oliver
(c) Raphael Kaplinsky
(d) Michael Porter

124. Who has given Different Contract Farming models?
(a) Clarence Danhof
(b) Eaton and Shepherd
(c) Richard Cantillon
(d) Margret et. al

125. Which of the following is the nodal agency for AEZ?
(a) APEDA (b) ICAR
(c) NIAM (d) DARE

126. Which among the following will suit the best as location for a chemical plant?
(a) Desert Area (b) Coastal Area
(c) Platen (d) Hill Area

127. What is location of Rural Entrepreneurship Development Institute?
(a) Kanpur (b) Bhubaneswar
(c) Ranchi (d) Pune

128. When has Coconut Development Board been established?
(a) 1975 (b) 1981
(c) 1979 (d) 1985

129. Who has written the book,"Innovation and Entrepreneurship" ?
(a) Philip Kotler
(b) Peter F Drucker
(c) Pradeep Khandewala
(d) C K Prahlad

130. Who has given the first report on corporate governance in India?
(a) Narayan Murthy committee
(b) Naresh Chandra committee
(c) Cadbury's committee
(d) Kumarmangalam Birla committee

131. Small and Medium enterprises are highly________
(a) Labour intensive
(b) Capital intensive
(c) Market leader
(d) Industry price determiner

132. Which organization provided Technology Development and Modernization Fund?
(a) SIDBI (b) NABARD
(c) KVIC (d) SIDO

133. When a marketer offers one brand with another brand of the same company or another company is called as ________.
(a) Bundling (b) Co-Branding
(c) Upselling (d) Cross Selling

134. The relationship between Profitability Index (PI), Net Present Value (NPV) and Initial Capital Expenditure (Co) is________
(a) PI= NPV/Co
(b) PI= NPV x Co
(c) PI= NPV + Co
(d) PI = NPV – Co

135. Nature fresh oilseed is the product of which company?
(a) Ruchisoya
(b) Kaneria Oils
(c) Cargil
(d) Chambal

136. The maximum number of constraints allowed in a linear programming is
(a) Less than 5
(b) Less than 1024
(c) Unlimited
(d) Less than 512

137. Which risk from the following reduces due to Crop insurance policy?
(a) Production risk
(b) Price risk
(c) Financial risk
(d) Interest risk

138. Slack variable is used in case of ____________ type of constraints in Linear Programming Problem
(a) ≤ (b) ≥
(c) Both ≤ and ≥ (d) Either ≤ or ≥

139. The repayment plan in which installment is directly proportional to amount of harvest is called as
(a) Balloon repayment Plan
(b) Amortised Even Repayment Plan
(c) Amortised Decreasing Repayment Plan
(d) Variable Repayment Plan

140. Which of the following is the sixth 'P' of credit?
(a) Principle of Proper Utilisation
(b) Principle of Payment
(c) Principle of Protection
(d) Principle of Personality

141. Which of the following is the third 'R' of credit?
(a) Rate of Returns
(b) Returns
(c) Repayment Capacity
(d) Risk Bearing Ability

142. Which of the following is the first 'C' of credit?
(a) Condition (b) Character
(c) Capacity (d) Capital

143. For a 4 variable and 5 constraint primal problem, the dual would be a ___ variable and _____ constraint problem
(a) 5,4 (b) 4,4
(c) 4,5 (d) 5,5

144. The cost differences in the Vogel's approximation method indicate the ________ for not using the respective least cost route.
(a) Bonus (b) Penalty
(c) Cost (d) Profit

145. What is the fulform of VAM?
(a) Vogeal's Approximation Method
(b) Vogel's Approximate Method
(c) Vangel's Approximation Method
(d) Vogel's Approximation Method

146. Which one of the following would be considered a potential cost of inventory?
(a) Shortage (b) Carrying
(c) Holding (d) Overtime

147. Which is the most popular SHG model in India?
(a) Direct Linkage Model
(b) Modified Direct Linkage Model 11
(c) IFAD Model
(d) Modified Indirect Model

148. What is the refinance limit for farm Mechanization through NABARD?
(a) 25% (b) 33.33%
(c) 50% (d) 75%

149. In the basic EOQ model, if annual demand doubles, the effect on the EOQ is
(a) It doubles
(b) It is half its previous amount
(c) It is about 70 percent of its previous amount
(d) It increases by about 40 percent

150. When ordering cost/order and holding cost/unit are both doubled then EOQ (Economic Order Quantity) will become
(a) Double
(b) Increases four times
(c) Becomes half
(d) No change

ANSWER KEY (Practice Set 2)

1.	(b)	**2.**	(c)	**3.**	(a)	**4.**	(b)	**5.**	(b)	**6.**	(a)	**7.**	(b)	**8.**	(b)	**9.**	(d)	**10.**	(c)
11.	(a)	**12.**	(a)	**13.**	(d)	**14.**	(b)	**15.**	(a)	**16.**	(d)	**17.**	(d)	**18.**	(d)	**19.**	(d)	**20.**	(d)
21.	(a)	**22.**	(b)	**23.**	(b)	**24.**	(c)	**25.**	(b)	**26.**	(c)	**27.**	(b)	**28.**	(b)	**29.**	(a)	**30.**	(d)
31.	(d)	**32.**	(b)	**33.**	(b)	**34.**	(a)	**35.**	(a)	**36.**	(a)	**37.**	(c)	**38.**	(b)	**39.**	(c)	**40.**	(c)
41.	(a)	**42.**	(d)	**43.**	(a)	**44.**	(a)	**45.**	(b)	**46.**	(d)	**47.**	(b)	**48.**	(b)	**49.**	(b)	**50.**	(b)
51.	(b)	**52.**	(d)	**53.**	(a)	**54.**	(c)	**55.**	(b)	**56.**	(c)	**57.**	(a)	**58.**	(c)	**59.**	(b)	**60.**	(d)
61.	(c)	**62.**	(b)	**63.**	(b)	**64.**	(d)	**65.**	(a)	**66.**	(b)	**67.**	(d)	**68.**	(d)	**69.**	(d)	**70.**	(a)
71.	(c)	**72.**	(b)	**73.**	(d)	**74.**	(c)	**75.**	(d)	**76.**	(a)	**77.**	(b)	**78.**	(b)	**79.**	(d)	**80.**	(c)
81.	(c)	**82.**	(c)	**83.**	(b)	**84.**	(c)	**85.**	(b)	**86.**	(c)	**87.**	(c)	**88.**	(d)	**89.**	(c)	**90.**	(a)
91.	(c)	**92.**	(c)	**93.**	(d)	**94.**	(d)	**95.**	(b)	**96.**	(b)	**97.**	(a)	**98.**	(c)	**99.**	(c)	**100.**	(a)

101. (c) **102.** (d) **103.** (b) **104.** (a) **105.** (a) **106.** (b) **107.** (d) **108.** (b) **109.** (b) **110.** (b)
111. (a) **112.** (c) **113.** (b) **114.** (a) **115.** (c) **116.** (d) **117.** (c) **118.** (d) **119.** (d) **120.** (c)
121. (b) **122.** (b) **123.** (b) **124.** (b) **125.** (a) **126.** (b) **127.** (c) **128.** (c) **129.** (b) **130.** (d)
131. (a) **132.** (a) **133.** (c) **134.** (a) **135.** (c) **136.** (c) **137.** (a) **138.** (a) **139.** (d) **140.** (b)
141. (d) **142.** (b) **143.** (a) **144.** (b) **145.** (d) **146.** (d) **147.** (b) **148.** (c) **149.** (d) **150.** (d)

Practice Set 3

1. Who is called as Father of Modern Economics?
(a) J.M. Keynes (b) Alfred Marshall
(c) Adam Smith (d) Lionel Robbins

2. Who has given welfare definition of economics?
(a) J.M. Keynes (b) Alfred Marshall
(c) Adam Smith (d) Lionel Robbins

3. Who has given growth definition of Economics?
(a) J.M. Keynes (b) Alfred Marshall
(c) Adam Smith (d) Lionel Robbins

4. What is the other name of microeconomics?
(a) Income theory
(b) Employment theory
(c) Price theory
(d) Demand theory

5. Who has coined 'Macro Economics'?
(a) J M Keynes
(b) Ragnar Frisch
(c) Alfred Marshall
(d) Lionel Robbins

6. What to infer from the meaning, 'things ought to be' in economics?
(a) Positive Science
(b) Negative Science
(c) Normative Science
(d) Applied Science

7. Which of the following is an external, material and non-transferable good?
(a) Table
(b) Good will of business
(c) Conduct Certificate of a student
(d) Skill of a doctor

8. Who has explained, 'Human wants are complementary'?
(a) LEMU
(b) LDMU
(c) Consumers' surplus
(d) Theory of Joint demand

9. Who has given the law of family expenditure?
(a) Adam Smith (b) J M Keynes
(c) Ernest Engel (d) Alfred Marshall

10. When NAFED came in to picture?
(a) 1958 (b) 1965
(c) 1963 (d) 1975

11. What is the utility of the first unit of commodity?
(a) Total Utility (b) Initial Utility
(c) Last utility (d) Average Utility

12. What is the marginal utility of free goods?
(a) Positive (b) Negative
(c) Infinity (d) Zero

13. What is the marginal utility of scarce goods?
(a) Positive (b) Negative
(c) Infinity (d) Zero

14. Who has given the concept of consumer surplus?
(a) Adam Smith
(b) J M Keynes
(c) Ernest Engel
(d) Alfred Marshall

15. Which is not comes under concept of consumer surplus?
(a) Comforts
(b) Necessaries
(c) Luxuries
(d) Comforts and Luxuries

16. The shape of indifference curve is ____________ to origin.
(a) Concave (b) Convex
(c) Hyperbola (d) U shape

17. Who was written the book 'Value and Capital'?
(a) Alfred Marshall
(b) J R Hicks
(c) J M Keynes
(d) Robbins

18. Which market situation best suits to Price Discrimination?
(a) Monopoly (b) Monopsony
(c) Oligopoly (d) Duopoly

19. Who has given the doctrine ' Conspicuous consumption'?
(a) Robert Giffen
(b) Alfred Marshall
(c) Thorstein Veblen
(d) Chamberlin

20. When the price of the commodity increases then demand__________.
(a) Decreases (b) Increases
(c) Extends (d) Contracts

21. Which commodity relates to joint demand?
(a) Tea (b) Coal
(c) Ink (d) Electricity

22. What is the value of price elasticity of demand?
(a) -1 to +1 (b) 0 to 1
(c) 0 to ∞ (d) -1 to ∞

23. What is the position of cross elasticity of demand curve at the time of perfect substitute?
(a) Convex
(b) Concave
(c) Parallel to x-Axis
(d) Parallel to y-Axis

24. Which of the following forms of business organisations is a step towards Socialism?
(a) Joint Stock Company
(b) Cooperative Society
(c) Partnership
(d) State enterprise

25. What is the shape of Average revenue curve in perfect competition?
(a) Parallel to y- Axis
(b) Parallel to x-Axis
(c) Convex
(d) Downward sloping curve from left to right

26. In very short period market, which of the following is relatively fixed?
(a) Supply (b) Demand
(c) Floor Price (d) Normal Price

27. What comes under long period permanent equilibrium between demand and supply forces?
(a) Market Price
(b) Normal Price
(c) Floor Price
(d) Procurement Price

28. Who has established the degree of price discrimination?
(a) Chamberlin (b) A C Pigou
(c) Cournot (d) Robinson

29. What comes under category of place price discrimination from the following list?
(a) Decentralization
(b) Centralization
(c) Empowerment
(d) Enlargement

30. Who has written the theory of monopolistic competition?
(a) Chamberlin (b) A C Pigou
(c) Cournot (d) Robinson

31. Which of the following is the sixth 'P' of credit?
(a) Principle of Payment
(b) Principle of Protection
(c) Principle of productive Purpose
(d) Principle of Productivity

32. Which of the following is the seventh 'P' of credit?
(a) Principle of Payment
(b) Principle of Protection
(c) Principle of productive Purpose
(d) Principle of Productivity

33. Which of the following 'C's of credit is considered as prima-facie for sanctioning the loan by the banker to the farmer borrower?
(a) Character (b) Capacity
(c) Condition (d) Commonsense

34. Which rate bank borrows from the RBI at the time of necessity?
(a) Bank rate
(b) Reverse repo rate
(c) Repo rate
(d) CRR

35. Which of the following is the third 'C' of credit?
(a) Condition (b) Capital
(c) Character (d) Capacity

36. Which of the following is the fourth 'C' of credit?
(a) Condition (b) Capital
(c) Character (d) Capacity

37. Which repayment plan the installment amount gradually decreases but principal amount remains same throughout the repayment period and in the last huge installment is paid?
(a) Straight end repayment plan
(b) Amortized decreasing repayment plan
(c) Amortized even repayment plan
(d) Partial repayment plan

38. What is the other term of reverse repayment plan?
(a) Amortized decreasing repayment plan
(b) Future repayment plan
(c) Amortized even repayment plan
(d) Partial repayment plan

39. Repayment of loans in a series of installment is called as______.
(a) Liquidity (b) Amortization
(c) Leverage (d) Hypothecation

40. Which of the following cases amortized even repayment plan is followed?
(a) Paddy Loan
(b) Tractor Loan
(c) Loan given for digging wells
(d) Machinery Loan

41. Which of the following government body recommended Crop loan scheme?
(a) AIRCSC (b) AIRCRC
(c) RBEC (d) CRAFICARD

42. When CADP (Command Area Development Programme) came into existence?
(a) 1961 (b) 1964
(c) 1971 (d) 1974

43. What is prepared to overcome the deficiencies in NWS and Income Statement?
(a) Net Worth Statement
(b) Profit and Loss Account
(c) Balance Sheet
(d) Fund Flow Statement

44. What is the financial statement prepared for the future?
(a) Net Worth Statement
(b) Profit and Loss Account
(c) Balance Sheet
(d) Fund Flow Statement

45. The financial ratio which indicates the liquidity position of the farm business in the shortest or immediate period is called as
(a) Intermediate ratio
(b) Working ratio
(c) Net capital ratio
(d) Current ratio

46. When SFDA and MFAL came into existence?
(a) 1961 (b) 1964
(c) 1971 (d) 1974

47. Who was recommended SFDA and MFAL in India?
(a) AIRCSC (b) AIRCRC
(c) RBEC (d) CRAFICARD

48. When IRDP (Integrated Rural Development Programme) came into existence?
(a) 1968-69
(b) 1978-79
(c) 1982-83
(d) 1985-86

49. When did Village adoption scheme come into existence?
(a) 1949 (b) 1959
(c) 1969 (d) 1979

50. When did Service Area Approach(SAA) come into existence?
(a) 1999 (b) 1989
(c) 1969 (d) 1979

51. Service Area Approach was recommended to overcome duplication of financing by the committee headed by
(a) F K F Nariman
(b) B Sivaraman
(c) T A Pai
(d) P D Ojha

52. In service area approach, Who prepares the credit plan of the village?
(a) Lead Bank Officer
(b) District Collector
(c) Branch Manager
(d) Agriculture Officer

53. Who was recommended General Insurance Company in India?
(a) A D Gorwala
(b) D R Gadgil
(c) Mohan Kanda
(d) V M Dandekar

54. Who was introduced Comprehensive Crop Insurance Scheme in India?
(a) LIC
(b) GIC
(c) DICGC
(d) NAIS

55. Which of the following affiliated agency of world Bank provides soft loans to developing countries
(a) MIGA (b) IMF
(c) IFC (d) IDA

56. Who coined the term, "Cooperation is self- help made effective by Organisation'?
(a) J M Keynes
(b) Mohan Kanda
(c) Lionel Robbins
(d) H Plunkett

57. What is way of borrowing money by RBI from the Banks?
(a) CRR
(b) Bank rate
(c) Repo rate
(d) Reverse Repo Rate

58. Which has the principle of, 'One Share-One Vote'?
(a) JSC
(b) Cooperation
(c) Indivisual Business
(d) Franchise

59. Where is Asian Development Bank (ADB) is located?
(a) India (b) USA
(c) Philippines (d) Geneva

60. The agency located at national level to fix the grading standards for the commodities on scientific lines is
(a) NIAM (b) CACP
(c) DMI (d) ICAR

61. When did Tea Board Act come in India?
(a) 1953 (b) 1940
(c) 1947 (d) 1976

62. When did Rubber Board Act come in India?
(a) 1953 (b) 1940
(c) 1947 (d) 1976

63. When did Tobacco Board Act come in India?
(a) 1953 (b) 1940
(c) 1947 (d) 1976

64. When did Decentralized procurement scheme come in India?
(a) 1995 (b) 2000
(c) 1997 (d) 2005

65. When did Jute Corporation of India come in?
(a) 1965 (b) 1970
(c) 1971 (d) 1975

66. For which commodity MSP is given statutory status?
(a) Paddy (b) Sugarcane
(c) Wheat (d) Mango

67. Keeping the rise in demand under check by allotting limited quantity per capita per time period is called as______________.
(a) Balancing (b) Neutralising
(c) Rationing (d) Buffering

68. When did Market Intervention scheme come in India?
(a) 2000-01 (b) 2010-11
(c) 2005-06 (d) 1995-96

69. The ISI mark standard design in a rectangle having width and Height ratio
(a) 3:4 (b) 4:3
(c) 3:3 (d) 4:4

70. Each SFDA is expected to cover about __________ number of small farmers
(a) 50000 (b) 5000
(c) 10000 (d) 1000

71. Which type of cooperatives first came in India?
(a) Credit (b) Farming
(c) Marketing (d) Housing

72. What is 'Taccavi'?
(a) Tax Return
(b) Tax and Levy by Government
(c) Government Loan
(d) Tribal Welfare

73. When did IDBI come into existence?
(a) 1955 (b) 1964
(c) 1977 (d) 1982

74. When did Rajiv Gandhi Vidyutikaran Yojana come in?
(a) 1995 (b) 2000
(c) 2005 (d) 2007

75. Which segment is the core of the tractor industry?
(a) Under 20HP (b) 21-30HP
(c) 31-40HP (d) 41-50HP

76. What is the total electricity consumption for agriculture in India?
(a) 10% (b) 17%
(c) 40% (d) 45%

77. The first state warehouse was set up inanother in tacit form is called
(a) Uttar Pradesh
(b) Bihar
(c) Odisha
(d) West Bengal

78. Which commodity stored maximum in cold storages?
(a) Tomato (b) Onion
(c) Brinjal (d) Potato

79. Where is the head quarter of WTO?
(a) Phillipines (b) Switzerland
(c) Washington (d) Newyork

80. 'Agricultural Situation in India' is published in India in every interval of
(a) 30 days (b) 7 days
(c) 365 days (d) 1 day

81. 'The monthly situation and outlook reports' are published by which organization?
(a) Ministry of Agriculture
(b) ICAR
(c) Ministry of Commerce
(d) DMI, GOI

82. Which is the company that introduced 'Sankat Haran Yojana' for the farmers?
(a) Monsanto (b) IFFCO
(c) KRIBHCO (d) CFCL

83. Which state cooperative is dominated in plantation crop segment?
(a) Gujarat
(b) Maharashtra
(c) Karnataka
(d) Uttar Pradesh

84. In 1976-77, NAFED undertook price support operation for which commodity?
(a) Moong
(b) Groundnut
(c) Toria
(d) Sunflower

85. In which year National Cooperative Development Corporation came?
(a) January, 1963
(b) March, 1963
(c) July, 1963
(d) October, 1963

86. 'Mahagrape', 'Mahabanana', 'Mahamango' is the marketing entity of which state?
(a) Madhya Pradesh
(b) Maharashtra
(c) Manipur
(d) Meghalaya

87. Which country is dominated in world for processed food product categories?
(a) India (b) Netherland
(c) Germany (d) USA

88. Market information in which we try to estimate the future price of the produce is
(a) Market News
(b) Market Intelligence
(c) Market Outlook
(d) Market Research

89. What is the minimum age for getting benefit under Annapurna Yojana?
(a) 60 (b) 62
(c) 65 (d) 70

90. Which committee proposed Contract farming for the interest of tomato farmers and PEPSI?
(a) The goals are broad while objectives are specific
(b) Goals are not quantified while objectives are quantified
(c) Goals are more influenced by external environment
(d) All are correct

91. Assembling of the produce comes under, which category of marketing function?
(a) Primary function of marketing
(b) Secondary function of marketing
(c) Tertiary function of marketing
(d) Multi-function of marketing

92. What is the working boundary of Project Kaladarsha of HUL?
(a) Marketing their product and providing self-sustainability
(b) Providing skill development in embroidery and garment designing
(c) Providing scholarships for girls
(d) Assisting career guidance for young girls

93. Petty merchant who moves from village to village is called as
(a) Broker
(b) Class A Trader
(c) Itinerant Trader
(d) Kachha Arhatiyas

94. Which company promote, 'Project Jagruti'?
(a) Dabur (b) HUL
(c) ITC (d) Colgate

95. An IT based project for dairy farmers' in Gujarat is__________.
(a) Tarahat (b) Akash Ganga
(c) Bhoomi (d) Gyandoot

96. What is the root product of Molasses?
(a) Paddy (b) Sugarcane
(c) Cotton (d) Turmeric

97. When did Kutir Jyoti programme come in India?
(a) 1982 (b) 1985
(c) 1989 (d) 1995

98. Where is National Institute of Rural Development (NIRD) located?
(a) Pune (b) Delhi
(c) Hyderabad (d) Kanpur

99. Consumer evaluation model in which a consumer selects a product that satisfies few attributes that he considers important is
(a) Expectancy value model
(b) Lexicographic model
(c) Conjunctive model
(d) Disjunctive model

100. What do you understand by a group of product possess the same brand name?
(a) Umbrella Brand
(b) Blanket Family Brand
(c) Either Umbrella Brand or Blanker Family Brand
(d) Neither Umbrella Brand nor Blanker Family Brand

101. What does it indicate by a stable dividend policy of an organization?
(a) Stable Earning/Price ratio
(b) Stable Pay-Out ratio
(c) Stable Price/Earnings ratio
(d) Stable earnings per share ratio

102. The short term securities in which a company is lending money to a stock market operator who wishes to carry forward his/her transaction from one settlement period to another is called as
(a) Treasury Bill
(b) Commercial Paper
(c) Badla Financing
(d) Bills Discounting

103. Which is the market where the short term securities are traded?
(a) Capital Market
(b) Secondary Market
(c) Money Market
(d) Black Market

104. What is not the decentralized method of cash collection from the following list?
(a) Payment by wire
(b) Concentration banking
(c) Payment in nearest bank
(d) Lock-box system

105. If credit terms is written as 2/10 net 30, then 30 signifies
(a) Credit period
(b) Cash discount
(c) Cash discount period
(d) Cash only

106. What is not the feature of capital structure?
(a) Rigidity in nature
(b) Risky
(c) Control Mechanism
(d) Time Bound

107. A portion of the outstanding bonds mature at regular intervals until eventually all of thebonds matured is called as____________.
(a) Income Bond
(b) Junk Bond
(c) Subordinate Debenture
(d) Serial Bond

108. Which is the following long term financing security is not a hybrid one?
(a) Preference Share
(b) Convertible Debenture
(c) Equity Share
(d) Income Note

109. When asset is used constantly over the years then method of calculation of depreciation is
(a) Straight Line Method
(b) Sum of the years digit Method
(c) Diminishing Balance Method
(d) Annual Revaluation Method

110. What is the difference between responsibility centres and cost centres?
(a) It deals with cost of the product
(b) It deals with cost of the process
(c) It deals with persons who may be managing
(d) It deals with cost of assembling items

111. A scale for measuring attitudes that consists of a single adjective in the middle of an even numbered range of values, from -5 to +5 without a neutral point is
(a) Likert Scale
(b) Semantic Differential Scale
(c) Stapel Scale
(d) Thurstone Scale

112. Which is the non-parametric test among the following?
(a) Z test (b) F test
(c) Run test (d) T Test

113. What is the shape of the normal distribution curve?
(a) Square Shape
(b) Curvilinear Shape
(c) Bell Shape
(d) Circular Shape

114. What you understand by Hypothesis? It is a ____________.
(a) Theory (b) Technique
(c) Assumption (d) Model

115. What is the usual presentation of an Index number?
(a) Square (b) Rate
(c) Percentage (d) Decimal

116. Where we find Icicle plot in a statistical analysis?
(a) Factor Analysis
(b) Cluster Analysis
(c) Conjoint Analysis
(d) Multidimensional Scaling

117. Where we major Eigen value in a statistical analysis for coming in to a conclusion?
(a) Factor Analysis
(b) Cluster Analysis
(c) Conjoint Analysis
(d) Multidimensional Scaling

118. Which tests speak about K-mean Clustering?
(a) Hierarchical Clustering technique
(b) Non Hierarchical Clustering technique
(c) Multi Linkage
(d) Double Linkage

119. Which is the shortest distance between two points which are elements of two different clusters?
(a) Single Linkage
(b) Double Linkage
(c) Complete Linkage
(d) Multi Linkage

120. What is the general formula of Eigen Value?
(a) Explained Variance/ Total Variance
(b) Unexplained Variance/ Explained Variance
(c) Explained Variance/ Unexplained Variance
(d) Unexplained Variance/ Total Variance

121. Which tests consider BOX's test of equality of Covariance matrices?
(a) Factor Analysis
(b) Cluster Analysis
(c) Conjoint Analysis
(d) Discriminant Analysis

122. What is the range of Wilk's Lambda?
(a) -1 to +1 (b) 0 to 1
(c) $-\infty$ to $+\infty$ (d) 0 to $+\infty$

123. Which tests consider KMO is an indicator for understanding data adequacy?
(a) Factor Analysis
(b) Cluster Analysis
(c) Conjoint Analysis
(d) Discriminant Analysis

124. The addition of more independent variables creates more relationships among them. So not only are the independent variables related to the dependent variable, they are also potentially related to each other. When this happens, it is called as
(a) Partial Collenearity
(b) Multi Collenearity
(c) Multiple Regression
(d) Overfitting

125. If sum of square within group is 42, sum of square between group is 6, then total sum of square is
(a) 7 (b) 48
(c) 36 (d) 0

126. Where is National Plant Protection Training Institute located?
(a) New Delhi (b) Pune
(c) Hyderabad (d) Chandigarh

127. What is the Strategies formulated to convert a sick unit to healthy one?
(a) Turnaround
(b) Expansion
(c) Diversification
(d) Stabilisation

128. What is the strategy of reducing the number of Organisational levels to get closure to the customer?
(a) Thinning
(b) Sliding
(c) Flattening
(d) Focusing

129. In the diffusion of innovation who are the people not willing to try out the product?
(a) Innovators
(b) Progressive People
(c) Early Adopter
(d) Early Majority

130. In channel distribution of Industrial products which level is popular?
(a) Zero level and one level
(b) One level and two level
(c) Two level only
(d) Multi-level

131. To ensure success of new products, which market segment should be targeted initially?
(a) Innovators
(b) Laggards
(c) Early Majority
(d) Early Adopters

132. Who copies something from the leader but maintains differentiation in terms of packaging, advertising, pricing or location?
(a) Cloner (b) Counterfeiter
(c) Imitator (d) Adapter

133. In organizational buying process, the person acts as a filter in the flow of information from buyer to seller or vice versa is called as__________.
(a) Buyer (b) User
(c) Gate Keeper (d) Manager

134. Producers of a strong brand sometimes sell it to dealers only if they will some or all of the rest of the line. This practice is called as
(a) Intensive Distribution
(b) Exclusive Distribution
(c) Exclusive dealing
(d) Full line forcing

135. Who has written the book ,"Positioning – the Battle for the Mind"?
(a) Philip Kotler
(b) Peter Drucker
(c) Friedman and Philips
(d) AlRice and Jack trout

136. Which type of cooperative farming both ownership and operation is individual?
(a) Cooperative better farming
(b) Cooperative joint farming
(c) Cooperative tenant farming
(d) Cooperative collective farming

137. Which type of cooperative farming ownership is indivisual but operation is collective?
(a) Cooperative better farming
(b) Cooperative joint farming
(c) Cooperative tenant farming
(d) Cooperative collective farming

138. Which type of cooperative farming both ownership and operation is collective?
(a) Cooperative better farming
(b) Cooperative joint farming
(c) Cooperative tenant farming
(d) Cooperative collective farming

139. Which type of wage is paid for completion of a specific task?
(a) Cash Wage
(b) Kind Wage
(c) Piece Wage
(d) Task Wage

140. Which is not a type of hired labour from the following list?
(a) Casual Labour
(b) Attached Labour
(c) Family Labour
(d) Contract Labour

141. What is the correct method for evaluating Land value for an organization?
(a) Cost less depreciation Method
(b) Market Price Method
(c) Income capitalization method
(d) Cost method

142. Which is the moderate degree of controlled inflation?
(a) Hyper Inflation
(b) Deflation
(c) Reflation
(d) Stagflation

143. Which one of the following is not an appropriate job-shop scheduling rule?
(a) Earliest due date
(b) Assignment method
(c) First come first served
(d) Johnson's rule

144. Relationship between firms in different stages of the supply chain such as processors and retailers is refereed as______________.
(a) Horizontal cooperation
(b) Vertical cooperation
(c) Conglomerate relation
(d) Concentric relation

145. Which of the following is an example of horizontal integration from the following models?
(a) Corporate farming
(b) Contract farming
(c) Cooperative farming
(d) Tenant farming

146. What is the fulform of NCDEX?
(a) National Commodity Derivatives Exchang
(b) National Commodity and Derivatives Exchange Limited
(c) National Commodity Derivatives Extension Limited
(d) National Common Derivatives Exchange

147. Forward Market commission was merged with SEBI in which year?
(a) 2000 (b) 2005
(c) 2010 (d) 2015

148. Which is the lead promoter of National Agricultural Market?
(a) ISAP (b) SFAC
(c) FICCI (d) CII

149. When did Government of India approve NAM?
(a) 2012 (b) 2014
(c) 2015 (d) 2016

150. How many markets are targeted to be covered under NAM by 2017-18
(a) 585 (b) 485
(c) 535 (d) 435

ANSWER KEY (Practice Set 3)

1. (a)	**2.** (c)	**3.** (a)	**4.** (c)	**5.** (b)	**6.** (c)	**7.** (c)	**8.** (d)	**9.** (c)	**10.** (a)
11. (b)	**12.** (d)	**13.** (a)	**14.** (d)	**15.** (b)	**16.** (b)	**17.** (b)	**18.** (a)	**19.** (c)	**20.** (d)
21. (a)	**22.** (c)	**23.** (c)	**24.** (b)	**25.** (b)	**26.** (a)	**27.** (b)	**28.** (b)	**29.** (b)	**30.** (a)
31. (a)	**32.** (b)	**33.** (d)	**34.** (a)	**35.** (b)	**36.** (a)	**37.** (b)	**38.** (b)	**39.** (b)	**40.** (c)
41. (a)	**42.** (d)	**43.** (d)	**44.** (d)	**45.** (d)	**46.** (c)	**47.** (b)	**48.** (b)	**49.** (c)	**50.** (b)
51. (d)	**52.** (c)	**53.** (d)	**54.** (b)	**55.** (d)	**56.** (d)	**57.** (d)	**58.** (a)	**59.** (c)	**60.** (c)
61. (a)	**62.** (c)	**63.** (c)	**64.** (c)	**65.** (c)	**66.** (b)	**67.** (c)	**68.** (b)	**69.** (b)	**70.** (a)
71. (a)	**72.** (c)	**73.** (b)	**74.** (c)	**75.** (c)	**76.** (b)	**77.** (b)	**78.** (d)	**79.** (b)	**80.** (a)
81. (d)	**82.** (b)	**83.** (c)	**84.** (b)	**85.** (b)	**86.** (b)	**87.** (d)	**88.** (c)	**89.** (c)	**90.** (c)
91. (a)	**92.** (b)	**93.** (c)	**94.** (d)	**95.** (b)	**96.** (b)	**97.** (c)	**98.** (c)	**99.** (c)	**100.** (c)
101. (b)	**102.** (c)	**103.** (c)	**104.** (c)	**105.** (a)	**106.** (a)	**107.** (d)	**108.** (c)	**109.** (a)	**110.** (c)
111. (c)	**112.** (c)	**113.** (c)	**114.** (c)	**115.** (c)	**116.** (b)	**117.** (a)	**118.** (b)	**119.** (a)	**120.** (c)
121. (d)	**122.** (b)	**123.** (a)	**124.** (b)	**125.** (b)	**126.** (c)	**127.** (a)	**128.** (c)	**129.** (b)	**130.** (d)
131. (a)	**132.** (c)	**133.** (c)	**134.** (d)	**135.** (b)	**136.** (a)	**137.** (b)	**138.** (d)	**139.** (d)	**140.** (c)
141. (c)	**142.** (c)	**143.** (b)	**144.** (b)	**145.** (c)	**146.** (b)	**147.** (d)	**148.** (b)	**149.** (c)	**150.** (a)

Practice Set 4

1. Third "P" in Marketing represents as:
(a) Product
(b) Promotion
(c) Price
(d) Place

2. Spontaneous Market of developed country like UK simulates:
(a) Secular Market
(b) Periodic Market
(c) Long Period Market
(d) Short Period Market

3. When a large number of sellers deal with heterogeneous and differentiated commodity is called as?
(a) Perfect Competition
(b) Oligopoly
(c) Monopoly
(d) Monopolistic Competition

4. The commission charges for fruit and vegetable is
(a) 1-2% (b) 2-5%
(c) 4-8% (d) 8-10%

5. Direct Marketing Model in Odisha is
(a) Apni Mandi
(b) Shetkari Bazar
(c) Krushak Bazar
(d) Kisan Mandi

6. What is regulated in regulated market?
(a) Price
(b) Marketing Practices
(c) Behaviour of Trader
(d) Buying and Selling

7. National Cooperative Development Corporation (NCDC) was established in the year
(a) 1963 (b) 1965
(c) 1957 (d) 1991

8. Present Chairman of Commission for Agriculture Costs and Prices (CACP)
(a) Dr. Ashok Gulati
(b) Dr. Montek Singh Ahluwalia
(c) Dr. Vijay Pal Sharma
(d) Dr. Trilochan Mohapatra

9. AAVIN and MAAVIN are milk cooperatives of
(a) Maharashtra
(b) Kerala
(c) Andhra Pradesh
(d) Tamil Nadu

10. National Dairy Development Board (NDDB)came in the year
(a) 1955 (b) 1965
(c) 1963 (d) 1975

11. Central Institute of Agricultural Engineering (CIAE) is at
(a) Ludhiana (b) Chandigarh
(c) Bhopal (d) New Delhi

12. Indian Institute of Packaging (IIP) is at
(a) New Delhi (b) Mumbai
(c) Kolkata (d) Chennai

13. Head quarter of Directorate of Marketing and Inspection (DMI) is at
(a) New Delhi (b) Nagpur
(c) Faridabad (d) Hyderabad

14. Total numbers of commodity are included under Minimum Support Price (MSP) are
(a) 23 (b) 25
(c) 24 (d) 22

15. Essential Commodity Act (ECA)came in the year
(a) 1965 (b) 1979
(c) 1955 (d) 1982

16. First Direct Marketing Model (Apni Mandi) was started in
(a) Punjab (b) Haryana
(c) Gujarat (d) Andhra Pradesh

17. Unza Market in Gujarat is popularized for
(a) Cotton (b) Cumin
(c) Coconut (d) Cardamom

18. Input market farmers mainly face
(a) Oligopoly Competition
(b) Perfect Competition
(c) Monopolistic Competition
(d) Cartel

19. Revamped Public Distribution System (RPDS) was launched in
(a) 1965 (b) 1944
(c) 1997 (d) 1992

20. Cover and Plinth (CAP) storage structure is developed by
(a) NAFED (b) CCI
(c) FCI (d) JCI

21. Types of sampling when an enumerator choose every item at a particular interval from the sampling frame is called as
(a) Stratified Sampling
(b) Cluster Sampling
(c) Systematic Sampling
(d) Quota Sampling

22. Non-comparative scaling technique, where respondents rate the object by placing a mark at the appropriate position is
(a) Stapel Scale
(b) Likert Scale
(c) Semantic Scale
(d) Thurstone Scale

23. Stapel scale is usually presented
(a) Horizontally (b) Vertically
(c) Circularly (d) Slanting Manner

24. Flexible and unstructured research process is the characteristics of which types of research?
(a) Causal Research
(b) Descriptive Research
(c) Explorative Research
(d) Conclusive Research

25. Which is the beginning part of research process?
(a) Formulation of Hypothesis
(b) Define Research Problem
(c) Design Research
(d) Collection of Data

26. The experimental design that is always balanced with both rows and column
(a) Completely Randomized Design
(b) Latin Square Design
(c) Randomised Block Design
(d) Longitudinal Design

27. Correlation Coefficient ranges from
(a) 0 to 1 (b) -1 to +1
(c) $-\infty$ to $+\infty$ (d) 0 to -1

28. Who has established the degree of price discrimination?
(a) 0 to 1 (b) -1 to +1
(c) $-\infty$ to $+\infty$ (d) 0 to -1

29. The value R1.23 lies between
(a) 0 to 1 (b) -1 to +1
(c) $-\infty$ to $+\infty$ (d) 0 to -1

30. In scatter diagram when the relationship between variables is curvilinear, then coefficient of correlation is
(a) 0 (b) 1
(c) -1 (d) 2

31. If sum of square within group is 42, sum of square between group is 6, then total sum of square is
(a) 48 (b) 36
(c) 7 (d) 252

32. If coefficient of correlation is 0.9 then the coefficient of determination is
(a) 1 (b) 0.3
(c) 0.18 (d) 0.81

33. Regression concept is discovered by
(a) Karl Pearson
(b) Francis Galton
(c) Robert Fisher
(d) Mendel

34. If two regression lines coincide then coefficient of correlation is
(a) -1 (b) 1
(c) 0 (d) Either -1 or +1

35. Bartlett's Test of Sphericity is an indicator to understand data adequacy in
(a) Discriminant Analysis
(b) Factor Analysis
(c) Conjoint Analysis
(d) Cluster Analysis

36. Country that is famous for its Six Sigma Concept is
(a) Japan (b) USA
(c) Korea (d) India

37. Sweep in 5S' Principle encapsulates
(a) Sustain (b) Standardise
(c) Shine (d) Sort

38. Type of Learning process when managers copy with the change is called as
(a) Generative Learning
(b) Adaptive Learning
(c) Single Loop Learning
(d) Double Loop Learning

39. A single use plan that covers relatively large set of activities
(a) Policy
(b) Rule
(c) Programme
(d) Procedure

40. The combining of various operations at a similar level into one job
(a) Paddy Loan
(b) Tractor Loan
(c) Loan given for digging wells
(d) Machinery Loan

41. Reality shock syndrome is stated by
(a) Hall (b) Harrison
(c) Hamburger (d) Freidman

42. Right inherent in a managerial position is called as
(a) Power
(b) Authority
(c) Administration
(d) Leadership

43. Communication that cuts across both work areas and organizational level
(a) Diagonal Communication
(b) Horizontal Communication
(c) Lateral Communication
(d) Vertical Communication

44. 14 principles of management is the contribution of
(a) Henry Fayol
(b) Max Weber
(c) Henry Gantt
(d) Chester Bernard

45. Judgmental shortcuts is otherwise called as
(a) Hypnotics (b) Rationality
(c) Heuristic (d) Fuzzy Logic

46. Concept of Value Chain Network is stated by
(a) Karl Roger
(b) Peter Drucker
(c) Michael Porter
(d) Philip Kotler

47. The tendency one has to react in a certain way to a given situation is called as
(a) Mental Set (b) Halo Effect
(c) Stereotyping (d) Attrition

48. The listing of employee skills, abilities and qualification for a job is called as
(a) Job Analysis
(b) Job description
(c) Job Specification
(d) Job Evaluation

49. Temporary separation of employee at the instance of the employer is called as
(a) Layout (b) Layoff
(c) Discharge (d) VRS

50. Factories Act came in the year
(a) 1955 (b) 1961
(c) 1948 (d) 1947

51. Minimum Support Price (MSP) was first announced by the government of India in 1966-67, for which of the following crop/s
(a) Paddy (b) Wheat
(c) Maize (d) Paddy, Wheat, Maize

52. Which was the first state to launch a policy to establish Special Agricultural Zones?
(a) Gujarat (b) Tamilnadu
(c) Uttarakhand (d) Andhra Pradesh

53. When was ICAR established?
(a) 16 July, 1929
(b) 19 July, 1926
(c) 16 July, 1926
(d) 16 July, 1949

54. When was ASRB established?
(a) LIC (b) GIC
(c) DICGC (d) NAIS

55. Which committee recommended ASRB as an independent recruitment agency?
(a) Randhawa Committee
(b) Gajendra Gadkar Committe
(c) Mohan Singh Mehta Committee
(d) Khusro Committe

56. KVK was established on the recommendation of
(a) Randhawa Committee
(b) Gajendra Gadkar Committe
(c) Mohan Singh Mehta Committee
(d) Khusro Committe

57. Central Institute of Fisheries Education is located in
(a) New Delhi (b) Mumbai
(c) Kolkatta (d) Chennai

58. World Food Day is celebrated on
(a) 15th October
(b) 19th October
(c) 16th October
(d) 14th October

59. CeRA stands for
(a) Consortium for electronic Reply in Agriculture
(b) Consortium for e-Resources in Agriculture
(c) Centre for Resource allocation
(d) Centre for Resilient Agriculture

60. Nodal Centre for Krishi Prabha is
(a) NDRI, Karnal
(b) IARI, New Delhi
(c) IVRI, Izzatnagar
(d) HAU, Hisar

61. Monopolist can decide
(a) Product
(b) Price
(c) Both Product and Price
(d) Either Product or Price

62. The wage which has to be paid by an employer to his workers, irrespective of his ability, is known as
(a) Minimum Wage
(b) Fair Wage
(c) Living Wage
(d) Basic Wage

63. In North-West Corner rule for solving a transportation problem
(a) The first assignment is made in the cell occupying the lower left- hand corner of the transportation table
(b) The last assignment is made in the cell occupying the upper left hand corner of the transportation table
(c) The last assignment is made in the cell occupying the lower left hand corner of the transportation table
(d) The first assignment is made in the cell occupying the upper left- hand corner of the transportation table

64. Minimum crop loan has been provided to farmers' at ______percent of interest
(a) 3
(b) 4
(c) 5
(d) 7

65. TRIPS covers the aspect related to
(a) Patents
(b) Trades in Services
(c) Globalisation of Business
(d) Labour Practices

66. "Supply creates its own demand" is popularly called as
(a) Keynesian Law
(b) Say's Law
(c) Veblen Law
(d) Ratchet Effect

67. The ABC analysis in Inventory management is carried out on the basis of
(a) 60/30/10 rule
(b) 80/20 rule
(c) 50/30/20 rule
(d) 50/50 rule

68. In growth share matrix 'cash cow' indicates
(a) High business growth-High market share
(b) High business growth-Low market share
(c) Low business growth-High market share
(d) Low business growth-Low market share

69. 'Global Compact' is related to
(a) Human Right
(b) Social Responsibility
(c) Environmental Protection
(d) Women Right

70. Standing Labour Conference (SLC) was constituted in line with the recommendations of Royal Commission on Labour in the year
(a) 1942 (b) 1946
(c) 1944 (d) 1948

71. The two lines of regression are 3x + 2y = 26 and 6x + y = 31 then the mean of x is
(a) 3 (b) 7
(c) 4 (d) 9

72. Two factors theory was propounded by
(a) Fredrick Herzberg
(b) David Mc Clelland
(c) Mc Cardhy
(d) Phillip Kotler

73. Dramatic reduction of manpower is called as
(a) Termination (b) Retrenchment
(c) Downsizing (d) Rightsizing

74. The technique used for classifying inventory according to the value of usage is called
(a) XYZ Analysis
(b) DEF Analysis
(c) ABC Analysis
(d) SOS Analysis

75. Sick enterprise is referred to
(a) World Bank (b) NABARD
(c) SIDBI (d) BIFR

76. In the case of Veblen goods, demand curve will slope
(a) Upwards (b) Downwards
(c) Horizontal (d) Veritcal

77. Who authored the book, 'positioning- the battle for the mind'?
(a) Philip Kotler and Gary Armstrong
(b) Theodore Levitt
(c) Peter F. Drucker
(d) Alrice and Jack Trout

78. The name,' Victor Vroom' is associated with which theory of motivation
(a) Two Factor Theory
(b) Equity Theory
(c) Expectancy Theory
(d) Contingency Theory

79. Market comes from the word
(a) Murcatus
(b) Marcatus
(c) Markat
(d) Marucat

80. The price below which farmer would find it uneconomical to sell the produce is
(a) Wholesale Price
(b) Floor Price
(c) Minimum Price
(d) Harvesting Price

81. Volume of output is very low in case of
(a) Batch Production
(b) Job Shop
(c) Repetitive Production
(d) Continuous Processing

82. Unit cost is very low in case of
(a) Job Shop Production
(b) Batch Production
(c) Mass Production
(d) Process Production

83. Which of the following layout uses standardized processing operations to achieve smooth, rapid, high-volume flow?
(a) Fixed Layout
(b) Product Layout
(c) Process Layout
(d) Combination Layout

84. Analysis of therbligs is popularly simulates to
(a) Method Analysis
(b) Simo Chart
(c) Learning Curve Effect
(d) Motion Study

85. A cause of variation in a process probably wouldn't be due to
(a) People (b) Standards
(c) Methods (d) Environment

86. What type of control chart would be used to monitor the number of defects in the output of a process?
(a) x-bar Chart (b) P chart
(c) c chart (d) R Chart

87. In the JIT philosophy, the ideal lot size is:
(a) The Economic Order Quantity
(b) The Economic run Size
(c) One Unit
(d) The Capacity of the Standard Container

88. The payback period for short term loan is
(a) Within 6 to 18 months
(b) Within 15 months to 5 Years
(c) Within 5 Years to 20 Years
(d) No Limit

89. Form of security in which borrower stands as the guarantor for loan advancement
(a) Mortgage
(b) Chattel Loan
(c) Personal Security
(d) Collateral Security

90. When job is very big and spread over long periods of time, the method of costing adopted is?
(a) Process
(b) Contract
(c) Operation
(d) Job

91. Continuous costing is also called as
(a) Operation Costing
(b) Batch Costing
(c) Process Costing
(d) Contract Costing

92. Cost accounting is based on the concept of
(a) Approximation
(b) Historical
(c) Estimation
(d) Empirical

93. The total of all direct expenses is known as
(a) Total Cost
(b) Work cost
(c) Overhead Cost
(d) Prime Cost

94. DU PONT Analysis deals with
(a) Analysis of Current Assets
(b) Capital Budgeting
(c) Analysis of Fixed Assets
(d) Analysis of Profit

95. In Net Profit Ratio, the denominator is
(a) Net Purchases
(b) Net Sales
(c) Credit Sales
(d) Cost of Goods sold

96. The term 'EVA' is used for
(a) Extra Value Analysis
(b) Economic Value Added
(c) Expected Value Analysis
(d) Engineering Value Analysis

97. The matching concept matches which of the following
(a) Asset with liabilities
(b) Capital with Income
(c) Revenues with Expenses
(d) Expenses with Capital

98. The accounting equation 'asset = capital + liability is part of which of the following concepts
(a) Realization Concept
(b) Materiality Concept
(c) Historical Record Concept
(d) Dual Aspect Concept

99. Revenue from sale of products, is generally, realized in the period in which
(a) Cash is collected
(b) Sale is made
(c) Products are manufactured
(d) None of the cases

100. Cost unit in a college may be
(a) Teacher
(b) Non-Teaching Staff
(c) Student
(d) Number of Department

101. Globalization has led to higher standards of living of
(a) Well off Consumers
(b) Poor Consumers
(c) Big Producers
(d) Small Producers

102. Where do MNCs choose to set up production?
(a) Cheap Goods
(b) Cheap Labour Resources
(c) Economic Sustainability
(d) None of the above

103. Cargill Foods, an MNC has bought over which indigenous Indian company
(a) AMUL (b) Brittania
(c) Parakh Foods (d) Dabur

104. One major factor that has stimulated the globalization process is
(a) Effective Utilisation of Resourecs
(b) Increase in Income and Wealth
(c) Willingness to Cooperate
(d) Rapid Improvement in Technology

105. If tax is imposed on Chinese toys, what will happen
(a) Chinese toy-makers will benefit
(b) Indian toy-makers will prosper
(c) Chinese toys will remain cheap
(d) Indian consumers will buy more Chinese toys

106. Removing barriers or restrictions set by the government is called
(a) Liberalisation
(b) Investment
(c) Favourable Trade
(d) Free Trade

107. Globalization has posed major challenges for
(a) Big Producers
(b) Small Producers
(c) Rural Poor
(d) None of the above

108. Globalization opportunities for paid work for women has denied them
(a) Good Health
(b) Good Education
(c) Permanent Job
(d) None of the above

109. International Trade
(a) Increases Choices of Goods
(b) Decreases prices of goods
(c) Increases competition in the market
(d) Decreases earnings

110. Globalization was stimulated by
(a) Money
(b) Transportation
(c) Population
(d) Computers

111. Small Scale industries face competition from
(a) Rising Prices
(b) Cheap Imports
(c) Export
(d) Subsidy

112. Longest path in a network analysis is also called as
(a) Critical Path
(b) Merge Point
(c) Dummy Path
(d) Crash Project

113. Game theory model is perfectly established in the market condition like
(a) Perfect Competition
(b) Oligopoly
(c) Monopolistic Competition
(d) Monopoly

114. The balancing point in game theory model is known as
(a) Neutral Point
(b) Saddle Point
(c) Hungarian Point
(d) Equilibrium Point

115. Pareto System best suits to
(a) XYZ Analysis of Inventory Control
(b) SOS Analysis of Inventory Control
(c) ABC Analysis of Inventory Control
(d) DEF Analysis of Inventory Control

116. The symbol that denotes as <u>an inspection</u> in the operation site
(a) Δ (b) □
(c) ○ (d) →

117. Agro-processing industry is mushrooming in rural areas due to
(a) Labour Facility
(b) Infrastructure facility
(c) Raw Material Facility
(d) Electricity Facility

118. Customer Operations Performance Centre (COPC) standard in quality management deals with
(a) Quality (b) Production
(c) Design (d) Aftersales Service

119. Assembling line and fabricated line are best described in
(a) Process Layout
(b) Product Layout
(c) Cellular Layout
(d) Fixed Layout

120. Kanban is a Japanese word simulates
(a) Paper (b) Card
(c) Colour (d) Paint

121. Z8101-1981 is a quality measurement standard of
(a) USA (b) JAPAN
(c) INDIA (d) EUROPE

122. ECR stands for
(a) Effective Customer Response
(b) Efficient Consumer Response
(c) Effective Consumer Response
(d) Effective Customer Reply

123. The term Supply Chain Management was coined by
(a) Peter Drucker
(b) Keith Oliver
(c) Raphael Kaplinsky
(d) Michael Porter

124. Different Contract Farming models have been given by
(a) Clarence Danhof
(b) Eaton and Shepherd
(c) Richard Cantillon
(d) Margret et. al

125. Which of the following is the nodal agency for AEZ (Agri Export Zone)
(a) APEDA (b) ICAR
(c) NIAM (d) DARE

126. The term 'Value Chain' was developed and popularized by
(a) Peter Drucker
(b) Keith Oliver
(c) Raphael Kaplinsky
(d) Michael Porter

127. Nestle's model of milk procurement in dairy industry is commonly known as
(a) Hub and Spoke Model
(b) Milk run Model
(c) Milk district Model
(d) Milk Center Model

128. Which of the following has maximum share in Indian pesticide industry
(a) Syngenta India
(b) Bayer Crop Science
(c) United Phosphorus
(d) Rallies India

129. Uttam Bandhan is the community welfare programme in which company trains rural youths as extension workers as Krishi Sewak. Name that compan
(a) HUL (b) DSCL
(c) CFCL (d) BPCL

130. Kutir Jyoti programme in the direction of rural sector electrification came in the year
(a) 1982 (b) 1989
(c) 1985 (d) 1995

131. 'Head Start' is the rural based educational programme of
(a) Andhra Pradesh
(b) Madhya Pradesh
(c) Kerala
(d) Rajasthan

132. National Institute of Rural Development (NIRD) is at
(a) Pune
(b) Delhi
(c) Hyderabad
(d) Kanpur

133. When a marketer offers one brand with another brand of the same company or another company is called
(a) Bundling
(b) Co-Branding
(c) Upselling
(d) Cross Selling

134. Samridhi is an initiative in rural farm sector segment is launched by
(a) TATA (b) Mahindra
(c) John Dheere (d) HMT

135. Asha is low cost snacks and beverages-especially for girls age between 12 to 16, is the product of
(a) Coca Cola (b) Pepsi
(c) Parle Agro (d) Dabur

136. State Bank of India (SBI) was set up in
(a) 1935 (b) 1955
(c) 1949 (d) 1960

137. Crop insurance policy reduces
(a) Production risk
(b) Price risk
(c) Financial risk
(d) Interest risk

138. In Net Worth statement, liability is placed on which side
(a) Right side
(b) Left side
(c) Middle
(d) Subjective Consideration

139. The repayment plan in which installment is directly proportional to amount of harvest is called as
(a) Balloon repayment Plan
(b) Amortised Even Repayment Plan
(c) Amortised Decreasing Repayment Plan
(d) Variable Repayment Plan

140. Which of the following is the sixth 'P' of credit
(a) Principle of Proper Utilisation
(b) Principle of Payment
(c) Principle of Protection
(d) Principle of Personality

141. Which of the following is the third 'R' of credit
(a) Rate of Returns
(b) Returns
(c) Repayment Capacity
(d) Risk Bearing Ability

142. Which of the following is the first 'C' of credit
(a) Condition (b) Character
(c) Capacity (d) Capital

143. Comprehensive Crop Insurance Scheme was introduced by GIC in
(a) 1979 (b) 1985
(c) 1989 (d) 1972

144. Which committee recommended the Pilot crop insurance scheme in India
(a) A D Gorwala Committe
(b) B Siva Raman Committee
(c) V M Dandekar Committee
(d) D R Gadgil Committee

145. General Insurance Corporation (GIC) was first administered in
(a) Maharashtra
(b) Gujarat
(c) Andhra Pradesh
(d) Punjab

146. The subsidy for SC/ST through Ganga Kalyan Yojana is
(a) 25% (b) 33.33%
(c) 50% (d) 75%

147. Most popular SHG (self Help Group) model in India is
(a) Direct Linkage Model
(b) Modified Direct Linkage Model 11
(c) IFAD Model
(d) Modified Indirect Model

148. The extent of refinance for farm Mechanisation through NABARD (National Bank for Agriculture and Rural Development) are
(a) 25%
(b) 33.33%
(c) 50%
(d) 75%

149. The extent of refinance for Rural Electrification Corporation through NABARD are
(a) 25%
(b) 33.33%
(c) 50%
(d) 75%

150. The convener of District Consultative Committee (DCC) is
(a) Lead Bank Officer
(b) Collector/Magistrate
(c) PD DRDA
(d) District Agricultural Officer

ANSWER KEY (Practice Set 4)

1. (d) **2.** (b) **3.** (d) **4.** (c) **5.** (c) **6.** (b) **7.** (a) **8.** (c) **9.** (d) **10.** (b)
11. (c) **12.** (b) **13.** (c) **14.** (b) **15.** (c) **16.** (a) **17.** (b) **18.** (c) **19.** (d) **20.** (c)
21. (c) **22.** (c) **23.** (b) **24.** (c) **25.** (b) **26.** (b) **27.** (b) **28.** (c) **29.** (a) **30.** (a)
31. (a) **32.** (d) **33.** (b) **34.** (d) **35.** (b) **36.** (b) **37.** (c) **38.** (b) **39.** (c) **40.** (b)
41. (a) **42.** (b) **43.** (a) **44.** (a) **45.** (c) **46.** (c) **47.** (a) **48.** (c) **49.** (b) **50.** (c)
51. (b) **52.** (a) **53.** (a) **54.** (c) **55.** (b) **56.** (c) **57.** (b) **58.** (c) **59.** (b) **60.** (d)
61. (c) **62.** (a) **63.** (d) **64.** (d) **65.** (b) **66.** (b) **67.** (b) **68.** (c) **69.** (a) **70.** (a)
71. (a) **72.** (a) **73.** (b) **74.** (c) **75.** (d) **76.** (a) **77.** (d) **78.** (c) **79.** (b) **80.** (b)
81. (b) **82.** (d) **83.** (b) **84.** (d) **85.** (b) **86.** (c) **87.** (c) **88.** (a) **89.** (c) **90.** (b)
91. (c) **92.** (c) **93.** (d) **94.** (d) **95.** (b) **96.** (b) **97.** (c) **98.** (d) **99.** (b) **100.** (c)
101. (a) **102.** (b) **103.** (c) **104.** (d) **105.** (b) **106.** (a) **107.** (b) **108.** (c) **109.** (d) **110.** (b)
111. (b) **112.** (a) **113.** (b) **114.** (b) **115.** (c) **116.** (b) **117.** (c) **118.** (d) **119.** (b) **120.** (b)
121. (b) **122.** (b) **123.** (b) **124.** (b) **125.** (a) **126.** (d) **127.** (c) **128.** (c) **129.** (c) **130.** (c)
131. (b) **132.** (b) **133.** (c) **134.** (b) **135.** (b) **136.** (b) **137.** (a) **138.** (a) **139.** (d) **140.** (b)
141. (d) **142.** (b) **143.** (b) **144.** (c) **145.** (b) **146.** (d) **147.** (b) **148.** (c) **149.** (c) **150.** (a)

Practice Set 5

1. Scientist who popularly known as 'Father of Economics'.
   (a) J.M. Keynes
   (b) Alfred Marshall
   (c) Adam Smith
   (d) Lionel Robbins

2. When economics comes into picture, it generates the flavor of
   (a) Unlimited (b) Limited
   (c) Free (d) Scarce

3. Marketing way of utility creation is through
   (a) Production
   (b) Exchange
   (c) Consumption
   (d) Distribution

4. 'Government with Bullets', popularly simulates with
   (a) Communism
   (b) Capitalism
   (c) Socialism
   (d) Mixed Economy

5. Direct Marketing Model in Maharashtra is
   (a) Uzhar Sandies
   (b) Shetkari Bazar
   (c) Rythubazar
   (d) Hadaspar Market

6. 'A Particular want is satiable', comes under the purview of
   (a) LEMU
   (b) LDMU
   (c) Consumer Surplus
   (d) Indifference Curve Analysis

7. Food Corporation of India (FCI) was established in the year
   (a) 1963 (b) 1965
   (c) 1957 (d) 1991

8. Present Chairman of Commission of Agricultural Costs and Prices (CACP)
   (a) Dr. Ashok Gulati
   (b) Dr. Montek Singh Ahluwalia
   (c) Dr. Vijay Pal Sharma
   (d) Dr. Trilochan Mohapatra

9. Vijaya milk cooperatives present in
   (a) Maharashtra (b) Kerala
   (c) Andhra Pradesh (d) Tamil Nadu

10. National Agriculture Cooperative Marketing Federation (NAFED) came in the year
    (a) 1958 (b) 1965
    (c) 1963 (d) 1975

11. When total utility is maximum, marginal utility will be
    (a) Maximum (b) Zero
    (c) One (d) 100

12. The shape of Indifference curve is ______ origin
    (a) Concave (b) Convex
    (c) Hyperbola (d) Sigmoid

13. The Cross Demand curve of substitute is
    (a) Concave (b) Convex
    (c) Hyperbola (d) Sigmoid

14. The price elasticity of demand, in general ranges from
    (a) -1 to +1 (b) 0 to 1
    (c) 0 to ∞ (d) $-\infty$ to $+\infty$

15. Cross elasticity of complement goods is
    (a) Positive (b) Negative
    (c) Zero (d) Infinity

16. Single buyer in a market creates a condition of
    (a) Monopoly (b) Monopsony
    (c) Oligopoly (d) Duopoly

17. Distress sale comes in a condition when
(a) Selling price < MSP
(b) Selling price < Cost of Production
(c) Selling price > Cost of Production
(d) Selling price > MSP

18. Price Discrimination is a market situation simulates with
(a) Monopoly
(b) Monopsony
(c) Oligopoly
(d) Duopoly

19. Targeted Public Distribution System (TPDS)was launched in
(a) 1965 (b) 1944
(c) 1997 (d) 1992

20. The type of unemployment arises due to lack of mobility on the part of labour refers to
(a) Structural unemployment
(b) Frictional unemployment
(c) Keynes unemployment
(d) Seasonal unemploment

21. When an enumerator makes group on the basis of variables then the sampling procedure is called as
(a) Stratified Sampling
(b) Cluster Sampling
(c) Systematic Sampling
(d) uota Sampling

22. Type of inflation when 'Too much money chasing too few goods' is
(a) Demand pull inflation
(b) Galloping inflation
(c) Cost push inflation
(d) Creeping inflation

23. The rate at which bank borrows from the RBI (Reserve Bank of India) is called as
(a) Repo rate
(b) Reverse Repo rate
(c) Bank rate
(d) SLR

24. Regional Rural Bank (RRB) came into existence in the year
(a) 1935 (b) 1969
(c) 1949 (d) 1975

25. Farmer Service Society (FSS) came into existence in the year
(a) 1971 (b) 1973
(c) 1975 (d) 1977

26. Farmer Service Society (FSS) has its jurisdiction
(a) 1 district (b) 1 Village
(c) 1 Mandal (d) 1 District

27. DIR Scheme was the concept of
(a) Mohan Kanda Committe
(b) (b) K. Hazare Committe
(c) V.L. Mehra Committe
(d) (b) Sivaraman Committe

28. Adangal Certificate that is produced by the farmer during acuisition of loan indicates
(a) Cropping Pattern
(b) Mortgage Details
(c) Ownership rights of Land
(d) Human right

29. The incremental income due to loan investment can be worked out by using
(a) Cobb-Douglas production function
(b) Linear Programming techniue
(c) Partial Budgeting techniue
(d) Compound growth rate

30. FSS on cooperative lines are recommended by
(a) NCA
(b) Bawa Team
(c) Royal Commission on Agriculture
(d) Gadgil Committee

31. Key and Open Loans are together called as
(a) Mortgage (b) Pledge
(c) Production (d) Term

32. If coefficient of correlation is 0.8 then the coefficient of determination is
(a) 1.6 (b) 0.46
(c) 0.64 (d) 0

33. Single window system in cooperative credit structure was introduced by
(a) Raiffiesen (b) Nicholson
(c) Narasimhan (d) Mohan Kanda

34. Loans granted by the Government during British rule is
(a) Chattal (b) Taccavi
(c) Term (d) Pledge

35. Kaiser-Meyer-Olkin (KMO) is an indicator to understand data adeuacy in
(a) Discriminant Analysis
(b) Factor Analysis
(c) Conjoint Analysis
(d) Cluster Analysis

36. Concept 'Kanban' came from
(a) Japan (b) USA
(c) Korea (d) India

37. Repayment plan in which the installment amount gradually decreases but principal amount remains same throughout the repayment period in the last, huge installment is paid is called as
(a) Straight end repayment plan
(b) Amortized decreasing repayment plan
(c) Amortized even repayment plan
(d) Partial repayment plan

38. Crop loan scheme is came into picture in
(a) 1960 (b) 1965
(c) 1970 (d) 1971

39. The component A and B in crop loan scheme indicates
(a) Kind and Cash
(b) Owned and Leased in Land
(c) Cash and Kind
(d) Leased in and Owned Land

40. The combining of similar activities into one job
(a) Job enrichment
(b) Job Enlargement
(c) Job rotation
(d) Job specification

41. The ratio that defines finance on the basis of cash and kind components is
(a) 1:3 (b) 1:2
(c) 3:1 (d) 2:1

42. Power is
(a) Unidirectional
(b) Bidirectional
(c) Multidirectional
(d) Tri-directional

43. 'One man-One vote' is the principle of
(a) Cooperation
(b) Individual Business
(c) JSC
(d) Corporation

44. Scientific management is proposed by
(a) Henry Fayol
(b) Max Weber
(c) Fredrick W Taylor
(d) Chester Bernard

45. Which of the following marketing intermediary is more common in regulated market
(a) Wholesaler
(b) Commission agent
(c) Retailer
(d) Broker

46. Three Generic Strategy is given by
(a) Karl Roger
(b) Peter Drucker
(c) Michael Porter
(d) Philip Kotler

47. The Indian Institute of Packaging is established in the year
(a) 1962 (b) 1965
(c) 1963 (d) 1966

48. In short period market, __________ is constant and plays no significant role in determining prices of the commodities.
(a) Demand
(b) Supply
(c) Previous price
(d) Competitor price

49. Support prices of Agricultural commodities are announced by
(a) FCI (b) CACP
(c) ICAR (d) SAU

50. Industrial Safety Act came in
(a) 1955 (b) 1961
(c) 1948 (d) 1947

51. Market in which future sale and purchase of commodities will take place at current time
(a) Future Market
(b) Spot Market
(c) Forward Market
(d) International Market

52. Market in which the buyers and sellers participate in large number
(a) Local Market
(b) Primary wholesale market
(c) Secondary wholesale market
(d) Terminal market

53. Market prices are settled by pressing the fingers under cover of piece of cloth is called as
(a) Open Auction
(b) Dara Sale
(c) Hatha System
(d) Closed Auction

54. Wholesalers and Retailers are considered in which categories of middlemen?
(a) Facilitative Middlemen
(b) Agent Middlemen
(c) Merchant Middlemen
(d) Speculative Middlemen

55. The difference between price paid by the consumer and price received by the producer for an euivalent uantity of commodity is known as
(a) Marketing Cost
(b) Marketing Margin
(c) Concurrent Margin
(d) Price Spread

56. The market that deals with durable goods is
(a) Wholesale market
(b) Regulated Market
(c) Sea Board Market
(d) Secular Market

57. Central Sheep and Wool Research Institute is located in
(a) Palampur (b) Makhdoom
(c) Avikanagar (d) Mukteswar

58. Earth Day is celebrated on
(a) 22nd March (b) 22nd April
(c) 22nd October(d) 22nd January

59. CeRA stands for
(a) Consortium for electronic Reply in Agriculture
(b) Consortium for e-Resources in Agriculture
(c) Centre for Resource allocation
(d) Centre for Resilient Agriculture

60. The agency located at national level to fix the grading standards for the commodities on scientific lines is
(a) NIAM (b) CACP
(c) DMI (d) ICAR

61. Which of the following correspond to perfect competitive market situation?
(a) Law of comparative advantage
(b) Law of One Price
(c) Law of Reciprocal demand
(d) Law of Absolute demand

62. Indian Grain Storage Management and Research Institute (IGSMRI) is located at
(a) Pune (b) Hapur
(c) Hyderabad (d) Lucknow

63. Kachha Arhatiyas function on behalf of
(a) Trader (b) Farmer
(c) Broker (d) Speculator

64. Which Country may overtake India in Tea Production in future?
(a) China (b) Srilanka
(c) Kenya (d) Turey

65. 'A Class Trader' is seen in
(a) Unregulated Market
(b) Local Market
(c) Terminal Market
(d) Regulated Market

66. Minimum Support Price (MSP) is given statutory status for
(a) Paddy (b) Sugarcane
(c) Wheat (d) Mango

67. In which year tea board of India came?
(a) 1963 (b) 1973
(c) 1953 (d) 1945

68. In growth share matrix 'Star' indicates
(a) High business growth-High market share
(b) High business growth-Low market share
(c) Low business growth-High market share
(d) Low business growth-Low market share

69. The members of NAFED include
(a) NCDC and RAJFED
(b) NCDC and TRIFED
(c) State level Marketing Federation and NCDC
(d) NCDC and CCI

70. The first Direct Marketing Model (Apni Mandi) was started in Punjab in the year
(a) 1963 (b) 1978
(c) 1987 (d) 1991

71. Who coined 4Ps of marketing mix?
(a) Lauterbern
(b) E. J. McCarthy
(c) Rogers
(d) Boom and Bitner

72. Who coined 7Ps of marketing mix?
(a) Lauterbern
(b) E. J. McCarthy
(c) Rogers
(d) Boom and Bitner

73. Who coined 4Cs of marketing mix?
(a) Lauterbern
(b) E. J. McCarthy
(c) Rogers
(d) Boom and Bitner

74. In which analysis, Inventory classification is mainly based 'on value of usage'?
(a) XYZ Analysis
(b) DEF Analysis
(c) ABC Analysis
(d) SOS Analysis

75. Which is the organization sick industry is referred?
(a) World Bank (b) NABARD
(c) SIDBI (d) BIFR

76. 'Regulation on treatment of imported fruit to prevent pests and diseases spreading' represents the trade barrier
(a) SPS (b) Patent
(c) TBT (d) Copyright

77. Regulation on 'permitted food safety: health warnings, use, dosage' represents the trade barrier
(a) SPS (b) Patent
(c) TBT (d) Copyright

78. The scientist,' Adam' is associated with which theory of motivation?
(a) Two Factor Theory
(b) Euity Theory
(c) Expectancy Theory
(d) Contingency Theory

79. Where is the head uarter of WTO based?
(a) Phillipines
(b) Switzerland
(c) Washington
(d) Newyork

80. Which of the following boxes of subsidies distort the trade?
(a) Green Box
(b) Blue Box
(c) Amber Box
(d) Red Box

81. In which manufacturing system, Volume of output is very high?
(a) Batch Production
(b) Job Shop
(c) Repetitive Production
(d) Process Production

82. In which manufacturing system, Unit cost is very high?
(a) Job Shop Production
(b) Batch Production
(c) Mass Production
(d) Process Production

83. Which of the following layout uses specialized processing operations to satisfy customer demand?
(a) Fixed Layout
(b) Product Layout
(c) Process Layout
(d) Combination Layout

84. Where first Krishi Vigyan Kendra (KVK) was established at?
(a) Madras
(b) Puducherry
(c) Andaman & Nicobar Island
(d) Goa

85. When National Agricultural Innovation Project (NATP) was launched?
(a) 2003 (b) 2005
(c) 2004 (d) 2006

86. National Academy of Agricultural Research & Management (NAARM) is located at
(a) New Delhi (b) Hyderabad
(c) Mumbai (d) Kolkatta

87. What is the ideal lot size in Just In Time inventory?
(a) The Economic Order uantity
(b) The Economic run Size
(c) One Unit
(d) The Capacity of the Standard Container

88. National Centre for Agricultural Economics & Policy Research is located in
(a) New Delhi
(b) Hyderabad
(c) Mumbai
(d) Kolkatta

89. What is the share of Co-operative Bank towards Agricultural Credit?
(a) 10% (b) 75%
(c) 15% (d) 80%

90. When job is very big and spread over long periods of time, the method of costing adopted is?
(a) Process (b) Contract
(c) Operation (d) Job

91. Where Directorate of Knowledge Management in Agriculture is located at?
(a) New Delhi (b) Mumbai
(c) Kolkatta (d) Chennai

92. What is the maximum limit of loan under microfinance?
(a) 5 Lakhs (b) 10 Lakhs
(c) 25 Lakhs (d) 50 Lakhs

93. What is the position of India in pulse production in world?
(a) First (b) Second
(c) Third (d) Fourth

94. Full Text database of Indian Agricultural Doctoral Dissertations is
(a) Krishi Kiran
(b) E- Prabha
(c) Krishi Prabha
(d) E- Krishikiran

95. When National Initiative on Climate Resilient Agriculture (NICRA) was launched?
(a) 2nd Feb, 2011
(b) 3rd March, 2012
(c) 2nd Feb, 2013
(d) 3rd March, 2011

96. Which is the main source of Molasses?
(a) Paddy (b) Sugarcane
(c) Cotton (d) Turmeric

97. Where we get Linamarin from the following sources?
(a) Maize (b) Tapioca Meal
(c) Turmeric (d) Groundnut Oil Cake

98. Where we get black soil from the following states?
(a) Gujarat (b) Maharashtra
(c) Tamil Nadu (d) West Bengal

99. Parent group of Hero company
(a) Mittal (b) Modi
(c) Jhawar (d) Munjal

100. Industrial Development Bank of India (IDBI) came into existence
(a) 1955 (b) 1964
(c) 1977 (d) 1989

101. In which country, organizations are uite hierarchical with family members, occupying key position?
(a) USA (b) Japan
(c) Korea (d) China

102. "A place for everything, everything in its place" is
(a) American Management
(b) 5S' Principle
(c) Peter Principle
(d) A Chinese Principle

103. The statement that unveils permanent part of organization's identity, its purpose, competencies and place in the world
(a) Policy (b) Rule
(c) Mission (d) Procedure

104. Who coined motivational job design in management?
(a) Henry Foyal
(b) J Richard Hackman
(c) Simon
(d) Maslow

105. The social approach in which organizations exhibit little environmental sensitivity
(a) Dark Green Approach
(b) Light Green Approach
(c) Market Approach
(d) Stakeholder Approach

106. What is the other name of Informal internal knowledge?
(a) Explicit Knowledge
(b) Tacit Knowledge
(c) Unstructured Knowledge
(d) External Knowledge

107. What is the end phase of group development stage?
(a) Performing
(b) Adjourning
(c) Norming
(d) Forming

108. Who highlighted sociometry concept of feelings acceptance and rejection among employee?
(a) Louis Kelso (b) Kaplan
(c) J L Marino (d) Porter

109. The term 'BARS' stands for
(a) Behavioural Abnormality Rating Scale
(b) Behavioural Abnormality Response Scale
(c) Behavioural Anchored Response Scale
(d) Behavioural Anchored Rating Scale

110. Who has given T-group training method?
(a) Louis Kelso (b) Kaplan
(c) J L Marino (d) Porter

111. What is the related area of Agenda 21?
(a) Human Right
(b) Child Labour
(c) Environmental Protection
(d) Business Ethics

112. Who coined Double entry system in accounting?
(a) Canadian Institute of Chartered Accountants
(b) International Accounting standard committee
(c) Lucas Pacioli
(d) Confederation of Indian Industry

113. What is the other name of Books of original entry in accounting?
(a) Ledger
(b) Trial Balance
(c) Journal
(d) Balance Sheet

114. What is name of the statement where assets and liabilities of the business on a particular date is revealed?
(a) Ledger
(b) Trial Balance
(c) Journal
(d) Balance Sheet

115. In personal account, the person who is receiving benefit is highlighted in which side of balance sheet?
(a) Debit
(b) Credit
(c) Both Debit and Credit
(d) Either Debit or Credit

116. What is the symbol that denotes an operation in operation site in organization?
(a) Δ (b) □
(c) ○ (d) →

117. What is the major area of agro-processing industry is mushrooming in rural areas?
(a) Labour Facility
(b) Infrastructure facility
(c) Raw Material Facility
(d) Electricity Facility

118. Which is the most commonly used method for evaluation of larger project?
(a) NPV
(b) IRR
(c) Pay Back Period
(d) Benefit to Cost Ratio

119. What is difference between current assets and current liabilities in accounting is called as?
(a) Capital Budgeting
(b) Capital Structure
(c) Gross Working Capital
(d) Net Working Capital

120. The financial statement not only show the absolute figures of various years but also provide for columns to indicate the changes in various figure is called as
(a) Cash Flow Statement
(b) Fund Flow Statement
(c) Comparative Statement
(d) Common Size Statement

121. Q91 is a uality measurement standard of which country?
(a) USA
(b) JAPAN
(c) INDIA
(d) EUROPE

122. The innovative science of Operations Research was discovered during which period?
(a) Civil War
(b) World War 1
(c) World War 1I
(d) Industrial Revolution

123. Inventory classification method that is based on the consumption figure of the items
(a) VED Analysis
(b) XYZ Analysis
(c) FSN Analysis
(d) HML Analysis

124. Which of the following is fixed-time period inventory model?
(a) The Least Cost Method
(b) The Model
(c) Periodic System Model
(d) Just-in-time Model

125. Which of the following is the nodal agency for AEZ
(a) APEDA
(b) ICAR
(c) NIAM
(d) DARE

126. Who developed and popularized the term, 'Value Chain'?
(a) Peter Drucker
(b) Keith Oliver
(c) Raphael Kaplinsky
(d) Michael Porter

127. Who has written the book,"Innovation and Entrepreneurship"?
(a) Philip Kotler
(b) C K Prahlad
(c) Pradeep Khandelwal
(d) Peter F Drucker

128. Which is the toll free telephone number for Kisan Call Center from the following?
(a) 18001801551
(b) 18001801555
(c) 18001801550
(d) 18001801559

129. What is the other name of, 'The future value of present sum'?
(a) Discounting
(b) Compounding
(c) Opportunity Cost of Capital
(d) Annuity

130. What is the other name of , 'The present value of future sum' ?
(a) Discounting
(b) Compounding
(c) Opportunity Cost of Capital
(d) Annuity

131. What is the limit for short term crop loan has been decided by RBI?
(a) 2,50,000 (b) 3,00,000
(c) 5,00,000 (d) 7,00,000

132. What is the contribution of Indian food processing industry in country's total food market in percentage?
(a) 20 (b) 32
(c) 45 (d) 55

133. What is the overall loss in farm produce in India?
(a) 50,000 Cr (b) 90,000 Cr
(c) 75,000 Cr (d) 1,00,000 cr

134. What is the average per capita consumption of Pulses in India annually?
(a) 10 Kilos (b) 14 Kilos
(c) 20 Kilos (d) 40Kilos

135. What is the percentage of losses estimated in poultry production?
(a) 0.8% (b) 3.7%
(c) 6.9% (d) 2.9%

136. What is the percentage of losses estimated in milk production?
(a) 0.8% (b) 3.7%
(c) 6.9% (d) 2.9%

137. What is the percentage of losses estimated in Inland Fish production?
(a) 0.8% (b) 3.7%
(c) 6.9% (d) 2.9%

138. What is the percentage of losses estimated in Marine Fish production?
(a) 0.8% (b) 3.7%
(c) 6.9% (d) 2.9%

139. What is the contribution of India in world Basmati rice cultivation in percentage?
(a) 50 (b) 75
(c) 60 (d) 85

140. What is the contribution of India in world milk production?
(a) 17 (b) 17.5
(c) 18.5 (d) 18

141. Which is the number one country in egg production?
(a) USA
(b) CHINA
(c) INDIA
(d) BRAZIL

142. Which in the number one country in chicken production?
(a) USA
(b) CHINA
(c) INDIA
(d) BRAZIL

143. What is the rank of India in egg production?
(a) First
(b) Second
(c) Third
(d) Fourth

144. What is the rank of India in chicken production?
(a) First
(b) Second
(c) Third
(d) Fourth

145. What is the per-capita egg consumption in India?
(a) 50
(b) 68
(c) 25
(d) 32

146. What is the per-capita chicken consumption in India?
(a) 2.5 Kg
(b) 3 Kg
(c) 5 Kg
(d) 7.5 Kg

147. Which of the following organisation is meant for promoting small scale industries at district level
(a) SIDBI
(b) NABARD
(c) DIC
(d) SFC

148. Which strategies formulated to convert a sick unit to healthy one?
(a) Turnaround
(b) Expansion
(c) Diversification
(d) Stabilisation

149. When Rastriya krishi Vikas Yojana (RKVY) came into existence?
(a) 2000-01
(b) 2005-06
(c) 2007-08
(d) 2008-09

150. Which ministry under central government control manufacturing of Cane Sugar and Khandsari?
(a) Ministry of Agriculture
(b) Ministry of Khadi and Village Industries
(c) Ministry of Commerce
(d) Ministry of Rural Development

ANSWER KEY (Practice Set 5)

1. (c) **2.** (d) **3.** (b) **4.** (a) **5.** (b) **6.** (b) **7.** (b) **8.** (d) **9.** (c) **10.** (a)
11. (b) **12.** (b) **13.** (a) **14.** (c) **15.** (b) **16.** (b) **17.** (b) **18.** (a) **19.** (c) **20.** (b)
21. (a) **22.** (a) **23.** (c) **24.** (d) **25.** (a) **26.** (c) **27.** (b) **28.** (a) **29.** (c) **30.** (a)
31. (b) **32.** (c) **33.** (d) **34.** (b) **35.** (b) **36.** (a) **37.** (b) **38.** (b) **39.** (c) **40.** (b)
41. (d) **42.** (d) **43.** (a) **44.** (c) **45.** (b) **46.** (c) **47.** (d) **48.** (b) **49.** (b) **50.** (d)
51. (c) **52.** (c) **53.** (c) **54.** (c) **55.** (d) **56.** (d) **57.** (c) **58.** (b) **59.** (b) **60.** (c)
61. (b) **62.** (b) **63.** (b) **64.** (c) **65.** (d) **66.** (b) **67.** (c) **68.** (a) **69.** (c) **70.** (c)
71. (b) **72.** (d) **73.** (a) **74.** (c) **75.** (d) **76.** (a) **77.** (a) **78.** (b) **79.** (b) **80.** (c)
81. (d) **82.** (d) **83.** (c) **84.** (b) **85.** (b) **86.** (b) **87.** (c) **88.** (a) **89.** (c) **90.** (b)
91. (a) **92.** (c) **93.** (a) **94.** (c) **95.** (a) **96.** (b) **97.** (b) **98.** (b) **99.** (d) **100.** (b)
101. (c) **102.** (b) **103.** (c) **104.** (b) **105.** (b) **106.** (b) **107.** (d) **108.** (c) **109.** (d) **110.** (c)
111. (b) **112.** (c) **113.** (c) **114.** (d) **115.** (a) **116.** (c) **117.** (c) **118.** (b) **119.** (d) **120.** (c)
121. (a) **122.** (d) **123.** (c) **124.** (c) **125.** (a) **126.** (d) **127.** (d) **128.** (a) **129.** (b) **130.** (a)
131. (b) **132.** (b) **133.** (b) **134.** (b) **135.** (b) **136.** (a) **137.** (c) **138.** (d) **139.** (d) **140.** (c)
141. (b) **142.** (b) **143.** (c) **144.** (d) **145.** (b) **146.** (b) **147.** (c) **148.** (a) **149.** (c) **150.** (b)